D0408416

DATE DUE

JA 28 '94			
MY 26 '95			
JY 2 '97			
JY 23 '97			
DE 17			
MY 31 '02			
NO 6 '02			
DE 6 '02			
MY 13 04			
DE 13 05			

Radon, Radium, and Other Radioactivity in Ground Water

Hydrogeologic Impact and Application to Indoor Airborne Contamination

Proceedings of the NWWA Conference
April 7–9, 1987
Somerset, New Jersey

Edited by Barbara Graves

Sponsors

Association of Ground Water Scientists and Engineers
(A Division of NWWA)
U.S. Environmental Protection Agency

Coordinated by

The National Water Well Association

LEWIS PUBLISHERS

national water well association

Library of Congress Cataloging-in-Publication Data

NWWA Conference (1987: Somerset, N.J.)
   Radon, radium, and other radioactivity in ground
water.

   "Coordinated by the National Water Well Association."
   1. Radioactive substances in rivers, lakes, etc. —
Congresses.   2. Water, Underground — Pollution —
Congresses.   3. Drinking water — Contamination — Congresses.
4. Radionuclides — Environmental aspects — Congresses.
5. Hydrogeology — Congresses.   6. Air — Pollution, Indoor —
Congresses.   I. Association of Ground Water Scientists
and Engineers (U.S.)   II. United States. Environmental
Protection Agency.   III. National Well Water Association.
IV. Title.
TD427.R3N85   1987   628.1′685   87-3624
ISBN 0-87371-117-3

A copublication with the National Water Well Association

LEWIS PUBLISHERS, INC.
121 South Main Street, Chelsea, Michigan 48118

PRINTED IN THE UNITED STATES OF AMERICA

Preface

The dangers associated with the inhalation of short-lived radon decay products have been well documented. The U.S. EPA estimates that radon contaminates one in eight U.S. homes and causes thousands of lung cancer deaths each year. The fact that ground water is one of the many sources of radon contamination is cause for concern, considering the growing number of persons who rely on this precious resource. This concern prompted the Association of Ground Water Scientists and Engineers and the U.S. EPA to host this first-of-its-kind conference on radon, radium, and other radioactivity in ground water in Somerset, New Jersey on April 7-9, 1987.

The conference sessions focused on the following topics: geologic and hydrogeologic controls influencing radon occurrence; monitoring radon, radium, and other radioactivity from geologic and hydrogeologic sources; mining impacts on the occurrence of radon, radium, and other radioactivity in ground water; sampling and analysis of radon, radium, and other radioactivity in ground water; radon and radium in water supply wells; predictive models for the occurrence of radon, radium, and other radioactivity; and remedial action for radon, radium, and other radioactivity.

The conference provided an opportunity for interested professionals to learn more about this important subject. These proceedings are a compilation of papers presented by the conference speakers.

The papers in this book have been reproduced exactly as submitted by the authors, without peer review, in order to ensure early publication and dissemination of this important work. It is the belief of the conference coordinating committee that these chapters have substantial technical merit, or they would not have been selected for presentation.

Acknowledgments

Without the special talents, services, and contributions of the following people, the conference and these proceedings would not have been possible:

Symposium Coordinating Committee
Barbara J. Graves
Jay H. Lehr
Kathy Butcher

NWWA Research/Education Department
Vaughn Shelton
Paula Williams
Dottie Semons

NWWA Conventions and Meetings Planning Department
Lisa Ammerman
Pat Behling

Conference Invited Speakers and Moderators
L. DeWayne Cecil, U.S. Geological Survey
C. Richard Cothern, U.S. Environmental Protection Agency
J. B. Cowart, Department of Geology, Florida State University
Harry LeGrand, Hydrogeologist
Henry F. Lucas, Argonne National Laboratory
Michael Weber, U.S. Nuclear Regulatory Commission

Author Index

Contents

SESSION III
MINING IMPACTS ON THE OCCURRENCE OF RADON,
RADIUM AND OTHER RADIOACTIVITY IN GROUND WATER

SESSION IV
SAMPLING AND ANALYSIS OF RADON, RADIUM AND
OTHER RADIOACTIVITY IN GROUND WATER

SESSION V
RADON AND RADIUM IN WATER SUPPLY WELLS

SESSION VI
PREDICTIVE MODELS FOR THE OCCURRENCE OF RADON AND OTHER RADIOACTIVITY

SESSION VII
REMEDIAL ACTIONS FOR RADON, RADIUM AND OTHER RADIOACTIVITY

Radon, Radium, and Other Radioactivity in Ground Water

Hydrogeologic Impact and Application to Indoor Airborne Contamination

DEVELOPMENT OF REGULATIONS FOR
RADIONUCLIDES IN DRINKING WATER

C. Richard Cothern, Office of Drinking Water,
U.S. Environmental Protection Agency

INTRODUCTION

The Office of Drinking Water in the U.S. Environmental Protection Agency (EPA) is currently reexamining
existing regulations for radionuclides in drinking
water and is considering the possibility of adding
maximum contaminant levels (MCLs) for uranium and radon.
Background analyses to support this activity include
evaluations of occurrence, exposure, health effects,
monitoring analytical methodology and treatment techniques. A summary of the information available was
published in the Federal Register, September 30, 1986.
This paper will discuss the scientific background data
and its possible contribution to regulation for natural
and manmade radionuclides in drinking water.

THE 1976 INTERIM REGULATIONS

The National Interim Primary Drinking Water regulations (NIPDWR) for radionuclides were promulgated
July 9, 1976 (41 FR 28404). The amendments to the Safe
Drinking Water Act of 1986 have dropped the designation
"interim" and these regulations are now simply called
the National Primary Drinking Water Regulations. The
interim drinking water regulations set maximum contaminant levels (MCLs) for manmade and naturally occurring
radionuclides in drinking water. The interim MCLs
were set at 5 pCi/L for radium-226 and -228 combined,
15 pCi/L for gross alpha particle activity (excluding
radon and uranium) and a dose equivalent of 4 mrem/y

for man-made radioactivity. Uranium and radon were
excluded because of the uncertainties about their
occurrence, toxicity and routes of exposure. A separate standard was set for radium because at the time
of promulgation the EPA believed that radium was the
most radiotoxic of the radionuclides.

The gross alpha particle activity and gross beta
particle activity standards were intended as screening
devices. If the gross beta particle activity is
greater than 50 pCi/L, then drinking water must be
analyzed to determine which particular manmade radionuclides are present. This dose level of 4 mrem/y was
chosen because it was felt that the corresponding
concentrations were as low as reasonably achievable
(ALARA) and well below the 170 mrem/yr maximum dose
recommended in 1961 by the Federal Radiation Council
for the general public.

In general, the interim regulation requires quarterly sampling for one year every four years. For
natural radionuclides, the monitoring strategy involves
testing for gross alpha particle activity. If the
gross alpha particle activity exceeds 5 pCi/L, then the
sample is analyzed for radium-226. If the radium-226
exceeds 3 pCi/L, the sample must be analyzed for
radium-228. Monitoring for manmade radionuclides is
only required of surface water systems serving more
than 100,000 people.

THE AMENDMENTS TO THE SAFE DRINKING WATER ACT OF 1986

Prior to the amendments of 1986, the revised
drinking water regulations were to be developed in two
steps. The first step was called a Recommended Maximum
Contaminant Level (RMCL). The RMCL was a non-enforce-
able goal based on health effects and set at the no
adverse effect level with an adequate margin of safety.
In the case where there is no threshold in the dose
response curve, the RMCL was to be set at zero. After
proposal of the RMCL, a public meeting was to be held
and public input would be sought. After consideration,
the RMCL was to be finalized and the MCL was to be
proposed. The MCL is the enforceable standard and is
to be set as close to the RMCL as feasible. It, too,
would go through a public comment period, followed by
finalization of the MCL. The amendments of 1986 to
the Safe Drinking Water Act have streamlined this
procedure. The goal is now to be called the Maximum
Contaminant Level Goal (MCLG). In addition, the MCLG
and the MCL are to be proposed simultaneously. The
Federal Register notice published on September 30,
1986, was an Advance Notice of Proposed Rule Making
(ANPRM). This advance advance notice summarized

information contained in four health criteria documents
for radium, uranium, radon and manmade radionuclides.
They also summarized summarized the available information on analytical methodology and treatment technology.
Current schedules are to propose MCLGs and MCLs for
radionuclides in drinking water in the late fall of
1987.

OCCURRENCE

Radium-226

The occurrence data available for radium-226 comes
from three sources: the compliance monitoring data
developed under the interim regulations, some limited
studies of East Coast aquifers and an EPA survey of
2,500 public water systems. EPA's Office of Radiation
Programs conducted a nationwide survey in 1980 and 1981
of 2,500 public water supplies in 27 states which
represented 45% of the drinking water consumed in the
United States. The survey had two limitations: it
included samples primarily from ground water systems
serving more than 1,000 people and it also employed a
gross alpha particle screening step of 5 pCi/L. As can
be seen in Table 1, the average population-weighted
concentration in U.S. community water supplies of
radium-226 is in the range of 0.3 to 0.8 pCi/L.

Table 1.   Population Weighted Average Concentration
of Natural Radionuclides in U.S. Community
Drinking Water (for Both Surface and Ground
Water Supplies)

Radionuclides	Average population-weighted concentrations in U.S. community water supplies (average of surface and ground water supplies) (pCi/L)
Radium-226	0.3-0.8
Radium-228	0.4-1.0
Natural uranium	0.3-2.0
Radon-222	50-300
Lead-210	<0.11
Polonium-210	<0.13
Thorium-230	<0.04
Thorium-232	<0.01

Between 300 and 3,000 public drinking water supplies
are estimated to have a Radium-226 concentration
exceeding 1 pCi/L, a level corresponding to projected
excess cancer risks of 1 in 100,000.

Radium-228

Because of the limited available data for
radium-228 occurrence in ground water, calculations of
population-weighted average cannot be made directly.
However, several studies have shown that natural waters
have approximately equal activities of radium-228 and
radium-226. From this and other available information
on radium-228, we estimated that the population-
weighted average radium-228 concentration in public
drinking water supplies is in the range of 0.4 to
1.0 pCi/L. The occurrence data for radium-228 are
skimpy but probably are similar to that of radium-226.
We analyzed the geochemical transport properties of
radium228 in order to determine which counties in the
United States might have high, medium or low concentration levels of radium-228. Those aquifers with
high levels of radium-228 are expected to be granite,
arcosic sand and quartose sandstone aquifers with high
total dissolved solids. Aquifers having low activity
of radium-228 are those whose geology is either carbonate metamorphic rock or quartose sand or sandstone
and basic igneous rocks.

Natural Uranium

Natural uranium contains three isotopes:
uranium-234, uranium-235 and uranium-238. The relative
occurrence of these isotopes are 0.006, 0.72 and 99.27%.
The occurrence of natural uranium in surface and ground
waters was estimated using a data base from the U.S.
Geological Survey's National Uranium Resource Evaluation (NURE) program. In this program, over 34,000
surface water and over 55,000 ground water samples were
analyzed during the late 1970s for natural uranium by
delayed neutron activation analysis. About 28,000 of
these samples were identified as possible drinking
water sources.
Natural uranium concentrations in ground water were
found to be generally higher than in surface water.
The largest reported concentration in either was about
600 pCi/L. Only a few supplies exceeded the level of
50 pCi/L. The arithmetic average for the ground and
surface water samples was 2 pCi/L. The average population-weighted concentration of natural uranium in
drinking water for individual states ranged from less

than 0.1 to 6.7 pCi/L. The expected range of nation-wide average population-weighted occurrence is in the range of 0.3 to 2.0 pCi/L. Between 100 and 2,000 public drinking water supplies are estimated to have concentrations exceeding 7 pCi/L, a level corresponding to a projected cancer risk of one in 100,000, as shown in Table 1.

Radon

Data on the concentrations of radon in ground water from large systems (greater than 1,000 people) has been collected by EPA's Office of Radiation Programs. About 2,500 systems were sampled across the country. In addition, studies have been made of radon occurrence in smaller systems and reported in the proceedings of the Workshop on Radioactive in Drinking Water held in Easton, Maryland in 1983 and published as the May 1985 issue of Health Physics. The average of Radon-222 concentrations in all ground water systems was esti-mated to be in the range of 200 to 600 pCi/L. The range of 50 to 300 pCi/L shown in Table 1 is for all systems -- ground and surface water. Being a gas, radon would not be expected to occur in surface waters. Available data, while limited, confirm that surface water has radon-222 concentrations less than the detectable level (5 to 10 pCi/L).

RELATIVE SOURCE CONTRIBUTION

Table 2 shows the average relative source contri-butions to the daily intake of naturally occurring radionuclides. As can be seen in the table, the contribution of radium-226 from food is about 1 pCi/L, which is about the same as that from drinking water. Similar daily intakes are found for radium-228. For uranium, the daily intake from food is about 1 pCi/day, while that from drinking water is in the range of 0.6 to 4 pCi/day. The primary source of indoor air radon levels is from soil and is roughly an order of magnitude greater than that from drinking water. It is estimated that the radon in indoor air from the drinking water source is in the range of one to seven percent of the total of indoor air levels of radon. Radon gets into indoor air from the drinking water source from showers, baths, dish washers, clothes washers, toilets, and other sources.

Table 2.   Average Relative Source Contribution to
           the Daily Intake of Natural Radionuclides

Radionuclide	Source	pCi/d
Radium-226...	Air	0.007 ....
	Food	1.1 - 1.7
	Drinking water	Generally small from surface supplies
		0.6-2
Radium-228...	Air	0.007 ....
	Food	1.1
	Drinking water	0.6-2
Uranium-234,. Uranium-236	Air	0.0007 ....
	Food	0.37 - 0.9
	Drinking water	0.6 - 4
Lead-210.....	Air	0.3 ....
	Food	1.2 - 3
	Drinking water	<0.02
Polonium-210.	Air	0.06 ....
	Food	1.2 - 3
	Drinking water	<0.02
Thorium-230..	Air	0.0007 ....
	Food	Probably negligible
	Drinking water	<0.06
Thorium-232..	Air	0.0007 ....
	Food	negligible
	Drinking water	<0.02
Radon 222....	Outdoors .... (1.8 Bq/m^3)	970 ....
	Indoors (15 Bq/m^3)(g)	8100
	Drinking water	100-800

HEALTH EFFECTS

At low and medium doses of internally deposited radium, the most severe biological damage is cancer arising in skeletal tissue. For radium-226, two types of malignancy are induced: bone sarcomas and head carcinomas. Among some 3,700 persons in the United States who were exposed to radium-226 and radium-228 in the process of painting watch dials in the early part of this century, from medical administration and other means, a total of 85 cases of bone sarcoma and 36 cases of head carcinoma have been observed.

Ingested uranium goes both to the bone and to the kidneys. There is no direct epidemiology study of the radiotoxicity of uranium. However, it is known that between one and five percent of ingested uranium goes to the bone and deposits in the bone in a similar way to which radium deposits there. Thus, the radio-toxicity of ingested uranium can be estimated using a dosimetric model.

The primary chemical toxic effect of natural uranium is on the kidneys. This has been seen from evidence for over a century from both medical admini-stration to humans and numerous animal studies. Nephritis (inflammation of the kidneys) and changes in urine consumption are clear symptoms. Based on this evidence, the Adjusted Acceptable Daily Intake for uranium is 60 micrograms per liter and is computed by allocation of the Acceptable Daily Intake for a 70-kg adult consuming two liters of water per day. This level is roughly equivalent to 40 pCi/L.

From studies of hard rock miners early in this century it was determined that inhaled radon led to lung cancer. Studies have been made producing dose-response data from four particular groups of miners: the Czechoslovakian uranium miners, the Newfoundland fluorospar miners, the Colorado uranium miners, and the Swedish metal miners. From this dose-response information, the risk and environmental levels can be estimated.

RISK ESTIMATES

Table 3 shows a summary of the risk levels and occurrence for radionuclides in drinking water. The interim MCL for radium is 5 pCi/L and this corresponds to a risk of about 5×10^{-5} per lifetime. The concen-tration of uranium in Table 3 corresponding to this risk level would be about 40 pCi/L, and a concentration for radon at this risk level would be of the order of a few hundred pCi/L. As noted earlier, the average concentration in ground water supplies in the United

Table 3. Summary of Risk Levels and Occurrence
for Radionuclides in Drinking Water*

Estimated lifetime risk level	pCi/L[1]			
	Radium	Radium	Uranium[2]	Radon-222
Risk levels				
10^{-3}	100	200	700	10,000
10^{-4}	10	20	70	1,000
10^{-5}	1	2	7	100
10^{-6}	0.1	0.2	0.7	10
Occurrence: Population- weighted concentration averages				
All supplies .	0.3-0.8	0.4-1.0	0.3-2.0	50-300
Ground water supplies	1.6	1.8	3	approx. 400
Surface water supplies	....	....	1	....
Actual concentration.	0-200	0-50	0-600	0-500,000

* The calculations in the table involve uncertainties of the order of 4 to 5.
[1] Rounded off to one significant figure. Note that the dose level for man-made radioactivity in drinking water under the interim regulations is 4 mrem/year, at the end of 70 years.
[2] Using $f_1 = 0.05$.

States is about a few hundred pCi/L. Thus, if the radon standard were to be set at the same risk level as the current radium standard, roughly 50% of the public ground water supplies in the United States would be out of compliance.

ANALYTICAL METHODS

Table 4 lists the analytical methods for deter- mining concentrations of radionuclides in drinking

Table 4. Analytical Methods for Radionuclides

Method	Has it been validated?*
Radium	
Alpha-emitting radium isotopes (method 903.0)	Yes
Radium-226-radon emanation techniques (method 903.1)	Yes
New York State Dept. of Health (Ra-226 and -228)	No
Total radium (method 304)	Yes
Radium-226 (method 305)	Yes
Coincidence spectrometry	No
Gamma ray spectrometry (Ra-226 and -228	No
Solid state nuclear track detector	Terradex is preparing equivalency test results
Radiochemical determination of Ra-226 in water samples (method Ra-03)	No
Radiochemical determination of Ra-228 in water samples (method Ra-05)	No
Ra-228 by liquid scintillation counting (method 904.1)	No
Radium-228 (method 904.0)	No
Gross Alpha Particle Activity	
Gross alpha and gross beta radioactivity (method 900.0)	Yes
Gross radium alpha screening (method 900.1)	Single lab tested
Gross alpha activity in drinking water by coprecipitation (method 00-02)	Single lab tested (being collaboratively tested
Gross alpha and beta (method 703)	Yes
Gross alpha particle activity (method D-1943)	Yes
Gross Beta Particle Activity	
Gross alpha and beta radioactivity (method 900.0)	Yes
Gross beta particle activity (method D-1890)	Yes

(continued on next page)

Table 4.   Analytical Methods for Radionuclides

(continued from previous page)

Method	Has it been validated?*
Uranium	
Radiochemical (method 906.0)	Yes
Fluorometric (method 966.1)	Yes
Laser induced fluorometry (method 906.2)	No
ASTM method D-2907	Yes
Radon	
Liquid scintillation (including modification using mineral oil so sample can be mailed)	Underway
Solid state nuclear track detector	Underway
Lucas cell	Underway
Man-Made Radionuclides	
Radioactive cesium (method 901.0)	Single lab tested
Gamma emitting radionuclides (method 901.1)	Yes
Radioactive iodine (method 902.0)	Yes
Radioactive strontium (method 905.0)	Yes
Tritium (method 906.0)	Yes
Strontium 89,90 (method 303)	Yes
Tritium (method 306)	Yes
Gamma ray spectroscopy (method D-2459)	Yes

* "Yes" means multi-lab validation.

water. There are several methods available for radium.
Some of these methods have not yet been validated as
officially sanctioned methods. The coincidence method
of McCurdy and Mellor in the gamma ray spectroscopy
methods may well be validated in the near future. The
gross alpha particle activity method for dealing with
high solids has been validated. The uranium laser-
induced fluorometric method will be validated in the
future. The three methods listed for radon, namely
liquid scintillation, solid state nuclear track
detector and Lucas cell methods, are currently being
validated.

TREATMENT TECHNOLOGY

Several methods are available for treating drinking water supplies for radium and have been discussed in previous papers. Recent studies indicate that the techniques available to remove uranium from drinking water supplies are anion exchange, lime softening at high pH and reverse osmosis. The two available methods for removing radon from drinking water are granular activated carbon and aeration.

The Agency has developed preliminary cost estimates for technologies that may feasibly remove radionuclides from drinking water. Depending on amount of water treated, estimated costs range from 30¢ to 80¢/1,000 gallons for cation ion exchange, 30¢ to 110¢/1,000 gallons for iron and manganese treatment and 160¢ to 320¢/1,000 gallons for reverse osmosis.

Preliminary cost estimates for aeration range from 10¢ to 75¢/1,000 gallons for systems serving about 100,000 people and 100-500 people, respectively.

Preliminary cost estimates for removing radon from household drinking water systems by point-of-entry treatment devices are a capital cost of $400 to $800 for granular activated carbon and about $900 for aeration. Operating costs are estimated at $20/year and $80/year, respectively.

CONCLUSION

The U.S. Environmental Protection Agency is in the process of collecting public comments from the September 30, 1986, Advance Notice of Proposed Rule Making. We are also assessing treatment technologies, the cost and feasibility of monitoring and treatment, and alternative monitoring strategies. These analyses will form the basis of the proposal of MCLGs and MCLs for radionuclides in drinking water currently scheduled in the late fall of 1987.

Session I: Geologic and
Hydrogeologic Controls
Influencing Radon
Occurrence

Moderator: Harry E. LeGrand,
Hydrogeologist

GEOLOGIC CONTROLS AND RADON OCCURRENCE IN NEW ENGLAND

Francis R. Hall, University of New Hampshire, Durham, New Hampshire

Eugene L. Boudette, University of New Hampshire, Durham, New Hampshire

William J. Olszewski, Jr., University of New Hampshire, Durham, New Hampshire

ABSTRACT

The predictability of radon production from bedrock is anticipated by a combination of uranium endowment and distribution of uranium into mineral phases. Olszewski and Boudette have synthesized a map of New England at 1:1,000,000-scale which shows 11 rock units. These units are discriminated by geologic factors including measured uranium distribution. The average uranium content varies from very low (< 1 ppm) to very high (> 29 ppm). Uranium in rocks and in soils derived from a rock and radon in air and ground water in the environment should show a direct correlation in most instances. This observation has been borne out by information assembled from air and water sampling particularly in Maine and New Hampshire. Other controls that influence radon production relate to mobility of uranium in ground water. Transport can led to a secondary enrichment of uranium (or radium) which contributes to local high levels of radon. Two-mica granite provides an excellent example because the uranium is nearly all labile, and high radon values occur in ground water associated either with pegmatite or deposits of secondary uranium minerals. Alkalic or calc-alkalic granite is also high in uranium, but it tends to be stabilized within accessory minerals. Thus, the radon levels associated with these rocks are not as high as in two-mica granite. Hydrologic response to a pumping well also has a role in determining radon content of well water. Altering geochemical conditions in the flow field and aquifer stress providing increased rates of radon transport with induced ground-water flow appear to be the primary controls.

INTRODUCTION

Radon is a radioactive, odorless, colorless, noble gas that is chemically inert. It's solubility in water is inversely proportional to temperature.

There are three isotopes of radon arising from the decay chains of uranium-238 (U-238), uranium-235, and thorium-232. They are radon-222, radon-219, and radon 220 respectively. The longer-lived radon-222 (Rn-222) with a half life of 3.82 days from the U-238 decay chain is the isotope of principal interest for epidemiological concerns. Radon has a coefficient of diffusivity in minerals and in non-porous rocks that is negligible. It's diffusivity in dry, porous materials on the other hand is measurable. Radon in standard, dry sand probably migrates less than 30 centimeters in it's lifetime in the absence of a pressure gradient or in a vacuum. The transport of large volumes of radon in natural systems is, therefore, dependent on appreciable ground water mobility and significant atmospheric or human induced pressure gradients. Radium-226 (Ra-226) the immediate parent of Rn-222 is also of considerable interest. Rn-222 decays rather rapidly down through intermediate products to lead-210 and then on to the isotopically stable lead-206. Rn-222 and these intermediate products pose a potential damage to health.

Radon gas (and its progeny) in high concentrations is known to be carcinogenic. Radon gas is dissolved in ground water used in homes and small public water supplies or comes from a direct gas flux into buildings from their geologic substrate. Ingestion of excessive radon and it's progeny from water and inhalation of radon after degassing from water in the environment or from the soil gas flux are, therefore, of concern to public health officials. Health standards have not been set for radon levels in potable water. One curie of radiation is equivalent to that level of radiation emitted from one gram of Ra-226, that is 3.700×10^{10} disintegrations per second. There is a general consensus among workers that 10,000 to 20,000 picocuries (10^{-12}) per liter (pCi/l) of radon gas in ground water is cause for concern and that 100,000 pCi/l or greater probably requires remedial action. Four pCi/l in air is the accepted upper limit of tolerance in homes and public buildings. The level of radon in water required to raise the level of radon in a building over 4 pCi/l has never been established, particularly because variables such as architecture, degassing efficiency, house/atmosphere pressure gradients, and ventilation are difficult to evaluate. We believe, however, that levels of radon approaching 100,000 pCi/l in water supplies are capable of being the principal contributors to excessive radon levels, especially in private homes and other dwelling units.

Past and ongoing studies particularly in Maine and New Hampshire have investigated the role of a variety of factors in the geologic-hydrologic substrate that might influence radon occurrence. These factors have included rock type, geochemical parameters, and the effect of the pumping well or so called pumping stress. The strongest relationship which has merged from these efforts is the one between rock type and radon in ground water. The radon in air from the direct soil gas flux typically into the basement of a dwelling is not always so strongly related to rock type. For example, in some cases radon measured in ground water supplied to a house is less than 10,000 pCi/l, yet the radon in household air is measured to exceed 4 pCi/l. Nevertheless, the general relationship of radon risk or production to rock type is clear. Olszewski and Boudette [1] prepared a 1:1,000,000-scale bedrock map of New England for the U.S. Environmental Protection Agency (EPA) to serve as a planning document for radon research. This map shows only major, regional lithogenetic domains classified according to their analysed or geochemically predictable uranium distribution. The map is not intended to serve any purpose for risk assessment of specific areas or cultural and demographic polygons.

The ultimate source of radon is the radioactive decay of uranium. The occurrence of radon will then, to a first approximation, be correlated with the uranium content of the bedrock. The distribution of uranium in the bedrock is dependent on a combination of the genesis, geochemistry, and mineral content of a rock unit.

URANIUM DISTRIBUTION IN NEW ENGLAND ROCKS

The areal distribution of the major igneous, metamorphic, and layered rock successions of New England
are shown on Figure 1. Olszewski and Boudette have
attempted to predict the uranium endowment of these
categories and thus the related potential for regional
radon production. This map is a derivative product
which combines formal geologic units into lithogenetic
domains. The definition of the domains then permits
application of geochemical models for uranium source
and mobility balanced against depletion and retention
events.

Synthesis of the map required simplification
dictated by the scale of the map (1:1,000,000).
Structural features such as faults, joints, and folds
could not be shown. It is important to point out,
however, that structures related to brittle deformation are, indeed important in the mobility of radon
especially in ground water. Uranium deposits are
known to occur along fissures, unconformities. faults,
and shatter zones in New England. Also, the map does
not accomodate important "young" surficial uranium
deposits (SUDS) in Holocene peat deposits, epithermal
uranium occurrences in Pleistocene deposits, and
Cretaceous regoliths over granitic rocks.

Eleven rock categories are discerned on the map
(Table 1). Any given category incorporates several
different lithologies, and from place-to-place the
ratio of lithologic types can vary greatly. The areal
extent of major scale plutonic bodies are shown accurately, but plutons too small to show at map scale are
blended in with the all-inclusive metamorphic rock
domain. Metamorphic facies cannot be shown accurately
at the scale of the map, but it is well established
that uranium endowment in rocks can be directly correlated with metamorphic grade, at least to a first
approximation. The level of radon in water from
drilled wells generally increases with metamorphic
grade. Extensive areas in Vermont and northern New
Hampshire extending to central and northern Maine are
characterized by relatively lower grade of metamorphism and, on balance, should be correlated with
lower levels of uranium dispersed in rocks. Higher
metamorphic grades in southwestern Maine, central and
southern New Hampshire, and all of southern New
England are, conversely, identified with relatively
higher levels of dispersed uranium.

Comprehensive analysis of uranium in New England
rocks has not been done. Some two-mica granite sheets

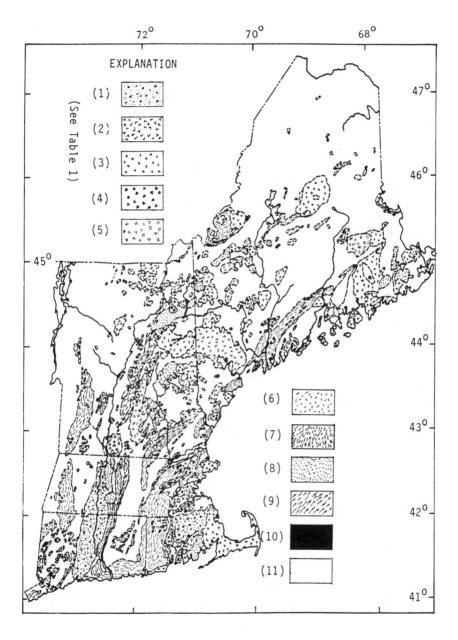

Figure 1.   Index map showing the principal lithogenetic domains in New England.

Table 1. The geologic context and uranium endowment of the principal regional lithogenetic domains in New England.

Domain	Tectonic or Geometric Context	Age	Lithologies	Metamorphism	Deformation	Uranium Endowment
(1) Sedimentary basins of southern New England	Boston basin	Precambrian	Terrestrial and marine clastic rocks with volcanics.	Low Grade.	Minor, if any.	Low to moderate may be high in conglomerate and sandstone.
	Narragansett and Norfolk basins	Carboniferous	" " " " " , coal.	" "		do.
	Hartford and Deerfield basins	Mesozoic	" " " " " with volcanics.	" "	None	do. Hydrothermal uranium deposits along western margins.
(2) Taconica granite	Sheet-like bodies in metamorphic terranes of medium to high grade. Most are < 2cm thick. Abundant minor bodies in Precambrian Z gneiss and migmatite.	Late Ordovician to Permian	Magnetite-type granite with primary muscovite.	Local retrograde.	Minor, if any.	Moderate to very high with local occurrences of uranium secondary minerals abundant. Other uranium occurs in places as dispersed primary oxides.
(3) Calc-alkalic granite	Widespread occurrence varying from thin sheets to batholiths.	Precambrian to Cretaceous	Ilmenite-type biotite or hornblende granite.	Retrograde or low.	Variable directly correlative with age.	Mostly moderate to high but very high in some places. Uranium occurs as: (a) dispersed oxides, (b) intragranular and along grain boundaries, and (c) in accessory mineral substitution sites
(4) Alkalic intrusive suites	Small stocks to batholiths of the White Mountain plutonic succession and in eastern Massachusetts.	Precambrian Z to Cretaceous	K and Na-rich riebeckite granite, syenite, and quartz syenite.	Mostly absent.	Mostly absent, locally moderate.	Moderate to very high, most of the uranium is distributed to refractory accessory minerals. Local uranium occurrences possible.
(5) Calcic to calc-alkalic intrusive suites of intermediate composition	Widespread stocks and sheet-like plutons typically associated with calc-alkalic granite and granitic gneiss.	Precambrian Y to Devonian	Granodiorite, tonalite, monzonite, and quartz monzonite of ilmenite-type.	Absent, locally low.	Variable.	Low to moderate with the uranium mostly distributed to accessories or is held in intragranular space.
(6) Calcic to calc-alkalic intrusive suites of mafic composition	Widespread stocks and sheet-like plutons typically associated with calc-alkalic granite and other crystalline rocks of intermediate composition.	Precambrian Y to Cretaceous	Gabbro diorite monzodiorite, and quartz diorite.	Absent, locally low.	Variable.	Very low to low with most uranium distributed to accessory minerals; some along grain boundaries.
(7) Gneiss and granofels unclassified	Mostly paragneiss with some ortho-gneiss that almost everywhere occurs as elongate bodies trending with the regional tectonic grain. Typically associated with migmatite and granitic to dioritic rocks and many bodies have boundaries that are either faults or an intrusive contact. Some bodies are either uplifted or remobilized basement.	Precambrian to Ordovician	Quartzo-feldspathic to hornblendic-biotitic locally alaskitic; some is dioritic; other variants are blastomylonite and granofels.	Polymetamorphism and retrograde overprints in some bodies.	Moderate to intense.	Variable low to high with a tendency for enrichment in higher metamorphic zones in quartzo-feldspathic rocks. Known uranium vein occurrences and some uranium secondary minerals. Disseminated uranium mostly along grain boundaries.

(8) Granitic gneiss	Irregular-shaped bodies and bodies elongated in the regional tectonic grain that are mostly interpreted to be orthogneiss; all in southern New England.	Precambrian Y to Ordovician	Includes undivided granite, granodiorite migmatite and gneiss of ferro-calcic or ferro-magnesian composition. Local segments of anorexite occur.		do.	Moderate to very high with local occurrences of uranium minerals in veins and secondary deposits possible.
(9) Quartzo-plagioclase amphibolitic and calcic/ferromagnesian gneiss	Mostly orthogneiss with some segments of paragneiss or granitic gneiss. Small to medium-sized bodies usually elongated in the regional tectonic grain.	Precambrian to Devonian	n.a.		do.	Variable; rocks of this domain define a Cu-Mo metallogenic province which could have uranium vein and other epigenetic uranium deposits.
(10) Ultramafic rocks	Semi-autochthonous and diapiric ophiolite associated with accretionary mélange. Vestiges of oceanic crust.	Cambrian to Ordovician	Serpentinite, websterite, and harzburgite with epidiorite and gabbro. Some cumulate, Cr-rich layers.		Mostly retrograde with hydration and propyllitization common.	Low to very depleted.
(11) Stratified metamorphic rocks undivided.	Successions of passive continental margins, extensional basins, island arcs, forearcs, and backarcs. Post-basement host rocks of the Caledonian-Hercynian orogen.	Precambrian Z to Lower Devonian	Protoliths were clastic and chemical sedimentary rocks with felsic to mafic volcanic rocks and epiclastic volcanics principally of marine estuarine or beach environments of deposition. Where metamorphic grades are highest in the domain there are abundant segments of granitic rocks, amphibolite, gneiss, and migmatite that cannot be mapped at scale.	Variable, but generally increases with metamorphic grade.	Variable, but regionally lowest in Maine increasing along the orogenic axis to southern New England (the orogen plunges northeasterly, concomitant with a rising paleocrustal level).	Variable, but directly correlative with increasing metamorphic grade. Uranium mineralization is known to occur in fissures and along faults and unconformities.

and some plutonic rocks in the White Mountain intrusive succession are known in relative detail. The uranium content of many rock units in New England is not known at all. The uranium content of any given lithologic unit can vary widely in space. Weathering generally depletes the uranium in a rock, and depth of weathering varies from unit to unit providing an additional variable. The prediction of uranium content in rock units is, therefore, risky business. Application of geochemical logic combined with knowledge of the genesis of rock domains has been an important influence in the synthesis of the map.

Discrimination of uranium endowment into five uranium content levels was adopted for the map especially in view of the geochemical knowledge (see Table 2). Some calc-alkalic to alkalic rocks of the Jurassic White Mountain succession of New Hampshire, that contain whole rock uranium levels in excess of 20 ppm, have been weighted downward to a rating of moderate to high because most of the uranium is documented to be locked-up in resistate or refractory minerals, where neither it, nor its daughter product radon is available for transport to ground water.

The map was synthesized from state geologic maps (2, 3, 4, 5, 6, 7). The map can be used as a first step in identifying areas of possible or existing radon hazard, and can define areas for future investigation. We emphasize, however, that many factors enter into making radon an environmental pollutant. Ground water flow, geologic structures (faults, folds, joints, etc.), surficial geology, water, and other factors are just as, or even more important than, bedrock geology in determining the potential for radon pollution.

Table   2.   Measured or Predicted Uranium Contents of
              Rock Units, and Relative Rating of Uranium
              Endowment and Radon Production.

Uranium Endowment and Radon Production.	Whole Rock Uranium Content.
(a) Very Low	<1
(b) Low	1-5
(c) Moderate	5-10
(d) High	10-20
(e) Very High	>20

CORRELATION OF RADON IN GROUND WATER WITH ROCK TYPE

Early evidence for high radon levels in ground water of northern New England was presented in 1961 by Smith et al. [8]. This was followed by more detailed studies of radon in ground water and air in Maine by Hess and coworkers [9, 10, 11, 12, 13]. Some sites in New Hampshire were included in this work. The findings in Maine implicated New Hampshire because of the regional geologic continuity. Complementary studies followed with some overlap into Maine by Hall and coworkers [14, 15, 16]. All of these studies demonstrate that high radon values are associated with granitic and high grade metamorphic terranes.

As field evidence accumulated, it became clear that high radon levels were correlated with rock type. In fact, the highest values are associated with two-mica granites (Figure 1, Group 2). Concentrations in ground water in the range of 50,000-100,000 pCi/l are fairly common and values in excess of 1,000,000 pCi/l are documented. Very high levels of radon commonly occur where pegmatite dikes are present in the granite [14, 17]. The two-mica granite is anatectic, formed by the partial melting of some protolith. If this protolith is enriched with uranium, then the partial melting can preferentially fractionate uranium into two-mica magma, thus further enriching it in uranium. Uranium in two-mica granite and associated pegmatite is commonly dispersed uranium oxides such as uraninite and coffinite, and it is nearly all labile. Labile uranium is readily transported by ground water, especially at a pH less than 7, over a wide range of redox conditions. Deposits of secondary uranium minerals are common in two-mica granite.

The alkalic or calc-alkalic granites (Figure 1, Group 3) can also be enriched with uranium, but in this case much of it is stablized in accessory minerals. Therefore, radon concentrations are not as high for the two-mica granites. High grade metamorphics particularly when associated with pegmatites also have fairly high radon. These are not so easily categorized however, because they occur in parts of Groups 7, 8, 9 and 11 (Figure 1). Most other igneous and metamorphic rocks, glacial deposits, and older sedimentary rock tend to have low values of radon, but there are local exceptions.

CORRELATION OF PUMPING STRESS AND RADON LEVEL

The influence of pumping stress on radon level in

well water is a matter for conjecture based upon studies completed to date [18]. These studies present data mainly for radon in drilled wells in bedrock with depths ranging from 100 feet to 500 feet or more. Most dug wells are shallow, and they are in glacial deposits that overlay bedrock. Thus, water from dug wells tends to be low in radon. Some of the earlier work suggested an inverse correlation between radon and both well depth and well yield [9, 15], but these relationships have not always been borne out by more recent work [16].

Six wells in three towns in New Hampshire were sampled at monthly intervals for seven months in the period 1984-85 [15]. The results, however, were somewhat confusing because of the variety of fluctuations that were observed. For example, three wells in one well field in Seabrook showed uniform radon concentrations except for a decrease in one well after an extended shutdown. Two wells in Newfields showed a seasonal recharge pattern with an increase in radon during the unusually dry fall and winter of that period. Radon in one well in Deerfield simply increased with time. The first five wells were in a metamorphic rock and the sixth in a two-mica granite. There were also variations in the type of overburden. Nevertheless, no obvious consistent relationships were found to explain what was happening.

Campisano [19] did short term sampling at a new, unused domestic well where radon first decreased at start up and then increased to a uniform level. Similar results were obtained by Smith et al. [8]. Campisano and Hall [14] report on results of sampling at two neighboring houses and a nearby office. The results were interesting in that radon in the office well was more erratic than for the other two. This may happen because the office well is not used as much as the domestic wells. Also, there seemed to be some kind of relationship between radon variations in the office well and one of the domestic wells. Nevertheless, the results did not provide a clear explanation of what was happening.

URANIUM DISEQUILIBRIUM IN GROUND WATER

Studies in New England [10, 14, 16] show that radon is normally grossly out of equilibrium with respect to Ra-226 and U-238 in solution in ground water. Therefore, radon must come from immobile radium probably stablized as coatings on fractures of the bedrock secondary porosity through which the water flows, possibly according to the model suggested by LeGrand [18]. Studies by Campisano and Hall [14] show positive correlation between Ra-226 and Rn-222 in one two-mica granite where radon was high and radium was rather low. These waters had little or no measurable uranium. Wathen and Hall [16] found measurable uranium in ground waters from four other two-mica granites, but there was no clear relationship to radon. There is some tendency for high uranium values to accompany high radon, but not in a consistent way.

The various findings described above indicate the need for further studies to determine the response of levels of uranium and its progeny within the dynamic regime of the pumping well. This might include investigation of features such as isotope build-up and depletion curves, distribution of dissolved versus adsorbed isotope, and the interrelationships of factors such as well depth, yield, water level, rock type, and the presence of a saline interface.

SUMMARY AND CONCLUSIONS

Geologic information is critical to the understanding of radon occurrence and mobility, in its ultimate delivery into environments at sufficient levels to become dangerous to human health. A comprehensive model for radon geochemistry requires knowledge of (1) the uranium content and distribution and bedrock, (2) the nature of deformation of the bedrock, (3) the textural and mineral characteristics of the bedrock, (4) the nature of unconsolidated deposits which mask the bedrock (including soils of the a & b zones), and (5) the regime of ground water in rocks and soils. Once radon is delivered into an environment devised by man, then physical and engineering, not geological, variables prevail. The rapidly evolving technology of drilling into bedrock to exploit the ground water resource has raised penetrating questions about accelerating radon mobility and raising its levels in ground water far beyond those predicted by natural systems.

Work to date has indicated that there is a strong correlation between rock-type and radon levels in exploited ground water. Research directed into radon abundance and distribution require a fundamental geologic and hydrogeologic components. The synthesis of primary and derivative geologic products that sense the fundamental control of the geologic substrate over the genesis and mobility of radon is essential to any further advances in radon research. The radon genetic model is a complex one; the interrelationships between uranium content in rock, ground water regime. regoliths, sediments, surficial deposits, and soils must be reconciled as an integrated system. The exchange of radon dissolved in ground water into the soil gas of the vadose zone is among several obvious controls that must be evaluated. Geologic-hydrogeologic research must simultaneously accommodate the constraints and boundary conditions dictated by the combined geochemistry of uranium, radium, and radon.

REFERENCES CITED

1. Olszewski, Wm. J., Jr. and E.L. Boudette. _Generalized Bedrock Geologic Map of New England with emphasis on uranium endowment and radon production_ (EPA Open file Map, 1986).

2. Doll, C.G., W.M. Cady, J.B. Thompson, Jr., and M.P. Billings. _Centennial Geologic Map of Vermont_ (Vermont Geological Survey, 1:250,000, 1961).

3. Lyons, J.B., W.A.Bothner, R.H. Moench, and J.B. Thompson, Jr. _Interim Geologic Map of New Hampshire_ (New Hampshire Department of Environmental Service, 1:250,000, 1986).

4. Osberg, P.H., A.M. Hussey, II, and G.M. Boone. _Bedrock Geologic Map of Maine_ (Maine Geological Survey, 1:500,000, 1985).

5. Quinn, A.W. _Bedrock Geology of Rhode Island_, USGS Bull. 1295:68p. with map 1:125,000 (1971).

6. Rodgers. J. _Bedrock Geological Map of Connecticut_ (Connecticut Geological and Natural History Survey, 1:125,000, 1985).

7. Zen, E. _Bedrock Geologic Map of Massachusetts_ (Mass. Dept. of Public Works, 1:125,000, 1983.

8. Smith, B.M., W.N. Grune, F.B. Higgins, Jr. and J.G. Terrill, Jr. "Natural Radioactivity in Ground Water Supplies in Maine and New Hampshire," _Jour. Amer. Water Works Assoc._, 53: 75-88 (1961).

9. Brutsaert, W.F., S.A. Norton, C.T. Hess, and J.S. Williams. "Geologic and Hydrologic Factors Controlling Radon-222 in Ground Water in Maine," _Ground Water_ 19: 407-417 (1981).

10. Hess, C.T., S.A. Norton, W.F. Brutsaert, R.E. Casparius, E.G. Combs, and A.L. Hess. "Radon-222 in Potable Water Supplies in Maine – The Geology. Hydrology, Physics, and Health Effects," Land and Water Resource Center, University of Maine at Orono (1979).

11. Hess, C.T., R.L. Fleischer, and L.G. Turner. "Measurements of Indoor Radon-222 in Maine, Summer-vs-Winter Variations and Effects of Draftiness of Homes," General Electric Corporate Research and Development Technical Series, Report-83CRD278, Schenectady, New York (1983).

12. Hess, C.T., C.V. Weiffenbach, and S.A. Norton. "Variations of Airborne and Waterborne, Radon-222 in Houses in Maine," Environ. Internatl. 8: 59-66 (1982).

13. Hess, C.T., C.V. Weiffenbach, and S.A. Norton. "Environmental Radon and Cancer Correlations in Maine," Health Physics 45: 339:348 (1983).

14. Campisano, C.D. and F.R. Hall. "Controls on Radon in Ground Water – A Small Scale Study in Southeastern New Hampshire," in Proceedings of Association of Ground Water Scientists and Engineers, Third Annual Eastern Regional Ground Water Conference (Dublin, Ohio: 1986) pp. 650-681.

15. Hall, R.R., P.M. Donahue, and A.L. Eldridge. "Radon gas in Ground Water of New Hampshire," in Proceedings of Association of Ground Water Scientists and Engineers, Second Annual Eastern Regional Ground Water Conference (Dublin, Ohio: 1985) pp. 86-101.

16. Wathen, J.B. and F.R. Hall. "Factors Affecting Levels of Rn-222 in Wells Drilled into Two-Mica Granites in Maine and New Hampshire," in Proceedings of Association of Ground Water Scientists and Engineers, Third Annual Eastern Regional Ground Water Conference (Dublin, Ohio: 1986) pp. 650-681.

17. Lanctot, E.M., A.L. Tolman, and M. Loiselle. "Hydrogeochemistry of Radon in Ground Water," in Proceedings of Association of Ground Water Scientists and Engineers, Second Annual Regional Ground Water Conference (Dublin, Ohio: 1985) pp. 66-85.

18. LeGrand, H.E. "Radon and Radon Emanations from
     Fractured Crystalline Rocks - A Conceptual Hydro-
     geological Model," Ground Water 25: 59-69 (1987).

19. Campisano, C.D.   Unpublished results (1986).

BIOGRAPHICAL SKETCHES:

Francis R. Hall received a B.S. in 1949 and an Ph.D. in 1960 in Geology from Stanford University and an M.A. in Geology in 1953 from the University of California at Los Angeles. He has worked for the U.S. Geological Survey, Stanford Research Institute, and the New Mexico Institute of Mining and Technology. He has been at the University of New Hampshire for 22 years where he is a Professor in the Department of Earth Sciences. His research interests are in chemical transport, ground water recharge, and isotope hydrology.

Eugene L. Boudette received a B.S. in Geology in 1951 from the University of New Hampshire and an M.A. in 1959 and Ph.D. in 1970 in Geology from Dartmouth College. He has worked for the U.S. Army Corps of Engineers, and he has spent over 33 years with the U.S. Geological Survey working in mineral resources and regional geology. He is now the New Hampshire State Geologist and an Adjunct Professor in the Department of Earth Sciences, University of New Hampshire. His primary research interests include the geochemistry of granites as an indicator of their genesis, emplacement, and resource endowment.

William J. Olszewski, Jr. received a B.A. in Geology from the University of Pennsylvania in 1973 and a Ph.D. in Geology in 1977 from the Massachusetts Institute of Technology. He is a Research Scientist in the Department of Earth Sciences, University of New Hampshire. His professional interests are in isotope geology geochronology, and geochemistry. He is currently engaged in Rb-Sr, U-Ph, and Sm-Nd geochronological studies of rocks from South America, New England, and Canada.

Acknowledgements

The work of Hall has been supported by the University of New Hampshire Water Resource Research Center and that of Boudette and Olszewski by the EPA and the New Hampshire Department of Environmental Services.

THE EFFECT OF URANIUM SITING IN TWO-MICA GRANITES
ON URANIUM CONCENTRATIONS AND
RADON ACTIVITY IN GROUND WATER

John B. Wathen
Shevenell Gallen and Associates, Inc., Portsmouth, New Hampshire

ABSTRACT

Two-mica granites have been identified as a lithology with the potential for yielding ground water containing elevated levels of Rn-222. Wide variations between individual wells and between average activities for wells drilled into apparently lithologically similar rock bodies suggest complex controls on factors predisposing high radon concentrations in ground water.

Research into factors affecting levels of Rn-222 was conducted on wells drilled into four two-mica plutons in Maine and New Hampshire. The study included the analysis of the well water samples for uranium concentration as well as autoradiographic studies of rock samples from outcrops spatially associated with the specific wells. Photomicrographs of the rocks and the exposed autoradiographs were compared. In all the two-mica rock samples studied, uranium was found to be sited along grain boundaries, in microcracks, and on alteration sites in the rock rather than in discrete mineral phases.

Ground water with the higher concentrations of uranium was found to be associated with the siting of the uranium on altered plagioclase feldspar grains in the granite. Lower uranium concentrations and lower and less variable activities of Rn-222 was associated with the siting of uranium on altered biotite grains. The distinction between these two modes of siting may be attributable to the relative strength with which the uranium is retained in the two types of sites. The uranium is retained on the ferric hydroxide alteration sites on the biotite approximately 10^6 more strongly than on the altered plagioclase.

The relatively loose retention of uranium on altered plagioclase grains provides the basis for mechanisms for the leaching, transport, and reconcentration of uranium. One implication of the existence of accumulations of uranium is that ground water in contact with secondary uranium mineralization will be in contact with the decay products of the uranium. This includes radon, which of all the isotopes in the U-238 decay chain, is the most soluble in ground water.

INTRODUCTION

Studies on the occurrence of Rn-222 in ground water from wells in Maine and New Hampshire (Hall *et al.*, 1985; Brutsaert *et al.*, 1981; Hess *et al.*, 1979) have been consistent in identifying two-mica or Concord granites, as they have been classified in New Hampshire, as a lithology which yields well water with relatively high average radon activities. Two-mica granites are highly felsic and highly differentiated rocks, and the high radon activities are explained in part by the geochemical behavior of uranium in a crystallizing rock body. The large ionic radius and high valence of uranium make it incompatible with the crystalline sites available in early-crystallizing minerals; therefore it remains in the liquid phase of a magma until the very last stages of crystallization (Adams *et al.*, 1959). Two-mica granites are formed in the last stages of crystallization or are the results of early remelting of other geologic materials and typically contain an amount of uranium consistent with this behavior. As radon is a radiogenic daughter product of the uranium decay, the presence of high radon activities in water drawn from wells drilled into two-mica granites is not surprising.

Not all granites containing as great or greater amounts of radioactive elements, however, have been found to yield well water with as high average radon activities. The results of limited sampling of wells drilled into the Conway granite are an example of this (Hall *et al.*, 1985). The Conway granite is a pink biotite granite of the White Mountain magma series, and contains and average of 25 ppm uranium (Boudette, 1987). The two mica granites in this study contain somewhat less than this amount of uranium, in the range of 15 ppm. This suggests that there may be particular qualities of two-mica granites which contribute to the high average levels of radon activity in well water drawn from them.

The Conway granite is much more richly endowed with accessory minerals than are two-mica granites, reflecting a much higher content of rare earth elements. Accessory minerals provide crystal sites for uranium, and the relative lack of such sites suggests a possible explanation for the elevated radon activities observed in ground water from the two-mica granite areas. The purpose of this study was to determine if the uranium in two-mica granites was, in fact, sited outside of mineral phases and to determine if this characteristic of two-mica granites was associated with the high radon activities which had been determined in previous work.

METHODS

Sampling of 109 wells drilled into the Fitzwilliam and Sunapee granites in New Hampshire and the Sebago and Waldoboro granites in Maine (Figure 1) was conducted in the summer and fall of 1985. Wells were selected for sampling in part because of their proximity to good outcrops of two-mica granite. The water samples were analyzed for radon activity using the method described by Prichard and Gessell (1977) and employed in previous work in Maine and New Hampshire (Hall *et al.*, 1985; Brutsaert *et al.*, 1981; Hess *et al.*, 1979). Uranium concentration was determined by the spectrophotometric PAR Method, with higher concentrations being redetermined by Atomic Absorption Spectrophotometry. Other parameters determined included pH, dissolved oxygen, specific conductivity, chloride, iron, and total phosphate.

Thin sections were cut from rock samples collected from the outcrops spatially associated with the well water samples. The preparation was performed in oil so as to prevent leaching of uranium from the samples. The uncovered thin sections were covered with Kodak AR 10 autoradiographic film, which was

Figure 1. Generalized geologic map of the areas studied.

allowed to develope for two weeks under refrigeration and in permanent contact with the thin sections. The autoradiographic film on the thin sections was examined and photographed in reflected light, and the underlying thin section in transmitted cross- or plane-polarized light on a petrographic microscope.

RESULTS

Water Analyses

As previous radon studies in New England have shown, variations of radon activities in wells within the same lithologies are great. The elevated average radon activities of wells drilled into two-mica granites were also consistent with earlier studies.   The average activity level of radon determined for ground water from the wells in the Fitzwilliam granite, although considerable when compared to average levels observed in other lithologies, was the lowest of the four areas studied and exhibited the least variability expressed as standard deviation (Table 1). Radon activities in the Fitzwilliam wells ranged from 1,109 pCi/L to 35,666 pCi/L. The Fitzwilliam wells had an average uranium concentration barely higher than the limit of determination for the PAR method initially employed. Many of the Fitzwilliam wells yielded water with essentially no uranium content, and there were no wells that yielded water with a uranium content of concern with respect to human consumption (Wrenn *et al*, 1985). Such was not the case for waters from wells in the other three granite bodies.

The average uranium concentrations of the water samples from the Sunapee, Waldoboro, and Sebago granites were found to be substantial, and, in individual wells, ranged to extreme levels. The highest average uranium concentrations were found in the well waters from the Sebago granite, as was the highest individual well concentration, 2700 ug/L, +/- 400 ug/L. The most pronounced aspect of the uranium concentrations from these areas was the wide variability. The highest average levels of radon were observed in wells in the Sunapee granite, 31,742 pCi/L, as compared to an average of 18,087 pCi/L for the Fitzwilliam wells. The Sunapee, Waldoboro, and Sebago granites all yielded water with radon activities that can be characterized as high and highly variable.

Table 1.   Summary of radon and uranium values.

Statistical Parameter:	Fitzwilliam	Sunapee	Waldoboro	Segago
Number of Samples:	30	29	24	28
Radon- (pCi/L):				
Mean	18087	31742	27979	26216
Median	18021	24401	15735	17761
Stan. Dev.	+/-11665	+/-27122	+/-33503	+/-24229
Minimum	1109	1454	2931	243
Maximum	35666	104758	145567	104139
Uranium- (ug/L):				
Mean	30	120	148	191
Median	27	59	42	43
Stan. Dev.	+/-15	+/-190	+/-283	+/-508
Minimum	7	0	2	6
Maximum	81	840	1050	2700

Rock Samples

Virtually all the radiation in the rock samples associated with the sampled wells was found to be emanating from areas outside the bodies of individual mineral grains. Development of the autoradiographic film was associated with grain boundaries (Plate 1), microfractures (Plate 2), and at alteration sites on mineral grains.

Greatest concentrations of radiation development in the samples of the Fitzwilliam granite were found to be associated with alteration sites on biotite grains (Plate 3). In the samples from the other three rock bodies, the predominant site of concentration was on alteration sites on plagioclase feldspar (Plate 4). In areas where the mineral grain had been altered to sericite, significant accumulations of uranium had apparently occurred, and was indicated by the presence of pronounced development on the autoradiographic film. In the Waldoboro granite, lesser amounts of the biotite siting were also observed, but siting of radiation development on plagioclase grains in the Fitzwilliam virtually was absent.

DISCUSSION AND INTERPRETATION

The siting of the uranium in two-mica granites along grain boundaries and in interstitial spaces rather than within accessory minerals is a phenomenon which has been observed in other granites (Guthrie and Kleeman, 1986; Andrews et al., 1982). The suggestion that this was the case for the two-mica granites of the New Hampshire had been made by J. B. Lyons (1964). This research was formulated based on Lyon's hypothesis as an explanation for the high radon activities observed in ground water from wells drilled into this lithology. The mode of siting of the uranium observed in the rocks studied is consistent with the high radon activities observed in this and other studies. The distinction between the sites of concentration in the various rock units, however, offers a new insight into the factors controlling uranium mobility and accumulation, with resultant implications for uranium concentrations and radon activities in well water.

The Effect of the Uranium Mobility Cycle on Radon Activities

The siting of uranium outside of mineral grains was characteristic of the four rock bodies studied, and may be associated with the high average radon activities characteristic of two-mica granites. Uranium concentrations on biotite grains where presumably the grain has been altered to ferric hydroxides as a results of a deuteric process are strongly bound in the alteration site. The strength with which uranium is bound on ferric hydroxides is approximately 10^6 greater than the strength with which uranium is bound on the clay-like sericite plagioclase alteration products. Where the uranium concentrations are bound in the stronger biotite sites, as in the Fitzwilliam granite, less variability in radon activities and uranium concentrations in ground water are observed than where uranium is bound in the easily leached plagioclase sites.

Uranium sited on altered plagioclase is subject to leaching by a number of chemical variables, *e.g.*, increased ionic strength and the presence of humic acids or other changes in pH. Once uranyl species are dissolved and mobile in ground water, they are available for subsequent reconcentration. It appears likely that reconcentrated uranium and the leaching of secondary reconcentrated deposits are the source of the high uranium concentrations detected in a significant number of the wells tested in this study.

Secondary uranium mineralization observed in the Interstate 89 roadcut near New London, New Hampshire, offers an opportunity to examine a likely

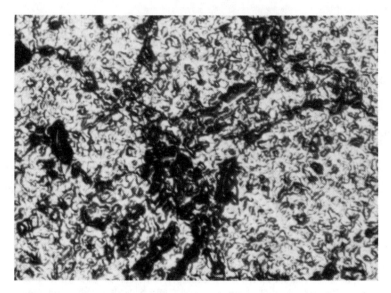

Plate 1A. Autoradiograph of radiation development along grain boundaries (Sunapee granite).

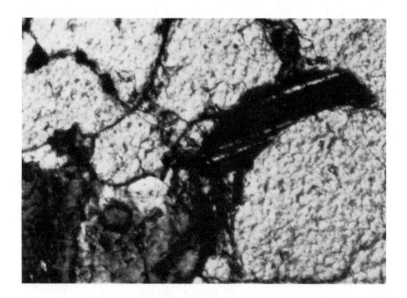

Plate 1B. Photomicrograph of underlying thin section.

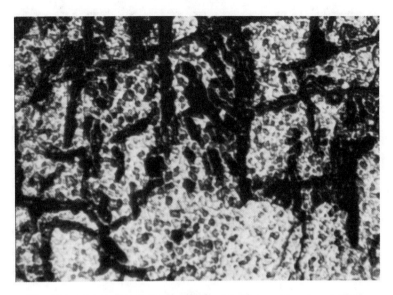

Plate 2A. Autoradiograph of radiation development in microcracks (Waldoboro granite).

Plate 2B. Photomicrograph of underlying thin section.

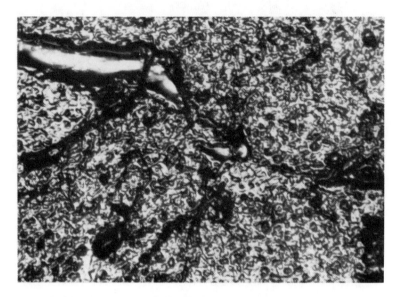

Plate 3A. Autoradioraph of radiation development on an altered
biotite grain (Fitzwilliam granite).

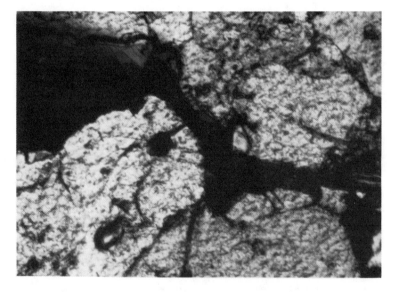

Plate 3B. Photomicrograph of underlying thin section.

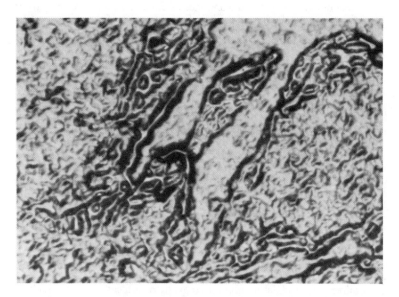

Plate 4A. Autoradiograph of radiation development on an altered
plagioclase feldspar grain (Sebago granite).

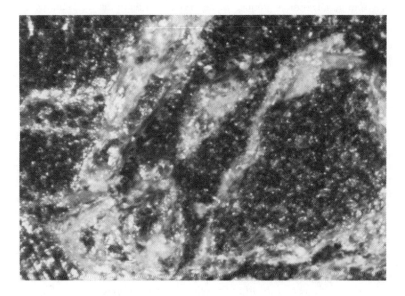

Plate 4B. Photomicrograph of underlying thin section.

sequence of conditions that led to the presence of uranium in water from wells in the Sunapee granite. At the time of the initial emplacement of the granite body, approximately 360 million years ago (Hayward, 1983), magmatic conditions may have been such that primary uranium minerals crystallized. Subsequent reheating and/or the concentration of volatile components such as CO_2 and water (Fehn et al., 1978), or other changes in magmatic conditions apparently caused the uranium to become mobile once again in the liquid phase of the magma. At this point the solid phases of the rock were essentially stewing in a soup of silica, volatiles, and incompatible elements, including uranium.

As the rock body cooled further, the uranium remained without specific mineral sites. Accumulations occurred preferentially where weak surface bonds were available on sericite, which formed as plagioclase grains were altered by the magmatic conditions which had become incompatible with their stability. The remaining uranium, finding no site at all, was essentially left as a coating on the outside of the mineral grains in the rock.

As the erosion of the approximately 11.5 km of originally overlying material brought the rock unit closer to the surface, oxygenated meteoric water circulated into the fractures in the rock, slowly dissolving that portion of the uranium which came in contact with the water. A substantial portion of this uranium was likely conveyed out of the rock unit, while that remaining within the rock unit may have reconcentrated at a redox barrier where the uranium was reduced and fixed as ground water descended. Relatively recent (Quaternary) unloading fracturing at depths close to the depth ranges of modern water supply wells would likely further mobilize intergranular uranium from exposed surfaces within the rock body, as well as from other secondary accumulations.

The depth of the secondary uranium mineralization in the I-89 road cut suggests that this accumulation occurred in the vicinity of the water table prior to the excavation of the road cut (Bothner, 1978), and it seems unlikely that this instance of mineralization is unique. Since the excavation of the road cut, most of the uranium mineralization at the site has dissolved, underscoring the fragile and leachable form that uranium takes under these conditions. The fate of releached material is that part will no doubt be transported to the sea; however, some will likely go through another process of redeposition (Johnson et al., 1987; Cameron and Boudette, 1984; Osmond and Cowart, 1976), and at least a portion of it will end up being consumed by thirsty well owners.

Of the wells sampled in this study, 18 % were found to contain over 100 ug/L of uranium. This is a clear indication that a substantial amount of dissolved uranium is circulating in the bedrock in these areas. High radon activities, however, do not occur as a result of the presence of parent radionuclides in the ground water, but as a direct result of a proximate accumulation of parent radionuclides. The transport distance of radon in the ground with respect to its half life and Darcian factors is very limited (Tanner, 1964). For a high level of radon to be observed in a water supply well, an accumulation of radium must be nearby. On the basis of information gathered in this study, no opinion can be formed as to whether the radium is a decay product of a nearby uranium accumulation, or whether the nearby radium is a function of the chemical or physical environment of the well itself. Metallic radium is co-precipitated with the ferric hydroxides that are often precipitated in the relatively oxygenated environment in the vicinity of well casings. Uranium is also co-precipitated under such conditions, but the amount of time required to approach secular equilibrium to a point at which high levels of radon would be produced is outside the range of the timespan of the life of a well. This would limit the role of uranium fixed in a well in producing radon anomalies unless very great amounts of uranium were so fixed and the U-234/U-238 were beyond the range of values observed in natural waters (Osmond and Cowart, 1976). This leaves radium accumulation at wells as a potential cause of radon anomalies.

Proximity of a well to a natural secondary uranium accumulation from any of the mechanisms described above, and some not described above, however, remains a viable potential source of radon anomalies. The two highest uranium concentrations in well water determined in the course of this research were from wells in Maine. A uranium concentration of 1050 ug/L was determined for a well in the Waldoboro granite, and a uranium concentration of 2700 ug/L was determined for a well in the Sebago granite. Radon activities for these two wells were both above 100,000 pCi/L, and a factor common to both wells may be proximity to the ocean. The Waldoboro well is located within a few hundred yards of an arm of the ocean, and the Sebago granite well is located on what was the shores of a cove at the height of the Holocene marine transgression.

The extreme uranium concentrations of water drawn from both these wells, as well as their highly-elevated radon activities, suggests two things. First, there is clearly an accumulation of uranium close to both these wells. This uranium is in such a form that it is highly leachable and, in fact, is being leached under the currently prevailing geochemical conditions near the wells. Secondly, the radon activities of the water from both these wells suggest that the uranium accumulations have existed for a considerable length of time, more on the order of geologic time than three-score-and-ten time, and that radon is therefore being produced in considerable quantities from the decay of intermediate isotopes.

The inferred association of these wells with proximity to the present or past seacoast invites speculation as to what quality of the sea water, or more likely the fresh water/salt water interface, may have contributed to the accumulations of uranium that apparently have occurred. Since the salinity of ocean water, and its higher ionic strength tend to increase leaching rather than accumulation of uranium minerals, the existence of another mechanism of uranium fixation in this area must be postulated. The dominant influence of redox potential on the mobility of uranium species suggests it as a possible factor relating to these accumulations. Perhaps microbial action associated with the presence of sea water in near-stagnant ground water conditions in the vicinity of the fresh water/salt water interface created a reducing environment that caused uranyl species to be reduced, fixed, and accumulated. The reader is encouraged to substitute an alternate hypothesis that explains the extreme uranium concentrations and radon activities determined for the water from these two wells.

Conceptual Model for Observed Radon Activities

As suggested earlier, and as suggested by a correlation coefficient between radon activity and uranium concentration of only slightly over 0.5 for the combined data, the presence of a high radon activity is indicative specifically of a radium concentration close to a well rather than the accompanying presence of the parent nuclides of radon in the well water. The very limited solubility of radium, however, has generally restrained workers from considering its transport and accumulation as a likely cause of the creation of radon anomalies (Lancetot *et al.*,1985). It must be considered that only 3.6×10^{-7} as much metallic radium as U-238 need be accumulated in the vicinity of a well to be equivalent in terms of radon production to the uranium under conditions of secular equilibrium, and secular equilibrium of the U-238 decay takes a very long time to be reached.

Nevertheless, the mobility of hexavalent uranium and, in the case of the Sunapee, Sebago, and Waldoboro granites, the apparent loose retention of uranium accumulations on alteration sites within the rock would seem to make unlucky proximity of a well to an area of secondary uranium mineralization a likely cause for a high radon activity in the well.

The fact that such a wide range of uranium concentrations was found to correlate so weakly with such a wide range of radon activities for the combined data from the four rock units suggests that locally varying conditions are

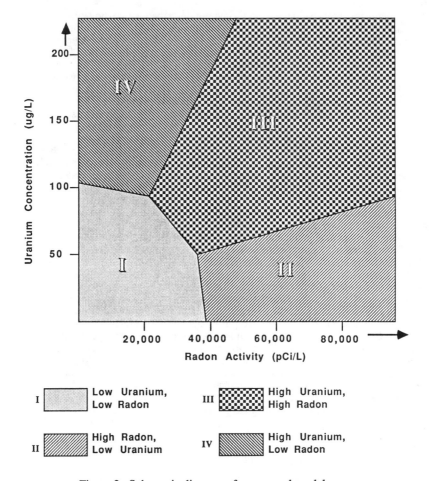

Figure 2. Schematic diagram of conceptual model.

controlling the observed levels of both parameters. As a conceptual model, Figure 2 depicts four fields, with nominally chosen limits. Field I defines a stable environment in which uranium is fixed and radon activities are relatively low. Field II is an area of elevated radon activities but low uranium concentration and defines a field of radon parent nuclide accumulation. Because of the time required to approach secular equilibrium, the accumulating parent nuclide may more likely be radium for wells in this field. Field III is an area characterized both by high radon activities and uranium concentrations. This implies the proximity of the well to an actively leaching uranium accumulation of geologic age. Field IV defines an area of high uranium concentration unaccompanied by a commensurate radon activity. Wells falling within this field are apparently down the hydraulic gradient from a leaching uranium accumulation, but are beyond the decay time/distance range, so to speak, of transported radon.

A comparison of these fields with the data from the four field areas reveals that the Fitzwilliam wells all lie within Field I. The wells from the other three

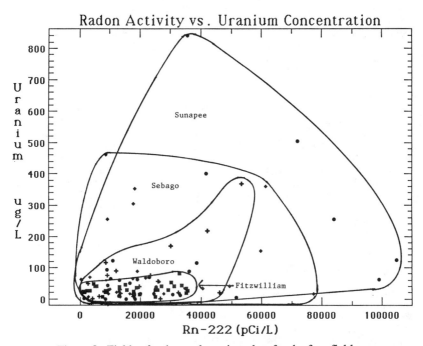

Figure 3. Fields of radon and uranium data for the four field areas.

areas are not similarly confined. The data field for the Sunapee granite is broad, but is skewed towards the accumulation side of the graph. The data field for the Sebago wells is also broad, but is offset towards the mobility side of the conceptual diagram. Although the two extreme wells in the Sebago and Waldoboro granites have been omitted from this plot for reasons of scaling, the trend of the Waldoboro field is clearly more constrained than the Sunapee or Sebago data, reflecting the high correlation coefficient between radon activity and uranium concentration for that area.

A correlation at greater than r=0.7 between chloride and both uranium and radon for the Waldoboro wells, along with the presence of some of the biotite siting in the rock samples of the Waldoboro granite, suggest an explanation for the more linear pattern exhibited by the Waldoboro uranium and radon data. Where the uranium is somewhat more tightly held, higher ionic strength in the form of an elevated chloride content may be mobilizing uranium. This would be consistent with the field of the Waldoboro data in Figure 3, which is midway between the tight field of the Fitzwilliam data and the broad variability exhibited by the Sunapee and Sebago data.

SUMMARY AND CONCLUSIONS

The siting of uranium in the two-mica granites studied was found to be outside of mineral grains, rather than within primary uranium minerals or accessory minerals, where uranium is sited in many granitic rocks. This is consistent with the elevated radon activities that previous studies have associated with this lithology. A distinction was observed between the siting of concentrations of what apparently was,or had been, mobile uranium in the Fitzwilliam granite and the other three rock units studied. Uranium was concentrated on altered biotite grains in the Fitzwilliam granite and on altered plagioclase feldspar grains in the Sunapee, Sebago, and Waldoboro granites. Clay-like materials such as the sericite alteration product of plagioclase only weakly adsorb uranium species, leaving uranium which had been so sited particularly available to a wide variety of remobilization and accumulation processes.

Two families of factors may be affecting these remobilization and accumulation processes. Natural factors include varying conditions brought about through erosion, sea level changes, and varying degrees of oxygenation. In addition, where water containing mobile uranium is tapped by a bedrock well, the geochemistry of the well environment may be influential in affecting the levels of radon activity resulting directly or indirectly from the presence of the uranium. The availability of uranium for processes of accumulation with which the higher and more variable radon activities and uranium concentrations are associated appears to be dependent on the nature of the original site of adsorption of the uranium within the rock.

List of References:

Adams, J.A.S., J.K. Osmond, & J.J.W. Rogers (1959) The geochemistry of thorium and uranium; in Physics and Chemistry of the Earth, L.H. Ahrens, F. Press, K. Rankama, & S.K. Runcorn, eds/ Pergamon Press.

Andrews, J.N., I.S. Giles, R.L.F. Kay, D.J. Lee, J.K. Osmond, J.B. Cowart. P. Fritz, J.F. Barker, and J. Gale (1982) Radioelements, radiogenic helium and age relationships for the granites at Stripa, Sweden. Geochim. et Cosmochim. Acta, v.46, p. 1533-1534.

Bothner, W.A. (1978) Selected uranium occurrences in New Hampshire. U.S. Geological Survey open file report 78-482.

Boudette, E.L. (1987) Personal communication.

Brutsaert, W.F., S.A. Norton, C.T. Hess, & J.S. Williams (1981) Geological and hydrologic factors controlling Radon-222 in ground water in Maine. Groundwater, v. 19, n. 4, p. 407-417.

Cameron, C.C., & E.L. Boudette (1984) Movement and concentration of uranium in peat deposits at New London, New Hampshire.

Fehn, U., L.M. Cathles, & H.D. Holland (1978) Hydrothermal convection and uranium deposits in abnormally radioactive plutons. Economic Geology, v.73, p. 1556-1566.

Guthrie, V.A., & J. D. Kleeman (1986) Changing uranium distirbutions during weathering of granite. Chemical Geology, v. 54, p. 113-126.

Hall, F.R., P.M. Donaue, & A.L. Eldridge (1985) Radon gas in ground water in New Hampshire. Proceedings of The Second Annual Eastern REgional Ground Water Conference, p. 86-100. National Water Well Association, Worthington, Ohio.

Hayward, J.A. (1983) Rb-Sr Geochronology and the evolution fo some "peraluminous" granites in New Hampshire. Master's thesis, University of New Hampshire.

Hess, C.T., S.A. Norton, W.F. Brutsaert, R.E., Casparius, E.g. Coombs, & A.C. Hess (1979) Radon-222 in Potable water suppleis in Maine: The geology, hydrology, physics and health effects. Land and Water Resoruces Center, University of Maine at Orono, 119 p.

Johnson, S. Y., J.K. Otton, D.L. Macke (1987) - Geology of the holocene surficial uranium deposit of the north fork of Flodelle Creek, northeastern Washington GSA Bulletin. V89, n. 1 p. 77-85.

Lancetot, E.M., A.L. Tolman & M. Loiselle (1985) Hydrogeochemistry of radon in ground water; in Proceedings of the Second Annual Eastern Regional Ground Water Conference, P.66-85. National Water Well Association, Worthington, Ohio.

Lyons, J.B. (1964) Distribution of thorium and uranium in three early Paleozoic plutonic series of New Hampshire. U.S. Geological Survey Bulletin 1144-F.

Osmond, J.K. & J. B. Cowart (1976) The theory and uses of natural uranium isotpoic variations in Hydrology. Atomic Energy Review, v.14, p.621-680.

Prichard, H.M. & T.F. Gessell (1977) Rapid measurements of 22Rn concentrations in water with a commerical liquid scintillation counter. Health Physics, v.33, p. 577-581.

Tanner, A.B. (1964) Radon migration in the ground: A Review; in The Natural Radiation Environment, Adams, J.A.S. and W.M. Lowder, eds., Univ. of Chicago Press, p. 161-190.

Wrenn, M.E., J. Lipsztein, P. Durbin, E. Still, D.L. Willis, B. Howard, & J. Rundo (1985) U and Ra metabolism. Health Physics, v.48, p.601-634.

The researched described in this report was conducted as a Master's thesis research project at the University of New Hampshire under a grant to the author from the Argonne National Laboratory / U.S. Department of Energy.

Other aspects of the research are described in "Factors affecting levels of Rn-222 in wells drilled into two-Mica granites in Maine and New Hampshire" (1986), Proceedings of the Third Annual Eastern Regional Ground Water Conference, Springfield, Ohio. National Water Well Association, Dublin, Ohio.

Biographical Sketch

John B. Wathen received a B.A. in Geology from Northeastern University in 1984 following 15 years in agriculture. He was awarded a M.S. in Geology in June of 1986 from the University of New Hampshire, where he conducted research on radon in granites under a grant from the Argonne National Laboratory and the U.S. Department of Energy.

He is currently an associate and hydrogeologist with Shevenell Gallen and Associates, Inc., of Portsmouth, New Hampshire.

Mailing Address:

John B. Wathen
Shevenell Gallen and Associates, Inc.
PO Box 433
Portsmouth, New Hampshire 03801

603-436-1490

SOURCE AND DISTRIBUTION OF NATURAL RADIOACTIVITY IN GROUND WATER IN THE NEWARK BASIN, NEW JERSEY

Otto S. Zapecza and Zoltan Szabo,

U.S. Geological Survey, West Trenton, New Jersey

ABSTRACT

Elevated levels of naturally occurring radionuclides in ground water are associated with uranium enrichment in the Newark Basin of New Jersey. The factors controlling the concentration, distribution, and migration of these radionuclides in ground water are being studied by the U.S. Geological Survey, in cooperation with the New Jersey Department of Environmental Protection, Division of Water Resources. Ground water from 260 sites in the basin was analyzed to determine the distribution of gross alpha-particle radiation. High levels of gross alpha radiation (greater than the 15 picocuries per liter maximum contaminant level established by the U.S. Environmental Protection Agency) were found predominantly in ground water near the contact of the Lockatong and Passaic Formations along the southeastern part of the basin, and in the Hopewell and Flemington fault blocks, where these formations are repeated. The source of the radioactivity has been determined by borehole geophysical testing and analysis of lithologic cores. Natural gamma-ray logs of wells near the Lockatong-Passaic contact depict thin but laterally extensive zones of high radioactivity. Analysis of lithologic cores of these zones indicates that uranium is concentrated in black mudstones that contain

abundant pyrite mineralization. The color and
mineralogy of the radioactive beds suggests that the
uranium was deposited in a reducing environment. This
is consistent with the geochemical behavior of uranium
precipitating in reducing environments and mobilizing
in oxidizing environments.

The uranium-bearing black mudstones, which are
common in the Lockatong and lower Passaic Formations,
are the primary source of radionuclides in the ground
water of the basin. The upper part of the Passaic
Formation, which is composed primarily of red shale and
sandstone, contains ground water with relatively low
concentrations of radionuclides. Localized and widely
scattered occurrences of elevated radioactivity in
ground water are found in the sandstone of the Stockton
Formation. Ground water from the basalt and diabase
aquifers in the basin contains low levels of
radionuclides.

INTRODUCTION

There is evidence that significant uranium
mineralization has occurred in the rocks of the Newark
Basin. Uranium enriched rocks can be a source of
radioactivity in ground water. Elevated activities of
radioactive isotopes, such as radon, radium, and
uranium, dissolved in ground water are known to be
carcinogenic. In addition, the isotopes are colorless,
odorless, tasteless, and cannot be detected by human
senses, unlike many water pollutants that impart
undesirable colors, odors, and tastes to the water.
The Newark Basin is a highly populated area in the
northeastern corridor of the United States with
significant ground-water use from fractured rock
aquifers. An estimated 90 million gallons per day of
ground water is pumped to serve part of a population of
more than 5 million.

Normally, radioactive substances are not analyzed
for in water-quality monitoring programs; they are
usually detected only during specific studies geared to
determine their presence. The U.S. Geological Survey
in cooperation with the New Jersey Department of
Environmental Protection, Division of Water Resources
is currently conducting a study to define the extent of
ground-water contamination from naturally occurring
radioactive substances in the New Jersey part of the
Newark Basin. The objectives of the study are to: (1)
define the occurrence and distribution of radionuclides
in ground water and identify areas, aquifers, or parts
of aquifers where elevated levels may be expected; (2)
identify nonradioactive constituents that might serve

as indicators of radionuclide presence in ground water; and (3) identify geological and geochemical factors controlling radionuclide distribution, migration, and concentrations in ground water.

The study area (fig. 1) was chosen based on previous evidence of uranium mineralization; known radioactive-ground-water anomalies; locally high, surficial radioactivity mapped by aeroradioactivity studies; and zones of high gamma activity in the subsurface, as recorded by borehole geophysical logs.

In 1985 and 1986, 260 ground-water samples were collected from the study area and analyzed for gross alpha- and beta-particle activity. Specific conductance, alkalinity, dissolved oxygen, pH, and Eh were measured on site. Selected samples were also analyzed for radium-226, radium-228, uranium, radon-222, and trace metals. In conjunction with the ground-water sampling, borehole geophysics and rock coring were used to identify specific zones of radioactivity in subsurface rocks.

This paper presents information on the geological factors contributing to the distribution of elevated levels of naturally occurring radioactivity in ground water in the Newark Basin, New Jersey. A companion paper in this volume [1] presents significant findings on some of the geochemical factors controlling specific radionuclides in ground water of the basin.

THE STUDY AREA

The Newark Basin in New Jersey (fig. 1) coincides with the Piedmont Physiographic Province and comprises a 10 county area of approximately 2,000 square miles. The basin is exposed at land surface in a band 16 to 32 miles wide trending southwestward from the New York-New Jersey border to the Delaware River. It is bordered on the northwest by the Precambrian and Paleozoic crystalline and sedimentary rocks of the New Jersey Highlands, also known as the Reading Prong, and on the southeast by the unconsolidated deposits of the Coastal Plain. The study area accounts for one-fifth of the State's land area as well as approximately two-thirds of its population.

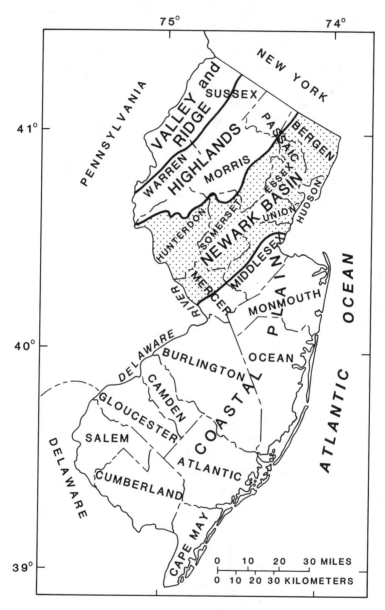

Figure 1. Location of study area, Newark Basin, New Jersey.

GEOLOGIC SETTING

The Newark Basin is the largest of 20 exposed fault-block basins that crop out in a sinuous belt more than 1,000 miles long from Nova Scotia to South Carolina (fig. 2). These basins were formed in Triassic and Jurassic time during initial rifting of the continents [2,3]. They subsequently were filled with thick sequences of continental sediments along with interbedded lava flows and intrusives, now termed the Newark Supergroup [4]. The Newark Basin contains the thickest sedimentary sequence of any exposures of the Newark Supergroup. According to Van Houten [5], Newark Supergroup in the basin attains maximum thicknesses of 16,000 to 20,000 feet. The source of sedimentation in the basin was erosion of Precambrian and Paleozoic highlands that surrounded the basin. Sediment dispersal was in all directions during basin filling [6]; however, the principal source of sediment was the crystalline highlands to the southeast [5].

A generalized geologic map of the New Jersey part of the Newark Basin is shown in figure 3. The structure of the basin is such that progressively younger rocks are exposed toward the northwest. Post-depositional faulting has repeated part of the stratigraphic sequence in outcrops in the Hopewell and Flemington fault blocks northwest of the faults in the southwestern part of the study area. These fault blocks are located approximately at the geographic center of the Newark Basin. The three major sedimentary units in the Newark Basin are the Stockton, and Lockatong Formations and the Passaic Formation of Olsen, 1980 [7]. In figure 3, the Jurassic sedimentary units that are interbedded with and overlie the basalt flows in the northwestern part of the basin have been termed "undifferentiated". The Passaic Formation and the "undifferentiated" unit were formerly called the Brunswick Formation [7].

The Triassic Stockton Formation crops out along the southeastern part of the study area and in the Hopewell and Flemington fault blocks. Its dominant lithology is gray- and buff-colored arkose and arkosic conglomerate, red arkosic sandstone, and siltstone [8]. The Stockton Formation, which is the oldest and most widespread unit in the subsurface, unconformably overlies Precambrian and Paleozoic basement rocks. Along the southeastern edge of the study area, the coarse fraction of the formation was deposited by streams with steep gradients [8]. Conglomerate and coarse sandstone generally grades into finer-grained sandstone and siltstone toward the center of the basin and may represent a facies change to a marginal lacustrine deposit [9]. The Stockton Formation has a maximum thickness of 5,000

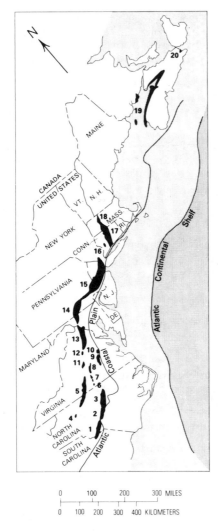

EXPLANATION

1. Wadesboro (N.C. – S.C.)
2. Sanford (N.C.)
3. Durham (N.C.)
4. Davie County (N.C.)
5. Dan River and
   Danville (N.C. – Va.)
6. Scottsburg (Va.)
7. Basins north of
   Scottsburg (Va.)
8. Farmville (Va.)
9. Richmond (Va.)
10. Taylorsville (Va.)
11. Scottsville (Va.)
12. Barboursville (Va.)
13. Culpeper (Va. – Md.)
14. Gettysburg (Md. – Pa.)
15. Newark (N.J. – Pa. – N.Y.)
16. Pomperaug (Conn.)
17. Hartford (Conn. – Mass.)
18. Deerfield (Mass.)
19. Fundy or Minas
    (Nova Scotia – Canada)
20. Chedabucto (Nova
    Scotia – Canada)

```
0        100       200       300 MILES
├──┬──┬──┬──┬──┬──┬──┤
0   100  200  300  400 KILOMETERS
```

Figure 2. Exposed basins of the Newark Supergroup
in eastern North America. (From
Froelich and Olsen [4])

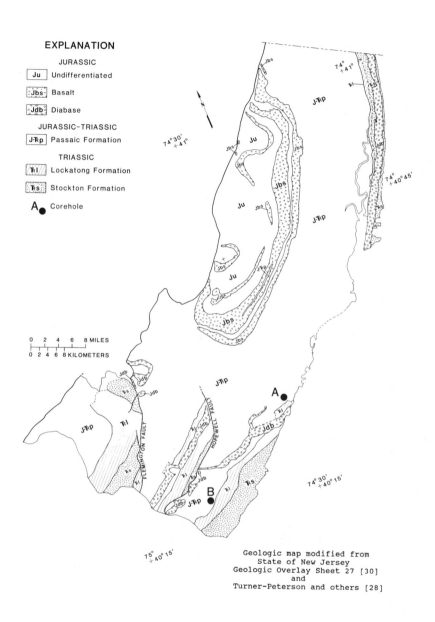

Figure 3. Generalized geologic map of the Newark Basin,
New Jersey.

feet in the Flemington fault-block in the southwestern part of the study area [5].

The Triassic Lockatong Formation conformably overlies and interfingers with the Stockton Formation over most of the Newark Basin. The Lockatong Formation is composed primarily of alternating sedimentary cycles of gray, green, and black mudstones, siltstones and minor carbonates that reflect the expansion and waning of an extensive lake [5]. The deepest part of the Lockatong lake was in the center of the basin. The formation is approximately 4,000 feet thick in the Flemington fault block and thins in all directions from the center of the basin [5]. In the northeastern end of the basin where the unit is thinnest, the Lockatong consists of darker colored mudstones that interfinger with beds of buff arkosic sandstone that are virtually indistinguishable from the sediments of the Stockton Formation [8]. Wider fluctuations of the Lockatong lake are noted in the upper part of the Lockatong Formation and continue into the lower part of the Passaic Formation. In this part of the stratigraphic section, extensive interfingering of both oxidizing and reducing environments are indicated by alternating deposits of nearshore red mudstones with offshore gray and black mudstones. The interfingering of these oxidation-reduction environments plays an important role in the formation of uranium enriched zones.

The boundary between the Triassic and Jurassic Passaic Formation of Olsen 1980 [7] and the Lockatong Formation is so transitional that it is defined both horizontally and vertically on the proportion of the red beds to the gray and black beds [8]. Overlying this transition zone, the Passaic Formation is composed predominantly of reddish-brown mudstone, siltstone, and sandstone. It is the most widespread unit cropping out in the basin. The depositional environment of the Passaic Formation was originally lacustrine and subsequently gave way to broad weakly drained mud flats and playas [5]. The thickness of the Passaic Formation in the southwestern part of the study area is approximately 6,000 feet [5].

During Early Jurassic time, continued crustal extension was accompanied by emplacement of diabase intrusives along the eastern and southwestern part of the study area and basaltic lava flows were extruded over the northwestern part [10].

Jurassic sedimentary deposits that are interbedded with and overlie the basalt flows record a return to cyclic lacustrine-nearshore deposits. Although these deposits are shown in figure 3 as being undifferentiated, Olsen [7, 8, 11] has divided these deposits into formal units based on lithology and biostratigraphy.

THE GROUND WATER SYSTEM

Ground water in formations within the Newark Basin
is stored and transmitted primarily through a complex
network of interconnected openings formed along joints,
fractures, faults, bedding plane partings, and, to a
much lesser extent, interstitial pore space and
solution channels. The intervening unfractured rock
has a negligible capacity to store and transmit water.
The primary porosity that originally existed in the
rocks has mostly been eliminated by compaction and
cementation. The openings, which contain ground water,
decrease in size and number with increasing depth below
land surface. Estimates of the thickness of the
ground-water producing zone in the Newark Basin
formations have been based typically on review of
drilling records. A well that has not successfully
tapped a water-yielding zone in the first 500 feet of
drilling is not likely to penetrate water-yielding
zones by deeper drilling.

Differences in hydraulic head in water-bearing
openings causes flow to occur from openings with higher
heads to openings with lower heads. In wells open to
multiple fractures and joints, water will flow within
the well under nonpumping conditions in response to
head differences between individual water-bearing
zones. In upland areas where recharge is dominant,
hydraulic heads generally decrease with increasing
depths and in unscreened nonpumping wells flow is
downward. In lowland areas where discharge is
dominant, heads increase with increasing depth and flow
is upward in such wells.

Secondary openings in consolidated rock aquifers
generally have some preferential alignment, giving some
degree of anisotropy to the ground-water system. In
wells open to the Passaic Formation, Herpers and
Barksdale [12] observed a drawdown in an observation
well located 2,400 feet from a pumped well in a
direction parallel to the strike of the beds, whereas
no drawdown was evident in observation wells 600 feet
from a pumped well in a direction transverse to the
strike. Similar observations of the anisotropy of the
Passaic Formation have been documented by Vecchioli
[13], and Vecchioli and others [14] and recently in the
Stockton Formation by Lewis [15].

Numerous county ground-water-resource reports [16-
17-18-19] have provided information on ground-water
yields from formations within the Newark Basin. The
highest yielding wells are the large diameter (10-
inches or greater), relatively deep wells (200 to 500
feet) used for public supply and industrial needs.
Yields of up to 500 gal/min (gallons per minute) are
common from these wells. The lowest yielding wells are

the small diameter (6 inches or less) domestic wells,
which are generally less than 250 feet deep. Yields of
10 to 20 gal/min are common from these wells. The most
productive water bearing units in the Newark Basin are
the Stockton and Passaic Formations. The Lockatong
Formation, basalt units, and diabase units consist of
finer-grained rocks with less interconnected openings
and are, therefore, poor aquifers. These units are
used primarily for domestic purposes. Well yields of
less than 5 gal/min are common, especially in the
fault-block areas in the southwestern part of the study
area.

URANIUM MINERALIZATION

    The crystalline rocks that flank the Newark Basin
are rich in uranium [20] and were the source of
sediments delivered to the basin during Triassic and
Jurassic time [5]. According to Grauch [21] the
crystalline rocks adjacent to the Newark Basin in the
New Jersey Highlands Province (Reading Prong) and in
eastern Pennsylvania form one of four uranium-rich
provinces of crystalline rocks in the eastern United
States. Uranium from these crystalline source rocks
was delivered along with clastic sediments to the
Newark Basin by sediment-laden streams and was
distributed by circulating ground water [20,22].
    In the early 1950's, work began to evaluate
domestic uranium resources in the United States. Since
that time, numerous uranium enriched zones have been
reported in formations of the Newark Basin [20,23-28].
Most of the uranium enriched zones are from the
Stockton and Lockatong outcrops in the Flemington fault
block. Turner-Peterson and others [20] report that
concentrations of 0.01 to 0.02 percent uranium oxide is
the usual grade for mineralized mudstones of the
Lockatong Formation. Locally, concentrations as high
as 0.29 percent (ore grade) are known. They also point
out that these concentrations are higher than uranium
concentrations of other black shales reported in the
literature. The middle Paleozoic Chattanooga Shale in
Tennessee and the Alum Shale in Sweden average 0.006
and 0.03 percent, respectively. Uranium concentrations
as high as 1.28 percent uranium oxide have been
reported for the Stockton Formation [9].
    Turner-Peterson [9,20] has provided a lacustrine-
humate model for uranium mineralization in the Newark
Basin for the offshore lacustrine black mudstone of the
Lockatong Formation and the marginal lacustrine
sandstone of the Stockton Formation. According to the
model, uranium mineralization in the Lockatong

Formation occurred during deposition. Humic acids and aqueous sulfide, capable of fixing uranium, were forming in offshore black muds at the time of deposition. Solubilization of humic acids was favored in the pore fluids of the muds because of an alkaline environment. Simultaneously, bacterial anaerobic respiration reduced sulfate to bisulfide within the bottom sediments. Uranium, which precipates from solution in reducing environments, was fixed near the sediment-water interface.

Mineralization in the sandstone beds of the Stockton Formation occurred shortly after deposition. Reducing agents (humic acids and aqueous sulfide) generated from the black lake bottom muds are believed to have been expelled during compaction into adjacent marginal lacustrine sands. Uranium-bearing ground water passed through the sands as it moved upward into the lake, and the reductants from the mud caused precipitation of uranium.

NATURAL RADIOACTIVITY IN GROUND WATER

Distribution

In order to determine the distribution of naturally occurring radioactivity in ground water within the basin, gross alpha- and gross beta-particle analysis were performed on 260 ground-water samples, including water from 257 wells and 3 springs. The wells sampled included approximately equal percentages of public supply, industrial, and domestic wells. The selection of the sites sampled was based on evaluation of existing radioactivity, geologic, hydrologic, well construction, and well availability data. Areas with known natural radioactive anomalies, both in rock and in ground water, were sampled more heavily; however, an effort was made to obtain a wide geographic distribution.

Gross alpha- and beta-particle activity were used as a screening tool to determine overall radioactivity in ground water of the basin. Gross alpha-particle activity measured in picocuries per liter (pCi/L) is an approximate measure of the alpha emitting radionuclide content of the sample. Uranium, radium-226, and radon are the most common alpha-emitting radionuclides found in ground water and all of these add to the total gross alpha-particle activity. Gross beta-particle activity is an indicator of beta emitting radionuclides, radium-228 being the most important to this study. Other alpha and beta emitting isotopes of the uranium and

thorium decay series are generally not present in
significant amounts in ground water, because most are
highly insoluble, and many have short half-lives that
preclude the buildup of large concentrations.

The gross alpha activities of ground water sites
sampled in this study are shown in figure 4. Eighty
percent of the sites sampled have gross alpha
activities less than 8 pCi/L; 15 percent have
activities between 8 and 15 pCi/L; and 5 percent have
activities greater than 15 pCi/L. Overall, gross-alpha
activities range from less than 0.1 to 124 pCi/L, with
a median value of 3.2 pCi/L. On the basis of current
U.S. Environmental Protection Agency standards, the
maximum contaminant level for gross alpha-particle
activity in drinking water is 15 pCi/L (including
radium but excluding uranium and radon) [29]. Based on
the 3.2 pCi/L median value, it is assumed that ground
water having a gross alpha activity more than 8 pCi/L
in the study area is influenced by some source of
elevated radioactivity. Elevated gross alpha
activities occur predominantly in ground water along
the southeastern border of the study area, in the upper
part of the Lockatong Formation and lower part of the
Passaic Formation, and in the Hopewell and Flemington
fault-blocks where these formations are repeated. Many
of the elevated values are in water from wells that are
located very near to the mapped stratigraphic contact
of these two formations.

Localized, elevated gross alpha activities are
found in ground water of the Stockton Formation.
Ground water sampled from the upper part of the Passaic
Formation, from the basalt units and diabase units, and
from the area shown as Jurassic undifferentiated (fig.
4) is characteristically very low in gross alpha
content.

The distribution of elevated gross alpha activities
in ground water from predominantly the same
stratigraphic interval, basin- wide, suggests that the
right conditions existed for uranium enrichment of
these formations during a specific period of basin
deposition.

Borehole Geophysics and Test Coring

Borehole geophysics, primarily gamma-ray logging,
and rock coring were used to determine the subsurface
source of radioactivity at specific sites within the
basin.

Gamma-ray-log data is consistent with the groundwater data presented here. Numerous zones of high
gamma levels appear in many of the logs from wells

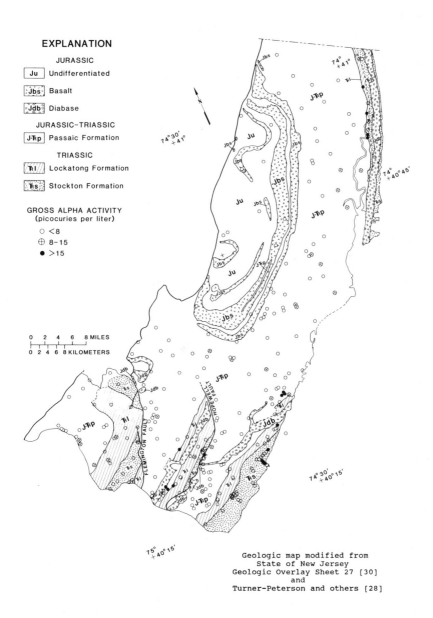

EXPLANATION

JURASSIC

Ju   Undifferentiated

Jbs   Basalt

Jdb   Diabase

JURASSIC-TRIASSIC

Jℝp   Passaic Formation

TRIASSIC

ℝl   Lockatong Formation

ℝs   Stockton Formation

GROSS ALPHA ACTIVITY
(picocuries per liter)

○   <8

⊕   8-15

●   >15

Geologic map modified from
State of New Jersey
Geologic Overlay Sheet 27 [30]
and
Turner-Peterson and others [28]

Figure 4. Gross alpha-particle activities in ground water,
Newark Basin, New Jersey.

drilled in the Lockatong and lower part of Olsen's [7] Passaic Formation. Measurement of gamma activity varies based on the sensitivity of the instrument used for measurement, changes in borehole diameter, and well-construction materials. Therefore, numbers reported for gamma activity in counts per second (cps) are relative numbers. The gamma tool used in this study records typical background readings in the range of 100 to 160 cps for sedimentary rocks of the basin. It is important to mention that the gamma tool used for this study is recording only the gamma radiation in the surrounding rock, and is not being influenced by the radioactive water in the well/borehole. In the lower part of the Passaic Formation, gamma levels from specific zones within boreholes were as high as 70 times background. Locally, high counts on gamma-ray logs of wells in the Stockton Formation also have been observed. Gamma-ray logs of numerous wells in the upper part of the Passaic Formation and in the area shown as undifferentiated (fig. 3) show no significant gamma levels above background. Gamma-ray logs of basalt and diabase wells recorded the lowest gamma counts of all the Newark Basin units. Background gamma levels recorded for the basalt and diabase units ranged from 20 to 40 cps.

Continuous cores were taken at two sites within the study area (fig. 3). The sites cored were adjacent to existing wells where gamma-ray logging indicated several large anomalies ranging from 1,500 cps to more than 7,000 cps. The unused well adjacent to corehole A is 409 feet in depth and is open to about 380 feet of the lower part of the Passaic Formation. Radiochemical analysis of water from this well showed concentrations of 44.9 pCi/L gross alpha, 18.1 pCi/L gross beta, and 9.0 pCi/L radium 226; both gross alpha and radium 226 exceed the maximum contaminant levels (15 pCi/L gross alpha, 5 pCi/L total radium) set by the U.S. Environmental Protection Agency. The source of radioactivity at this site was determined by corehole A.

Gamma-ray logs, a generalized lithologic description, and the interpreted environment of deposition for corehole A are shown in figure 5. Sixty-eight feet of continuous 2-inch-diameter core was obtained at this site. The alternating red, gray, and black mudstones are indicative of the expansion and contraction of an extensive lake. The red beds were deposited in a mudflat/playa environment near the margin of the lake where more oxidizing conditions prevailed. The progressively darker colored beds represent deepening of the lake. The black mudstones formed farthest offshore under anoxic conditions. The black color of these lake bottom sediments and the

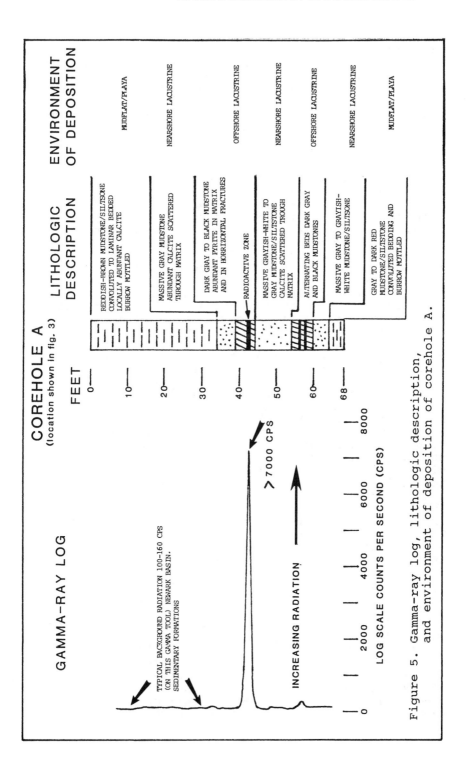

Figure 5. Gamma-ray log, lithologic description, and environment of deposition of corehole A.

presence of pyrite indicate a highly reducing
environment ideal for the precipitation of uranium.

The radioactive anomaly on the gamma-ray log (fig.
5) indicates that the radioactive rock interval is
approximately 42 to 43 feet below land surface. In
order to define the radioactive zone within this
interval, a hand-held scintillometer was used to scan
the core material. The radioactive zone was
approximately 0.5 feet thick and was composed of a dark
black mudstone with abundant pyrite mineralization.
(Analysis of the percentage of uranium, by weight,
within the core was not received by press time.) A
horizontal fracture zone, which is assumed to be water-
producing, is in direct contact with the top of the
radioactive zone in the black mudstone. On completion
of coring, the 68-foot corehole was flushed out and
pumped for 3 hours at 40 gal/min to obtain a
representative sample of aquifer water. Radiochemical
analysis of the sample shows concentrations of 66.4
pCi/L gross alpha, 25.2 pCi/L gross beta, and 22.5
pCi/L radium 226. The radiochemical analysis of ground
water from the shallow corehole and the adjacent deep
well indicate vertical changes in ground-water
radioactivity at the site. Concentrations of
radioactivity in ground water from the 409-foot-deep
well are less than those of the shallower corehole
because the radioactive water in the deeper well is
diluted by the mixing of less radioactive water from
deeper producing zones within the well.

At corehole B, approximately 20 miles to the
southwest, continuous core of the lower part of the
Passaic Formation was collected to a depth of about 187
feet. The lithologic sequence encountered was
strikingly similar to that of corehole A. At this
site, two radioactive zones, each approximately 0.5
feet thick, were encountered; both are black mudstones
with abundant pyrite mineralization in the matrix, and
locally, along bedding planes. Although these zones
are thin, they do appear to have great lateral extent.
The radioactive zones functioned as stratigraphic
marker beds and were recorded on a gamma log from
another well more than 1,100 feet from the corehole.

Radiochemical analysis of ground water from a well
700 feet deep, adjacent to corehole B, shows
concentrations of 0.89 pCi/L gross alpha, 1.86 pCi/L
gross beta. Concentrations of 2.50 pCi/L gross alpha,
1.90 pCi/L gross beta, and 0.19 pCi/L radium 226 were
measured in a water sample from a well 300 feet deep
located 1,100 feet from the corehole. At corehole B,
the ground water was not sampled because the water in
the hole did not clear of sediment after numerous hours
of pumping. However, because of the very low
concentrations found in the deep-well ground-water
samples, significantly higher concentrations in the

corehole are unlikely. It is also important to note that, at this site, core material for the two radioactive zones is more competent than at site A, and no open fracturing intercepted these two zones.

Discussion

Most elevated gross alpha activities in ground water of the Newark Basin, New Jersey, were found at approximately the same stratigraphic position throughout the basin. These high activities result from thin, laterally extensive, radioactive, black mudstones in the Lockatong Formation and upper part of Olsen's [7] Passaic Formation. This distribution is consistent with uranium mineralization models proposed by Turner-Peterson [9,20] and Turner-Peterson and others [28].

Turner-Peterson and others [28] report that uranium-rich, black mudstone occurrences have been located in outcrops at precisely the same stratigraphic interval in the Lockatong Formation at three widely spaced localities in the basin, spanning a lateral distance of approximately 37 miles. They believe that uranium enrichment occurred during deposition and relate it to uranium precipitation in response to reducing conditions near the sediment-water interface of the lake bottom.

Uranium enriched arkosic sandstones [23] may be the source of elevated radioactivity in ground water from wells in the Stockton Formation and along the Lockatong-Passaic geologic contact in the northeastern part of the basin.

It is important to note that water from many wells sampled, which are open to the Lockatong Formation and the lower part of the Passaic Formation, contains low levels of radioactivity. Elevated levels of radioactivity in ground water are not only a function of the geology, they are a combination of the specific conditions as suggested by interpretation of data from two coreholes within the basin. The following factors are necessary for elevated levels of natural radioactivity to occur in ground water: (1) radioactive rock must be present, (2) sufficient amounts of ground water must come in contact with the radioactive source, (3) the radioactive minerals must be soluble in the ground water, and (4) no significant dilution must occur.

SUMMARY AND CONCLUSIONS

     Elevated levels of naturally occurring
radionuclides in ground water are associated with
uranium enrichment in the rocks of the Newark Basin in
New Jersey. Gross alpha- and beta-particle activities
were measured in 260 ground-water samples from the
study area. Five percent of these samples have gross
alpha activities in excess of the 15 pCi/L standard set
by the U.S. Environmental Protection Agency.
Apprixomately 20 percent of the wells sampled contained
ground water that is influenced by some source of
elevated radioactivity.
     Elevated gross alpha-particle activities are
present mainly in ground water along the southeastern
border of the study area, in the upper part of the
Lockatong Formation and lower part of the Passaic
Formation of Olsen, 1980 [7], and in the Hopewell and
Flemington fault-blocks where these formations are
repeated. Many of the elevated values are from wells
located very near to the mapped stratigraphic contact
of these two formations.
     The source of radioactivity has been determined by
borehole geophysical testing and analysis of lithologic
cores. Thin, laterally extensive, uranium-enriched
black mudstones within the Lockatong Formation and
lower part of the Passaic Formation are the primary
source of natural radioactivity in ground water in the
study area.
     Elevated gross alpha-particle activity in ground
water also occurs locally in wells open to uranium
enriched zones of arkosic sandstone of the Stockton
Formation and in the Lockatong and lower part of the
Passaic Formation in the northeastern part of the
basin. Ground water from the upper part of the Passaic
Formation, basalt and diabase units, and from the
undifferentiated sedimentary units interbedded and
overlying the basalt flows is characteristically low in
radionuclides.

REFERENCES

1. Szabo, Z., and O.S. Zapecza. "Relation between
    Radionuclide Activities and Chemical Constituents
    in Ground Water of the Newark Basin, New Jersey,"
    in Proceedings of Radon, Radium and Other
    Radioactivity in Ground Water Conference:
    Hydrogeologic Impact and Application to Indoor
    Airborne Contamination.

2. Van Houten, F. "Triassic-Liassic Deposits of
    Morocco and Eastern North America: Comparison,"
    Amer. Assoc. Petrol. Geol. Bull. 61:79-99 (1977).

3. Manspeizer, W., J.H. Puffer, and H.L. Cousminer.
    "Separation of Morocco and Eastern North America:
    A Triassic-Liassic Stratigraphic Record," Geol.
    Soc. Amer. Bull. 89:901-920 (1978).

4. Froelich, A.J., and P.E. Olsen. "Newark Supergroup,
    a Revision of the Newark Group in Eastern North
    America," in Robinson, G.R. and A.J. Froelich
    editors, Proceedings of the Second U.S. Geological
    Survey Workshop on the Early Mesozoic Basins of the
    Eastern United States, U.S. Geological Survey
    Circular 946 (1985), pp. 1-3.

5. Van Houten, F. "Late Triassic Newark Group, North
    Central New Jersey and Adjacent Pennsylvania and
    New York," in, Geology of Selected Areas in New
    Jersey and Eastern Pennsylvania, S. Subitzky, Ed.
    (New Brunswick, New Jersey: Rutgers Univ. Press,
    1969) pp. 314-347.

6. Abdel-Monem, A.A., and J.L. Kulp. "Paleogeography
    and the Source of Sediments of the Triassic Basin,
    New Jersey, by K-Ar Dating," Geol. Soc. Amer. Bull.
    79:1231-1242 (1968).

7. Olsen, P.E. "The Latest Triassic and Early Jurassic
    Formations of the Newark Basin (Eastern North
    America, Newark Supergroup): Stratigraphy,
    Structure and Correlation", New Jersey Acad. of
    Sci. Bull. 25:25-51 (1980).

8. Olsen, P.E. "Triassic and Jurassic Formations of
    the Newark Basin" in Field Studies of New Jersey
    Geology and Guide to Field Trips: 52nd annual
    meeting of the New York State Geological
    Association, Warren Manspeizer, Ed. (Newark,
    New Jersey: Rutgers University Press, 1980) pp. 2-
    41.

9. Turner-Peterson, C.E. "Uranium Mineralization
    During Early Burial, Newark Basin, Pennsylvania-New
    Jersey", U.S. Geological Survey Uranium-Thorium
    Symposium, J.A. Campbell, Ed. U.S. Geological
    Survey Circular 753 (1977) pp. 3-4.

10. Manspeizer, W. "Rift Tectonics Inferred from
    Volcanic and Clastic Structures" in Field Studies
    of New Jersey Geology and Guide to Field Trips:
    52nd annual meeting of the New York State
    Geological Association, Warren Manspeizer, Ed.
    (Newark, New Jersey: Rugers University Press,
    1980) pp. 314-350.

11. Olsen, P.E. "Fossil Great Lakes of the Newark
    Supergroup in New Jersey" in Field Studies of New
    Jersey Geology and Guide to Field Trips: 52nd
    annual meeting of the New York State Geological
    Association, Warren Manspeizer, Ed. (Newark,
    New Jersey: Rugers University, 1980), pp. 314-350.

12. Herpers, H. and H.C. Barksdale. "Preliminary Report
    on the Geology and Ground-Water Supplies of the
    Newark, New Jersey Area", New Jersey Dept. Conserv.
    Econ. Devel., Div. Water Policy and Supply, Spec.
    Rept. 10 (1951), p. 52.

13. Vecchioli, J. "Directional Behavior of a Fractured-
    shale Aquifer in New Jersey," in Proceedings of the
    Dubrovnik Symposium, Hydrology of Fractured Rocks,
    Vol. I, (Internat. Assoc. Sci. Hydrology, Pub. no.
    73, 1967), pp. 318-326.

14. Vecchioli, J. and L.D. Carswell, and H.F. Kasabach.
    "Occurrence and Movement of Ground Water in the
    Brunswick Shale at a site near Trenton, New
    Jersey," U.S. Geological Survey Prof. Paper 650-B
    (1969) pp. 154-157.

15. Lewis, J.C. U.S. Geological Survey, Unpublished
    results (1986).

16. Nemickas, B. "Geology and Ground-Water Resources of
    Union County, New Jersey," U.S. Geological Survey
    Water-Resources Investigations 76-73 (1976), p.
    103.

17. Carswell, L.D. and J.G. Rooney. "Summary of Geology
    and Ground-Water Resources of Passaic County, New
    Jersey," U.S. Geological Survey Water-Resources
    Investigations 76-75 (1976), p. 47.

18. Kasabach, H.F. "Geology and Ground Water Resources of Hunterdon County, N.J.," New Jersey Bureau of Geology Spec. Rept. 24 (1966), p. 128.

19. Vecchioli, J. and M.M. Palmer. "Ground Water Resources of Mercer County, New Jersey," New Jersey Dept. of Cons. and Econ. Develop. Spec. Rept. 19 (1962), p. 71.

20. Turner-Peterson, C.E. "Uranium in Sedimentary Rocks: Application of the Facies Concept to Exploration," Society of Economic Paleontologists and Mineralogists, Rocky Mountain Section (1980), pp. 149-175.

21. Grauch, R.I. "Uranium Deposits in crystalline Rocks of the Eastern United States--a preliminary report," Geol. Soc. of Amer. Abstracts with Programs, VIII (2) 184-185 (1976).

22. Popper, G.H.P. and T.S. Martin. "National Uranium Resource Evaluation, Newark Quadrangle, Pennsylvania and New Jersey," Bendix Field Engineering Corp. Report PGJ/F-123 (82) for the U.S. Dept of Energy (1982), p. 73.

23. Stewart, R.H. "Radiometric Reconnaissance Examination in Southeastern Pennsylvania and Western New Jersey," U.S. Geological Survey Trace Elements Memoranda Report TEM-255 (1952), p. 10.

24. McKeown, F.A., P.W. Choquette and R.C. Baker. "Uranium Occurrences in Bucks County, Pennsylvania and Hunterdon County, New Jersey," U.S. Geological Survey Trace Elements Investigations Report TEI-414 (1954), p. 37.

25. McKeown, F.A., H. Klemic and P.W. Choquette. "Occurrence of Uranium at Clinton and Hunterdon Counties, New Jersey," U.S. Geological Survey Trace Elements Investigations Report TEI-382 (1954), p. 18.

26. Popper, G.H.P., and R.P. Blauvelt. "Work Plan, Newark Quadrangle Covering Parts of Pennsylvania and New Jersey," Bendix Field Engineering Corp., (in conjunction with the National Uranium Resource Evaluation Program, 1980), p. 45.

27. Bell, C. "Radioactive Mineral Occurrences in New Jersey," New Jersey Geological Survey Open-File Report 83-5 (1983), p. 22.

28. Turner-Peterson, C.E., P.E. Olsen, and V.F. Nuccio. "Modes of Uranium Occurrence in Black Mudstones in the Newark Basin, New Jersey and Pennsylvania," in Robinson, G.R. and A.J. Froelich editors, Proceedings of the Second U.S. Geological Survey Workshop on the Early Mesozoic Basins of the Eastern United States, U.S. Geological Survey Circular 946 (1985), pp. 120-124.

29. Lappenbusch, W.L. and C.R. Cothern. "Regulatory Development of the Interim and Revised Regulations for Radioactivity in Drinking Water-Past and Present Issues and Problems, "Health Physics 48 (5): 535-551 (1985).

30. State of New Jersey Geologic Overlay Sheet 27.

Session II: Monitoring Radon, Radium and Other Radioactivity from Geologic and Hydrogeologic Sources

Moderators: Harry E. LeGrand
Hydrogeologist

Rick Cothern

ELEVATED LEVELS OF RADIOACTIVITY IN WATER WELLS IN LOS ANGELES AND ORANGE COUNTIES, CALIFORNIA

Jeffery Wiegand, Alton Geoscience, Irvine, California

Gary Yamamoto, California Department of
  Health Services, Los Angeles, California

Wilbert Gaston, Alton Geoscience, Irvine, California

INTRODUCTION

Levels of gross alpha particle radioactivity nearly
three times the maximum contamination levels (MCL)
have been detected for several years in well waters
and related surface waters in Los Angeles and Orange
Counties, California. A few elevated levels of
uranium have also been recorded. The affected wells
and related surface waters represent only a minor
fraction of the water sampled and tested in this area.
None of the excessive radioactivity is believed to
persist in the municipal waters sold to the public,
due to the customary blending of waters from several
wells or sources which water purveyors practice. This
paper is a preliminary survey of the occurrence,
possible sources, fate, and implications of these
elevated radioactivity levels.

OCCURRENCE

Maximum levels of gross alpha particle activity of
44.2, 44.7, and 30 pico Curies per liter (pC/l) have
been detected in the Raymond Basin, Central Los
Angeles Basin, and Santa Ana Plain, respectively, of
the South Coast Hydrologic Basin. Please see Figure
1, Index Map, for the locations of these areas.

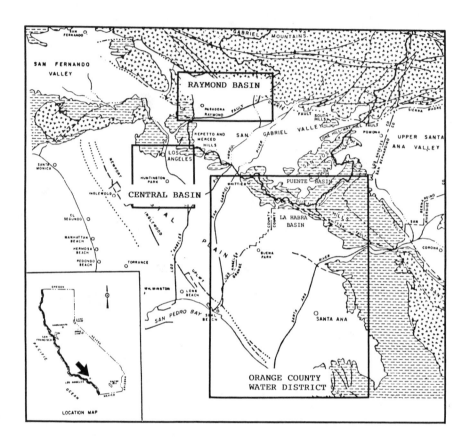

Figure 1. Hydrogeologic basins and physiographic
features of the Los Angeles area.
From P.K. Saint, "Hydrogeology of
Southern California," 1986.

Laboratory analyses have been performed on water
samples taken from different wells and related surface
waters in these areas since 1980, in compliance with
the laws of the State of California and the Federal
Government. Sections 64441 and 64443 of Title 22 of
the California Administrative Code mandate monitoring
of all public water supplies for radium-226 and
radium-228 at least once every four years. Gross
alpha particle activity is typically substituted for
measurement of radium-226 and -228. The water supply
is considered to be in compliance if the gross alpha
particle activity does not exceed 5 pC/l. The maximum
contaminant level (MCL) of each pertinent constituent
is shown in the following table.

Table 1. Maximum Contaminant Levels For Drinking
         Water, State of California (CAC Title 22)

Constituent	Maximum Contaminant Level-MCL, pC/l
Combined Radium-226 and -228	5
Gross Alpha particle activity (including Radium-226 but excluding Radon and Uranium)	15
Tritium	20,000
Strontium 90	8
Gross Beta particle activity	50

Pending an MCL for uranium from the USEPA, a
recommended action level of 20 pC/l has been adopted
for drinking water supplies in the State of
California.

The identity of the well or sampling point and the
highest reported level of gross alpha particle
activity are shown in Table 2 and 3. Ten pC/l was
arbitrarily selected as the minimal value, to
differentiate from lower levels for this summary.

Some levels of uranium were reported for the Raymond
Basin Area; these are also presented in Table 2.

TABLE 2: Alpha level radioactivity from sampling
points in Raymond Basin. Map identification number
refers to location of sampling points of Figure 1.
Data from California Department of Water Resources,
"Water Quality Monitoring in Raymond Basin", October
1981, January 1985, March 1986, January 1987 and
unpublished data.

RAYMOND BASIN SAMPLING POINTS

MAP ID#:	NAME:	ID_#:	ALPHA (pC/l)	URANIUM (pC/l)	LEVEL DATE
1	H.Prss.Tnnl	1N12W12H1S	12.4		
2	Tunnel 8	2N12W34Q1S	44.3	43.7	7/84
3	Hugo Reid	1N11W3or3S	10.12	10.77	7/85
4	Winston	1N12W26R1S	26.2	5.31	7/85
5	Arroyo Seco	Z-6-2930.00	9.6		
6	Ventura	1N12W5N1S	5.3		
7	Millard Cnyn	Z-6-2951.00	5.7		
8	Eaton Wash	Z-7-2950.10	14.0	14.9	7/84
9	Copelin #3	1N12W20B01S	14.6	12.71	7/85
10	WELL #4	1N12W08H03S	16.0	11.94	7/85
11	WELL #3 G.T.	1N12W03G01S	18.4		
12	WELL K3	1N12W13E03S	10.1		
13	WELL #5	1N12W05Q1S	12.8		
14	WELL #2	1N12W8H025	12.2		
15	WELL #1	1N12W06M6S	15.3	10.3	7/84
16	WELL #2	1N12W09R01S	13.9		
17	WELL #5	1N12W08H01S	18.7		

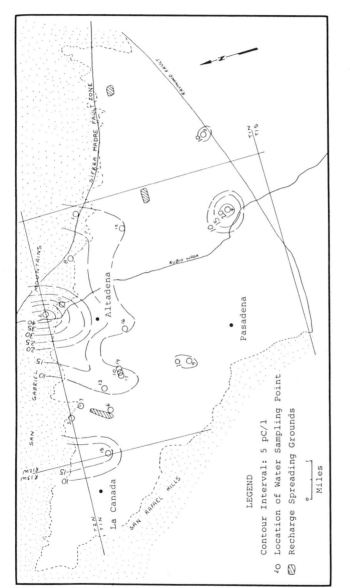

Figure 2. Elevated levels of gross alpha radioactivity in Raymond
Basin. Base map from Raymond Basin Watermaster Service.

Table 3:   Alpha radioactivity from sampling points
           in Los Angeles Central Basin and the
           Orange County Water District. Map
           identification number refers to location
           of sampling points on respective figures.
           Data from Los Angeles Central Basin Water
           Master and Orange County Water District
           unpublished data.

LOS ANGELES CENTRAL BASIN SAMPLING POINT

MAP ID#:	NAME:	ID #:	ALPHA LEVEL IN IN pC/l & DATE:	
1	Nadeau #2	2513W28H01	44.7	3/86

(See Fig. 3)

ORANGE COUNTY WATER DISTRICT SAMPLING POINTS

1	PLACENTIA	3S9234J05	28.1	10/86
2	PLACENTIA	3S10W36N02	10.6	6/86
3	STANTON	4S10W19B01	16.3	4/86
4	PLACENTIA	3S9W3SM01	14.1	4/86
5	TUSTIN	4S9W33J1	30.2	4/84
6	ANAHEIM	3S9W32K8	17.8	1/84
7	GARDEN GROVE	4S11W26N1	10.2	4/84
8	GARDEN GROVE	5S11W1H2	13.6	1/84
9	HUNT. BEACH	5S11W24Q	13.2	1/84
10	FULLERTON	3S10W35R1	12.2	1/84
11	BUENA PARK	3S11W34R1	11.7	1/84
12	ORANGE	4S9W7P1	12.8	1/84
13	ANAHEIM	4S10W7E1	15.1	4/84
14	ANAHEIM	4S10W21L	17.4	1/84
15	GARDEN GROVE	4S10W29K1	11.3	5/84
16	SO.CA. H2O	4S11W16H2	10.5	4/84

(See Fig. 4)

These data were obtained from the regional offices of the California Department of Health Services in Los Angeles and Santa Ana, California. The Raymond Basin data was included in: Water Quality Monitoring in Raymond Basin, by the California Department of Water Resources, 1980 through 1985.

Discussion

The Raymond Basin occurrences appear to be related to water sources issuing from tunnels in the adjacent mountains of the San Gabriel Range to the north, part of the Transverse Ranges, please see Figure 2. Indeed, the highest level reported in the Raymond Basin--44.2 pC/l--was in Tunnel Number 8, a water supply tunnel. Some of the other tunnels in the area also exhibited elevated levels of gross alpha activity. All but one of the wells that contained water with elevated levels of gross alpha also are located close to the mountain range front. Screened sections of these wells are typically from a depth of approximately 160 feet to 800 feet.

The plume of the elevated levels clearly depicts this in Figure 2, Isoconcentration Contours of Elevated Gross Alpha Activity in the Raymond Basin, California.

The Los Angeles Basin occurrence is restricted to a single well---State Well Number 2S13W28H01---in the vicinity of Huntington Park, approximately 5 miles south of downtown Los Angeles. The location of this occurrence, which reached 44.7 pC/l, is shown in Figure 3, Isoconcentration Contours of Elevated Levels of Gross Alpha activity in the Central Los Angeles Basin, California. Since there is but one data point, this occurrence may be regarded as anomalous, and the isoconcentration contours are presented, therefore, for comparative purposes only.

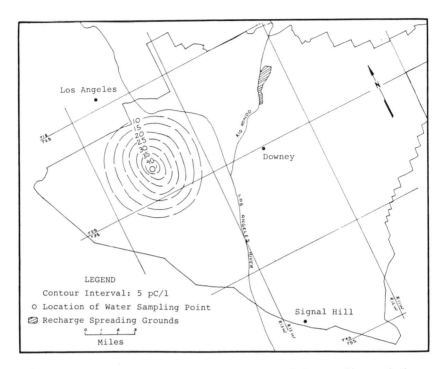

Figure 3. Elevated levels of gross alpha radioactivity
          in the Los Angeles Central Basin. Base map
          from Central Basin Watermaster Service.

The occurrences in Orange County are located in the
Santa Ana Plain southwest of the Chino Hills and Santa
Ana Mountains. The affected area is shown in Figure
4, Isoconcentrations of Elevated Levels of Gross Alpha
Activity in the Santa Ana Plain, Orange County,
California. The highest level encountered was 30.0
pC/l in a well slightly south of the point where the
Santa Ana River emerges from the mountains. An
associated plume of elevated levels of gross alpha
activity lies west of the Santa Ana River, with the
higher concentrations toward the north.

Here, there appears to be a potential correlation
between the wells exhibiting elevated gross alpha
levels and the aquifer recharge spreading grounds.
The latter are also shown in Figure 4.

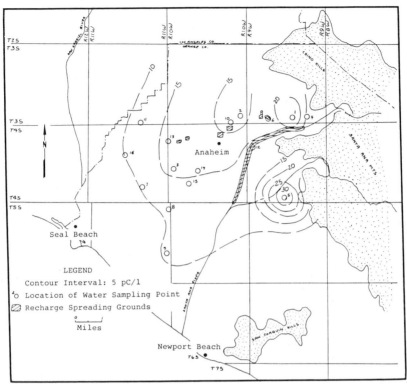

Figure 4. Elevated levels of gross alpha radioactivity in Orange County Water District. Base map from Orange County Water District.

POSSIBLE SOURCES

The shape of the isoconcentration contour plume for the Raymond Basin points to the foothills of the San Gabriel Mountains to the North as a possible source. Tunnel 8, the Kinneloa High Pressure Tunnel, and the Number 3 Gravity Tunnel penetrate these mountains. The tunnels were hand excavated into the mountain bedrock near the end of the last century to provide gravity-flow irrigation water.

The mountains are composed of Cretaceous granitic rocks with some undivided Pre-Cambrian granite inclusions. The presence of uraniniferous secondary mineralization in this portion of these mountains certainly cannot be ruled out.

Other portions of the Transverse Ranges reportedly
contain numerous uranium and rare earth deposits (Fife
and Minch, eds., 1982), which may be related to the
radioactivity reported here.

The well in the more southerly portion of the Raymond
Basin---State Well Number 1N12W26R1S---is close to the
Rubio Wash. This drainage comes from near Tunnel
Number 8, suggesting a possible causal pathway between
the two sites.

Possible sources of the solitary occurrence reported
in the Central Los Angeles Basin are not apparent
without further research. The possibility of an
accidental discharge, though remote, cannot be ruled
out.

As in the Raymond Basin, the occurrences in Orange
County appear to be related to the low mountains to
the east and north, and possibly to rocks traversed by
the stream course of the Santa Ana River. The stream
flow of the river consists of natural runoff as well
as waters imported from Northern California and the
Colorado River. (The highest level of gross alpha
found in the well waters of the Southern California
area in this survey---100 pC/l---occurs in a portion
of the Transverse Ranges in San Bernardino County to
the east of the Orange County area.)

In both the Raymond Basin and Orange County
occurrences, the presence of spreading grounds may
have had a significant impact on facilitating the
migration of the ground water with elevated levels of
gross alpha activity throughout the well fields. This
possible relationship is shown in Figure 5.

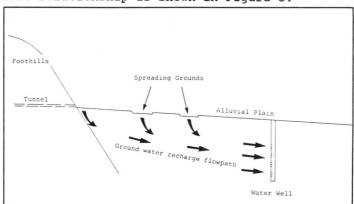

Figure 5. Cross section showing relationship between
         suspected alpha radioactivity source,
         spreading grounds, and water wells.

An analysis of aeroradiometric survey data of the
study areas, currently in progress, should help to
relate high radiation levels in the source rock areas
to the ground water phenomena under discussion.

FATE OF THESE CONTAMINANTS

Several unknowns cloud the picture presented so far.
Very little is known about the radionuclides
derivative from the gross alpha and uranium detected,
although no elevated levels of radium 226 were
reported in Los Angeles County. The presence of
radium-228, and/or radon-222 in significant
quantities cannot be ruled out. Also, although these
data are from groundwater, virtually nothing is known
about the presence and concentration of radioactivity
in the surface soil; in the adsorbed phase in the soil
grains in the aquifer media; in the distribution
systems such as tunnels, stream courses, and spreading
grounds; or in the rocks in the source areas. The
behavior of the radon gas is a separate issue. The
fate of these contaminants can therefore not yet be
addressed.

Regarding the well water, if the blending of water
supplies from several wells maintains the water
offered to the public within the MCLs, then no problem
can be said to exist. Similarly, if the trend over
time is for a relative decrease in the levels of gross
alpha as a high water throughput disperses and dilutes
the radioactive contaminants, then health risks should
diminish. The converse of these conditions also might
occur.

IMPLICATIONS

Mitigating the effects of elevated levels of
radioactivity at the sources, in the pathways from the
sources to the pertinent aquifers, and in the specific
wells, is a natural goal, provided that such action
does not result in even higher health hazards
elsewhere.

Also, an analysis of the short and long term health
risks related to the various aspects of the system
here described would seem to be indispensable. This
might include an analysis of the radionuclides present
and their concentrations, throughout the life-cycle of
these contaminants.

Are there some tasks that can be taken in the relatively immediate future to mitigate to a practical degree the most serious of the threats that confront the affected public?

1. Based on an analysis of available data, rank the geographic areas by risk/consequence level, either because of high levels of radioactive contaminants or due to sole reliance on the affected ground water, or both.

2. Evaluate the practicality as well as cost-effectiveness of various measures, such as: pursuing different water management practices in order to decrease contamination levels; plugging and abandoning water supplies such as Tunnel Number 8, if appropriate, etc.

Finally, an evaluation is needed to quantify the significance of the phenomenon of elevated levels of radioactive contaminants in the ground water and wells and their sources and pathways, to the larger picture of naturally-occurring radioactivity and public health.

ACKNOWLEDGMENT

Mr. Frank Hamamura of the Orange County office of the State of California Department of Health Services provided valuable and essential material and guidance to this study. Mr. Eldon Gath of Leighton and Associates helped plot early versions of the data.

REFERENCES

"Water Quality Monitoring in Raymond Basin," California Department of Water Resources, Southern Section, separate vols, 1980-81, 81-82, 82-83, 83-84, 84-85.

"Geology and Mineral Wealth of the California Transverse Ranges," Donald Fife, and John Minch, Eds. (Santa Ana, South Coast Geological Society, 1982.)

Albert E. Ogden, Tennessee Technological University,
Water Research Center, Cookeville, Tennessee

William B. Welling, Groundwater Technology Incorporated,
Albany, New York

Robert D. Funderburg and Larry C. Boschult, Idaho
Division of Environment, Boise, Idaho

INTRODUCTION

With rising energy prices in the 1970's, a nationwide surge of home weatherization began. Unfortunately,
these efforts can cause indoor air pollutants, such
as radon, to be trapped and concentrate in homes.
Increased public concern about the relationship between
radon and lung cancer has brought the Environmental
Protection Agency to encourage state governments to
perform widespread screening surveys to determine
the extent of the problem. Much of Idaho is underlain
by granitic and related rocks associated with the
Idaho Batholith. Since granitic-type rocks commonly
have higher concentrations of radium which decays
to radon, the Idaho Division of Environment obtained
a grant from the Idaho Cancer Coordinating Committee
to survey radon levels throughout the state. This
paper presents the results of the first phase of this
state-wide survey and examines the possible factors
affecting these levels in two Idaho counties.

LOCATION AND GEOLOGY

Figure 1 shows the principal aquifers in Idaho
as depicted by rock type (1,2). Most of the areas
shown in white are Cretaceous Age granite of the Idaho
Batholith or Precambrian metasediments and gneiss.

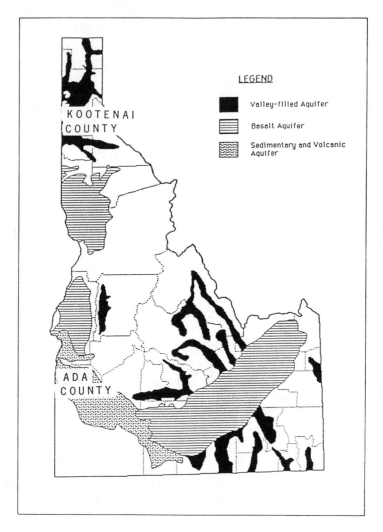

Figure 1.   Principal aquifers in Idaho as depicted
            by rock type after Shook [1] and Graham
            and Campbell [2].

The batholith comprises much of central Idaho while
the metasediments are found in northern Idaho. Some
of the unshaded areas in the southeastern portion
of the figure are mountain ranges of sedimentary rock.
In this rugged basin and range topography, homes are
built in the valleys upon transported sands and gravels.
The basalt aquifers are of two types: younger basalts
comprising the Snake Plains Aquifer of southeastern
Idaho, and the older, Columbian River basalts of

west-central Idaho. The basin fills of southwestern
Idaho are composed of a mixture of basalt flows and
fluvial and lacustrine deposits. Areas of suspected
higher radon levels include homes built on granite,
gneiss, or valleys filled with erosional products
of these rock types. Also, suspect are homes built
on phosphate-rich rocks and processed ore of the
Phosphoria Formation in southeastern Idaho, as well
as the geothermal areas of central Idaho.

Figure 1 also shows the location of Ada and Kootenai
counties where a more in-depth study was made of the
factors affecting radon levels. Ada County is the
most populous Idaho county and contains the capitol
city of Boise as well as the towns of Garden City,
Meridian, Eagle, and Kuna. Ada County lies on the
northwestern edge of the Snake River Plain and the
southern tip of the Idaho Batholith. Geologic
formations at house locations include: 1) recent
stream alluvium, 2) Quaternary sedimentary sequences
of the Idaho Group, primarily the Caldwell-Nampa
sediments which consist of clay, silt, sand, and gravel,
3) the Tenmile Gravel which is silt, sand, and gravel,
and 4) the Idaho Formation composed of clay, silt,
sand, volcanic ash, and fine gravel. All four of
these groups are composed of various amounts of
weathered and eroded granitic and basaltic rock.

Kootenai County lies 245 miles north of Ada County
and 70 miles south of the Canadian border. Larger
cities include Coeur d'Alene, Post Falls, Hayden Lake,
and Spirt Lake. Most homes are located on the Rathdrum
Prairie which is composed of glacial outwash sands
and boulders primarily of granite, gneiss, and basalt
lithologies. The Rathdrum Praire Aquifer was the
second EPA designated sole-source aquifer in the
country. Some homes are also located on benches
composed of Miocene Columbian River flood basalts
and hills of Precambrian schist, gneiss, and quartzite.

METHODS

With the help of the Idaho Health Districts,
charcoal canisters were distributed throughout the
counties on a volunteer basis during the winter months
of 1986 and 1987. Residents were instructed to place
the canister in the room occupied the most hours each
day, but not near drafts such as from doors or windows.
The intent was to sample still air during winter months
which is believed to represent a worst case scenario.
Basements were not chosen since most people spend
little time there. Canisters were left open for one
week and were then sealed and sent to the University

of Pittsburgh for analysis. Home owners were also asked to fill out a questionaire regarding: 1) age of the house, 2) degree of weatherization (yes/no), 3) substructure of the house (i.e. basement, crawl space, earth cover), 4) type of heating, and 5) presence or absence of an electric air cleaner. A total of 544 analyses were made in 35 counties. Sampling is ongoing at this time of writing.

Results from Ada and Kootenai counties were examined in more detail to determine the relationship between radon levels and various geologic, hydrologic, and soil parameters. Water table data for the two counties were obtained from well drillers' reports on file with the Idaho Department of Water Resources and from water table maps published by the Ada Planning Association [3] and Jones and Lustig [4]. Soil data and published maps were utilized for Ada [5] and Kootenai [6] counties. Geologic data, including fault locations were taken from Mitchell and Bennett [7] and Jones and Lustig [4].

RESULTS

Table 1 presents the median, mean, and range of radon levels in 35 Idaho counties. The median values are considered more representative for comparative purposes since just a few high values can significantly skew the mean. Insignificant sample sizes occur in many counties, thus making it difficult to attribute causes for some higher levels. Four counties with the highest median levels are Shoshone (2.9), Ada (3.1), Kootenai (4.0), and Blaine (7.9). Only the latter two equal or exceed the 4.0 pC/l health limit set by the Environmental Protection Agency. Custer County is not included due to small sample size. Table 2 presents a compilation of the radon data versus foundation or home type. Some home owners did not fully complete the questionaire; thus an "unknown" category exists. The data show that no significant difference exists for median radon levels among the different house foundation types except for mobile homes and earth covered homes. These two represent the low and high as would be expected.

A review of just the data for Ada and Kootenai counties shows conflicting results. In Ada County, homes that are earth covered or have basements have the highest radon levels, whereas in Kootenai County, homes with crawl spaces have the highest levels followed by earth covered homes and homes with basements.

Surprisingly, it does not appear that weatherizing

Table 1.   Median, Mean, and Range of
           Radon Values versus County.

COUNTY	N	MEDIAN	MEAN	RANGE
		pCi/l	pCi/l	pCi/l
Ada	82	3.1	4.1	0.0-31.4
Adams	11	1.6	2.5	0.6-5.9
Bannock	5	1.4	1.5	0.0-2.8
Benewah	9	1.5	2.9	1.0-12.7
Bingham	1		1.1	
Blaine	39	7.9	9.2	0.4-42.7
Boise	1		7.0	
Bonner	25	1.4	2.6	0.6-15.3
Bonneville	13	1.7	1.9	0.6-3.3
Boundary	11	1.2	3.9	0.9-28.3
Canyon	47	1.8	2.3	0.5-8.1
Caribou	2	0.9	0.9	0.8-0.9
Cassia	6	1.1	3.1	0-14.0
Clearwater	25	2.1	3.1	0.2-13.8
Custer	2	9.0	9.0	2.0-15.9
Elmore	13	1.7	5.8	0.3-28.3
Gem	16	2.2	2.3	1.1-4.7
Gooding	3	3.0	2.9	1.8-3.8
Idaho	20	1.9	2.5	0.2-7.3
Jefferson	1		1.8	
Jerome	7	1.3	1.6	0.2-3.5
Kootenai	67	4.0	8.2	0.2-134.9
Latah	22	1.9	2.4	0.1-10.1
Lemhi	1		1.9	
Lewis	3	2.3	2.1	0.9-3.3
Lincoln	4	1.5	2.4	1.4-5.4
Madison	3	1.0	3.7	0.9-9.2
Nez Perce	33	1.8	2.6	0.0-25.6
Oneida	1		0	
Owyhee	11	1.5	1.9	1.1-4.9
Payette	15	1.5	1.9	0.6-7.8
Shoshone	12	2.9	4.3	0.2-14.0
Twin Falls	16	2.4	3.2	1.2-10.2
Valley	1		1.4	
Washington	16	2.1	2.2	1.4-4.5
STATEWIDE	544	4.1	2.3	

a home has a large effect on increasing radon levels
based on the state-wide data (Table 3).   The data
show a difference in the median of only 0.3 pC/l.
The radon levels for Ada County show a similar small
difference, but there is nearly a 1.0 pC/l difference
in Kootenai County.

The type of heating in the home also does not
appear to significantly influence radon levels.   When
the data for the entire state is reviewed, there are

Table 2.   Mean and Median Radon Levels
versus Type of House Construction.

STATEWIDE RADON DATA AS OF 1/22/87

TYPE	N	MEAN	MEDIAN
		pCi/l	pCi/l
Basement	249	3.9	2.3
Crawl Space	116	4.2	2.1
Mobile Home	15	0.9	0.8
Condo	31	2.9	2.5
Bsmt. & Crawl	7	4.3	2.3
Apt.	6	2.9	2.9
Slab	10	4.2	2.4
Earth covered	12	11.5	7.8
Unknown	149	4.1	2.2

ADA COUNTY RADON   DATA AS OF 1/22/87

Basement	35	4.9	4.1
Crawl space	19	3.8	2.9
Mobile home	2	1.1	1.1
Condo	10	2.8	2.6
Bsmt. & crawl	2	1.3	1.3
Apt.	3	3.4	3.1
Slab	0		
Earth covered	3	8.9	6.8
Unknown	8	3.2	3.5

KOOTENAI COUNTY RADON DATA AS OF 1/22/87

Basement	28	7.9	4.8
Crawl space	13	16.3	6.6
Mobile home	2	0.9	0.9
Condo	0		
Bsmt. & Crawl	0		
Apt.	0		
Slab	0		
Earth covered	3	6.1	4.9
Unknown	21	4.7	2.8

only small differences in radon levels versus home
heating method (Table 4).   Homes with wood heat have
the highest levels of all types in Ada County, but
in Kootenai County homes with gas heat are highest.
It was hypothesized that homes with air cleaner systems
might have lower radon concentrations, but there were
only slightly lower levels based on the entire state's
data (Table 5).   Levels were also slightly lower in
homes with air cleaners in Kootenai County, but in
Ada County radon levels were significantly higher.
These contrasts in results suggest that geologic,
soil, and/or hydrologic factors may be more important
in affecting radon concentrations in homes.   To
investigate this further, radon levels in Ada and

Table 3. Mean and Median Radon Levels
versus Weatherization of Homes.

STATEWIDE RADON AS OF 1/22/87

	N	MEAN	MEDIAN
		pCi/l	pCi/l
Yes	317	4.0	2.4
No	133	3.9	2.1
Unknown	145	4.2	2.1

ADA COUNTY RADON DATA AS OF 1/22/87

Yes	44	4.5	3.2
No	28	4.0	3.0
Unknown	10	2.7	3.1

KOOTENAI COUNTY RADON DATA AS OF 1/27/87

Yes	30	7.9	5.0
No	16	13.1	4.0
Unknown	21	4.9	2.3

Table 4. Mean and Median Radon Levels
versus Type of Heating System.

STATEWIDE RADON DATA AS OF 1/22/87

	N	MEAN	MEDIAN
		pCi/l	pCi/l
Electric	188	4.6	2.4
Gas	101	3.5	2.6
Wood	108	3.8	1.9
Other	55	3.0	2.4
Unknown	143	4.2	2.1

ADA COUNTY RADON DATA AS OF 1/22/88

Electric	29	4.7	2.9
Gas	38	3.5	3.0
Wood	2	6.5	6.5
Other	4	5.6	4.2
Unknown	9	3.4	3.6

KOOTENAI COUNTY RADON DATA AS OF 1/22/87

Electric	24	13.0	4.1
Gas	9	7.3	7.1
Wood	11	5.7	4.9
Other	3	4.5	3.7
Unknown	20	4.8	4.0

Table 5.   Mean and Median Radon Levels versus
Presence or Absence of a Home Air
Cleaner.

STATEWIDE RADON DATA AS OF 1/22/87

	N	MEAN	MEDIAN
		PCi/l	pCi/l
Yes	94	4.0	2.1
No	339	4.1	2.2
Unknown	162	3.9	1.6

ADA COUNTY RADON DATA AS OF 1/22/87

Yes	24	5.3	4.7
No	46	3.8	3.0
Unknown	12	2.9	3.1

KOOTENAI COUNTY RADON DATA AS OF 1/22/87

Yes	9	5.0	4.5
No	37	11.2	5.1
Unknown	21	4.3	2.8

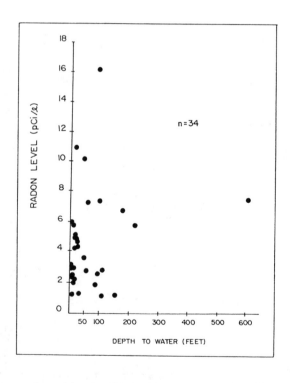

Figure 2.   Radon level versus depth to water, Ada
County.

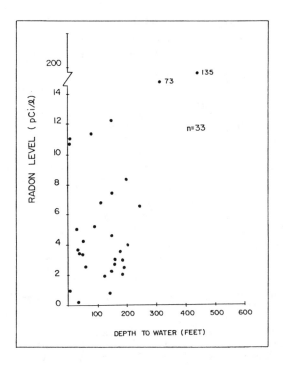

Figure 3.   Radon level versus depth to water, Kootenai
             County.

Kootenai counties were compared to: 1) depth to the
water table,   2) underlying rock type,   3) underlying
soil type, and 4) topographic setting.

    Figures 2 and 3 show the relationship between
radon concentration and depth to water.   Both graphs
show a positive correlation between the two variables.
It is possible that a thicker vadose zone allows for
a greater accumulation of radon gas.   A fluctuating
water table could cause a pumping action to aid in
the upward migration of the gas.

    It is suspected that the soil and rock type on
which a house is built could substantially affect
radon levels.   In Ada County, most homes are built
on unconsolidated materials derived from weathered
and eroded granite and basalt.   Therefore, few
differences in lithology exist.   Figure 4 shows a
breakdown of the four substrates versus median radon
level.   The only significant difference occurs for
houses built on Caldwell-Nampa sediments.   These
sediments are composed primarily of arkosic granitic
material that is possibly of greater granitic

consistency than the other three groups. It is important to note that homes in counties underlain exclusively by basalt such as Jerome and Lincoln have lower radon levels than those where granite or its eroded products are exposed at the surface. Figure 5 shows the median radon levels versus soil group and topographic setting. Homes on the Purdham-Abo-Power soil group have the highest radon levels. These soils form on low alluvial terraces and commonly have caliche layers. The soils are derived primarily from weathering of the Caldwell-Nampa sediments and wind blown deposits. They have relatively high permeability which allows the upward movement of radon gas into homes. The homes on the Quincy-Lankbosh-Brent soil group have the lowest radon levels. They are usually sandy soils near the surface and are found on the foothills around Boise. Clay layers in the subsoils of this group may inhibit upward movement of radon causing the lower concentrations found in homes.

Radon levels were also compared to rock type, soil group, and topographic setting in Kootenai County. Figure 6 shows the distribution of median radon levels versus rock type. Similar radon levels are found in houses built on Precambrian schist and gneiss as built on the glacial outwash composed of the same lithologies. Homes built on granite have slightly higher concentrations. Too few samples were obtained from homes built on recent alluvium and lake deposits for comparative purposes.

Figure 7 shows median radon levels compared to soil group and topographic setting. As in Ada County, soil types generally mimic topographic setting. Both the Kootenai-Bonner and Avonville-Garrison-McGuire groups are deep, well-drained soils formed from gravelly, glacial outwash; but there is a significant difference in the median radon levels. There does not appear to be any hydrogeologic difference between the two groups, therefore, house construction style is a suspected cause. The lowest radon concentrations are from homes built on Lenz-Schumacher-Skalam soils which occur on mountains and foothills generally composed of gneiss, schist, or basalt.

Other Factors

Other factors that possibly affect radon concentrations in Idaho homes are: 1) proximity to faults, 2) occurrence of geothermal waters, and 3) occurrence of phosphate-rich rocks. The radon levels for homes in Ada and Kootenai counties were compared to the distance to faults. The data suggest that

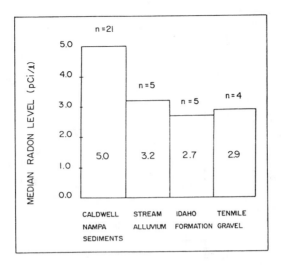

Figure 4. Median radon levels versus rock type, Ada County.

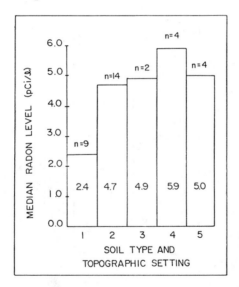

Figure 5. Median radon levels versus general soil type and topographic setting, Ada County: 1) Quincy-Lankbush-Brent (lacustrine fathills), 2) Notus-Moulten-Falk (flood plains and low terraces, 3) Power-Aeric Haplaquepts Jenness--drainage ways and low terraces, 4) Purdham-Abo-Power--low alluvial terraces, and 5) Colthorp-Elijah-Purdham--intermediate alluvial terraces and basalt plains.

Figure 6.   Median radon levels versus rock type, Kootenai County.

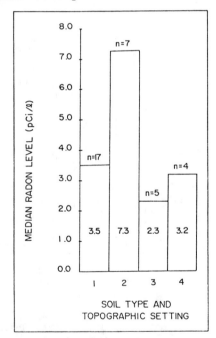

Figure 7.   Median radon levels versus general soil type and topographic setting, Kootenai County: 1) Avonville-Garrison-McGuire (glacial outwash plain, 2) Kootenai-Bonner (glacial outwash plain), and 3) Lenz-Schumacher-Skalan (mountains and foothills), and 4) Chatcolet Mokins-Selle (lake terraces)

radon levels may be higher in homes located on or near faults. A similar conclusion was found by Lloyd [8] for the Butte, Montana area where extensive mineralization has ocurred along numerous faults in the city area. As noted previously, radon levels in Shoshone County, Idaho are relatively high. The "Silver Valley" of Idaho occurs in this county. Extensive lead and silver mining associated with mineralization along faults has occurred in and around the towns of Kellog and Wallace. Adding to the problem, many homes in Kellog are built on lead-rich mine tailings and processed ore wastes. This has caused abnormally high levels of lead found in the blood of the city's children. As a result, the town has become a major Superfund site.

The Idaho county with the highest radon levels is Blaine County, home of Sun Valley. Most of the homes in the Sun Valley/Hailey/Ketchum area are built on stream alluvium with a shallow water table. Along the edges of the alluvium and beneath the alluvium at the bedrock contact are numerous hot springs and seeps. Deep circulation of ground water through the fractured granite has caused the water to be heated and enriched with radon gas. As a result, homes in the area have some of the highest radon concentrations found in the state.

A last factor that may affect radon levels in Idaho homes is the widespread occurrence of phosphate bearing rocks in Bingham and Caribou counties of southeastern Idaho. These rocks also have relatively high concentrations of radioactive minerals. Some homes may be built on mine tailing as well as waste from the fertilizer factories which process the rock. So far, these counties have not been surveyed; but they will be in the next phase of the project.

CONCLUSIONS

The results from this preliminary survey show that earth covered homes have the highest radon levels, but the presence or absence of a basement or crawl space does not appear to significantly affect radon levels. Also, there was no significant difference between homes that are weatherized versus those that are not. The type of heating in the homes and presence or absence of an air cleaning system also seem to have little influence on the radon levels.

A more detailed look at radon levels in Ada and Kootenai counties suggests that levels are higher if the home is near a fault and also where the water

table is deeper. The greater the granitic composition
of the alluvium and outwash on which homes are built,
the greater the possibility of higher radon levels.
In general, homes built in counties where the Snake
Plains basalt crops out have lower radon levels than
homes built on granite or granitic rich alluvium.
Soils with high permeability appear to produce more
radon than tighter soils. Finally, homes built in
geothermal areas have higher radon levels.

The radon survey in Idaho is ongoing. With the
new data and further analyses, many of the questions
raised by the results of this first phase of the project
will hopefully be answered.

REFERENCES

1. Schook, Gary. "Groundwater Quality Management
   Plan for Idaho," Idaho Division of Environment
   (1986),   p. 23.

2. Graham, W. G. and L. J. Campbell. "Groundwater
   Resources of Idaho," Idaho Department of Water
   Resources (1981), p. 100.

3. "Southwest Community Waste Management Study
   Groundwater Subtask, Ada County, Idaho," Ada
   Planning Association Technical Memorandum 308.04 g
   (1979), p. 69.

4. Jones, F.O. and K. W. Lustig. "Gound-water Quality
   Monitoring, Rathdrum Prarie Aquifer," Panhandle
   Health District No. 1 Technical Report (1977),
   p. 94.

5. Collett, R. S. "Soil Survey of Ada County Area,
   Idaho," National Cooperative Soil Survey,
   Washington, D.C. (1980), p. 286.

6. Weisel, C. J. "Soil Survey of Kootenai County
   Area, Idaho," National Cooperative Soil Survey,
   Washington, D.C. (1984), p. 255.

7. Mitchell, V.E., and E. H. Bennett. "Geologic Map
   of the Boise Quadrangle, Idaho," Idaho Bureau
   of Mines and Geology, Moscow, Idaho (1979).

8. Lloyd, L. L. "Evaluation of radon sources and
   phosphate slag in Butte, Montana," Environmental
   Protection Agency - Office of Radiation Program
   No. 520/6-83-026, p. 75.

NATURAL RADIOACTIVITY IN SOME
GROUNDWATERS OF THE CANADIAN SHIELD

Alberta E. Lemire and Melvyn Gascoyne,
Atomic Energy of Canada Limited, Manitoba, Canada

INTRODUCTION

The Canadian Nuclear Fuel Waste Management Program (NFWMP) is
presently assessing the concept of the disposal of nuclear fuel
waste in deep, stable geological formations [1]. As part of the
research, Atomic Energy of Canada Limited (AECL) has constructed
an Underground Research Laboratory (URL) in a granitic batholith
near Lac du Bonnet, Manitoba, to provide a representative geological environment in which to pursue experiments related to the
program.

In 1982, Whiteshell Nuclear Research Establishment (WNRE)[1]
instituted an environmental monitoring program to ensure that any
adverse environmental effects arising from the construction or
operation of the URL were identified and mitigated. As part of
the pre-operational environmental survey, measurements of natural
radioactivity were made in surface and well waters of the vicinity. A similar survey was later done at a second hydrogeological
research site near Atikokan, in northwestern Ontario. The results
of these surveys and of the ongoing environmental monitoring
program form the body of the present paper.

METHODS

Analyses on split samples were done in-house and by outside
laboratories under contract. Methods of analyses used by the
various laboratories are summarized in Table 1.

Locations of AECL research areas are shown in Figure 1.

---

[1]WNRE, located 12 km southwest of the URL, is one of two AECL
research laboratories. The other is located at Chalk River,
Ontario.

Table 1. Summary of Analytical Techniques.

	Uranium	Thorium	Radium	Radon
Lab 1	Alpha counting U on disc after co-precipitation with $Fe(OH)_3$ and ion exchange	Same as U method	^{222}Rn emanation & scintillation counting after storage for $\geq$ 7 days	Degassing - scintillation counting method
Lab 2	Neutron activation analysis using gamma ray from ^{239}U	NA	Gamma counting of 0.352 MeV ray from ^{214}Pb daughter after storage for 21 days	NA
Lab 3	Neutron activation analysis using ^{239}Np gamma ray	Neutron activation using ^{233}Pa gamma ray	Chemical separation of radium followed by co-precipitation with $BaSO_4$, thin source preparation and alpha counting	NA
Lab 4	Solvent extraction of U. Fluorometric analysis	NA	Co-precipitation with $BaSO_4$, thin source preparation, alpha spectrometry	NA
Lab 5	Details not provided	NA	Details not provided	Degassing - scintillation counting method

Lab 1 = WNRE Geochemistry and Applied Chemistry Branch
Lab 2 = WNRE Analytical Science Branch
Lab 3 = Saskatchewan Research Council
Lab 4 = Monenco Laboratories
Lab 5 = Radiation Protection Bureau
NA   = not applicable

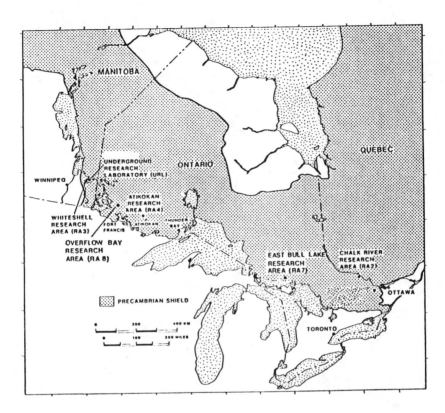

Figure 1.   AECL Research Areas.

RESULTS AND DISCUSSION

(1) Underground Research Laboratory, Pre-Operational Survey

The pre-operational environmental survey in June 1983 at the
URL was a cooperative program, conducted jointly by the WNRE and
the Manitoba Department of the Environment and Workplace Safety
and Health.   Samples were collected from surface water bodies
adjacent to the URL, from boreholes on the URL site, and from
private and public wells adjacent to the property (Figure 2).   The
wells were situated within an approximate 5 km radius of the URL
site and had been drilled into bedrock to varying depths.
One of the initial aims of the study was to establish consistency of methods and analyses between the WNRE environmental
monitoring program and the independent environmental monitoring
program conducted by the Manitoba Department of the Environment
and Workplace Safety and Health.
To this end, independent analyses were conducted on split
samples both in-house by WNRE and by outside laboratories under
contract to WNRE or the Manitoba Department of the Environment and

Workplace Safety and Health. Samples were analysed for uranium,
thorium, radium and radon. However, not all of the analyses were
done by all of the laboratories on all of the samples.

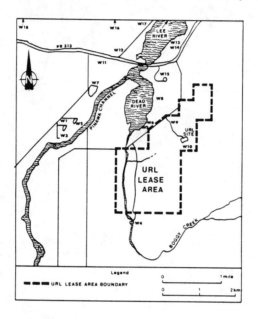

Figure 2. Location of Wells.

(a) Uranium, Thorium and Radium

     Results of thorium and uranium analyses are listed in Tables 2
and 3. Levels of thorium in both surface and well waters were
uniformly low. Levels of uranium in surface waters were low but
varied considerably in the well waters. (Overall, the range was
from 0.7 to 585 µg/L). Agreement of analytical results between
laboratories was reasonable, although some anomolies were
evident.
     Present federal Canadian guidelines for uranium in drinking
water recommend a maximum concentration of 20 µg/L. The guideline
was based on chemical rather than radiological toxicity [2]. If
the results of the analyses of each well are averaged, it is found
that six wells have uranium concentrations of less than 20 µg/L,
seven are between 20 µg/L and 100 µg/L* and six are in excess of
100 µg/L. The results illustrate the point recently made by Hess
et al. [4] that generally small population water supplies have the
highest radionuclide concentrations. The variability between
wells illustrates the need for individual well monitoring even

---

*100 µg/L was the recommended Manitoba provincial guideline at
 the time of the study [3].

within a small geographic area and is analogous to the variation in radium levels in adjacent wells in Illinois reported by Lucas [5].

Most of the well waters had ratios of ^{234}U to ^{238}U in excess of the expected 1:1. This phenomenon is well known (indeed ratios as high as 28 [4] have been reported), and has been variously ascribed to α-particle recoil or to preferential leaching of the radiation damaged site [6,7]. For high-ratio values, the radiological toxicity of the dissolved uranium also becomes more significant.

The study of uranium concentrations in groundwaters is now being extended to cover a larger region of southeastern Manitoba. Two hundred and eighty wells within a 15,500 km² area have presently been studied. The highest concentration recorded to date in an individual well is 2020 µg/L [8].

Table 2.   Underground Research Laboratory Pre-operational
           Survey 1983:   Results of Thorium Analyses.

		THORIUM[1]	
	CODE	LAB 1 (µg/L)	LAB 3 (µg/L)
Surface Water			
Pinawa Channel #5	F1	<5	<3.5
Pinawa Channel #5.5	F2	<5	<3.5
Lee River #3	F3	<5	<3.5
Lee River #8	F4	<5	<3.5
Well Water	W1	<5	<3.5
	W2	<5	<3.5
	W3	<5	<3.5
	W4	<5	<3.5
	W5	<5	<3.5
	W6	<5	<3.5
	W7	<5	<3.5
	W8	<5	<3.5
	W9-1[2]	<5	<3.5
	W9-2	<5	<3.5
	W9-3	<5	<3.5
	W10	<5	<3.5
	W11	<5	−
	W12	<5	−
	W13	<5	−
	W14	<5	−
	W15	<5	−

[1]Thorium Analyses: Measurement Accuracies reported were:
           Lab 1, ±8%; Lab 3, not reported.
[2]W9-1, W9-2, W9-3 refer to samples taken from different depths
  (43 m, 39 m, 150 m) in the same borehole.

Table 3. Underground Research Laboratory Pre-operational
          Survey 1983: Results of Uranium Analyses.

	CODE	URANIUM[1]					$^{234}U/^{238}U$[2]
		LAB 1 (µg/L)	LAB 2 (µg/L)	LAB 3 (µg/L)	LAB 4 (µg/L)	LAB 5 (µg/L)	LAB 1
Surface Water							
Pinawa Channel #5	F1	1.0	3	<3.0	5.7	–	1.4
Pinawa Channel #5.5	F2	0.4	9	<3.0	2.6	–	9
Lee River #3	F3	0.9	5	<3.0	6.0	–	3.3
Lee River #8	F4	0.3	4	<3.0	3.0	–	3.5
Well Water	W1	313	335	268.6	355	299	4.4
	W2	277	305	319	345	229	4.5
	W3	255	285	232	365	271	4.7
	W4	87	100	81.0	140	81	4.2
	W5	17	14	15.1	26	15.5	2.6
	W6	41	45	38.3	39.5	30.1	2.4
	W7	573	550	490	585	562	1.4
	W8	0.7	<5	<3.0	2.8	4.7	1.4
	W9-1	10	26	6.3	15	–	6.1
	W9-2	5	30	5.4	14	–	6.8
	W9-3	12	50	10.1	17	–	4.6
	W10	2	14	<3.0	14	–	1.6
	W11	86	–	–	–	63.1	2.7
	W12	348	–	–	–	410	1.4
	W13	85	–	–	–	65	1.4
	W14	338	–	–	–	333	1.6
	W15	51	–	–	–	51.0	1.2
	W16	–	–	–	–	46.5	–
	W17	–	–	–	–	10.2	–
	W18	–	–	–	–	9.0	–

[1]Uranium Analyses: Measurement Accuracies reported were:
                    Lab 1, ±8%; Lab 2, ±20%; Lab 3, not reported;
                    Lab 4, ±2%, Lab 5, ±9-27%.
[2]Isotope Activity Ratio

    Radium-226 concentrations (Table 4) were generally low in both
surface waters and groundwaters. In one well, however, the radium
concentration exceeded the 1 Bq/L, Canadian federal guideline for
drinking water [2].
    The presence of relatively high levels of uranium in some of
the wells aroused concern among local residents about the safety
of using such water as a long-term drinking water source. Because
of these concerns, personnel of the Whiteshell Nuclear Research
Establishment designed and tested a uranium and radium removal
system that could be used by local residents to treat their well
water [9].

Table 4. Underground Research Laboratory Pre-operational
Survey 1983: Results of Radium Analyses.

	CODE	LAB 1 (Bq/L)	LAB 2 (Bq/L)	LAB 3 (Bq/L)	LAB 4 (Bq/L)	LAB 5 (Bq/L)
			RADIUM-226[1]			
Surface Water						
Pinawa Channel #5	F1	<0.02	<0.7	<0.03	0.027	<0.005
Pinawa Channel #5.5	F2	-	<0.7	<0.03	<0.005	<0.005
Lee River #3	F3	<0.02	<0.7	<0.03	0.015	-
Lee River #8	F4		<0.7	<0.03	0.007	-
Well Water	W1	0.45	0.59	0.7	0.83	0.729
	W2	3.52	3.89	2.5	3.78	4.09
	W3	0.15	<0.5	0.2	0.38	0.277
	W4	0.16	<0.5	0.15	0.21	0.202
	W5	<0.02	<0.5	0.02	0.06	0.041
	W6	-	<0.5	0.11	0.13	0.145
	W7	0.23	<0.5	0.35	0.46	0.383
	W8	0.02	<0.5	<0.03	0.02	0.004
	W9-1	0.21	<0.7	<0.03	0.05	-
	W9-2	-	-	<0.03	0.016	-
	W9-3	0.05	<0.7	0.09	0.079	-
	W10	<0.02	<0.5	<0.03	≤0.005	-
	W11	-	-	-	-	0.07
	W12	-	-	-	-	0.20
	W13	-	-	-	-	0.20
	W14	-	-	-	-	0.04
	W15	-	-	-	-	0.006
	W16	-	-	-	-	0.05
	W17	-	-	-	-	0.007
	W18	-	-	-	-	0.04

[1]Measurement Accuracies were:
Lab 1, ±5% at 1 Bq/L; Lab 2, ±8% at 4 Bq/L; Lab 3, not reported;
Lab 4, ±2% at 3 Bq/L; Lab 5, not reported.

Figure 3 shows the result of a test of the system in a local
household in which the well water supply had high concentrations
of both uranium and radium. For the test, the units were interconnected in series and installed in the cold water line of the
kitchen sink tap. Over 5000 L of water were passed through the
system at a rate of 2 L/min over a 2 month period before the
materials were renewed. For the first 4000 L of water, removal
efficiencies of radium and uranium were 70% and 90%, respectively.
Radium removal was somewhat less consistent than uranium removal,
possibly because of a compaction of the fibre absorbant and
channelling of flow.

The test was continued at lower water demand rates for a period of over a year to confirm the results and demonstrate the suitability of the system for long-term use. A commercial version, designed to last for about one year in a household with average drinking water demand, has now been developed by WNRE in conjunction with Water Conditioning of Canada Limited, Regina.

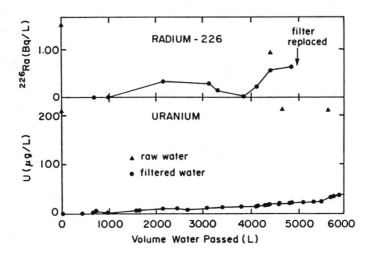

Figure 3.   Effectiveness of the system to remove uranium and radium from drinking water.
Note: Difference between untreated (▲) and treated water (●).

(b) Radon-222

During the pre-operational survey, measurements of ^{222}Rn concentrations were also made on water from several wells. As may be seen in Table 5, considerable variation in radon levels (from < 1 to 390 Bq/L) was observed among wells.

Radon in water leads to exposure of the individual in two ways: the first is by ingestion by direct water consumption and the second is by inhalation when radon degasses from the water.

At present, there are no federal or provincial guidelines in Canada for maximum radon levels in drinking water. The problem of setting such levels has been addressed by several workers and has been reviewed recently by Cross et al. [10] and Dundulis et al. [11]. Proposed standards for maximum radon concentrations have ranged from 67 to 3700 Bq/L depending on the dose models used. Recently Cross et al. [10] have suggested that a rounded ^{222}Rn-in-water concentration limit of 370 Bq/L can be supported by health-effect considerations alone. Three of the wells studied in this survey had radon levels close to this limit.

As in the case of uranium, even the highest radon levels are consistent with levels previously observed in granitic rock [4] and are unusual only in the variation of levels seen within such a small geographic region.

Because the number of wells sampled was small, deriving any general correlations is difficult. The presence of high radon levels did not appear to correlate with any particular geographic location (Figure 2); and, indeed, wells located quite close to each other (specifically W1,W2,W3, and W5,W6) differed markedly in radon concentrations.

No correlation was observed between radium and radon concentrations in the wells studied.

Table 5. Underground Research Laboratory Pre-operational
          Survey 1983: Results of Radon Analyses.

CODE	RADON-222 (Bq/L)	WELL DEPTH m
W1	170	124.0
W2	370	90.5
W3	33	145.7
W4	230	33.5
W5	75	53.3
W6	390	43.6
W7	100-250	90.2
W8	0.1-0.2	27.7
W9-1	-	43
W9-2	-	39
W9-3	-	150
W10	-	-
W11	37	39.3
W12	33	91.4
W13	55	95.7
W14	310	-
W15	2	-
W16	-	-
W17	7	-

(2) Underground Research Laboratory, Operational Monitoring

Shaft sinking for the URL began in May 1984 and was completed to a depth of 270 m by March 1985. During construction and later operations, the groundwater that drained into the shaft was pumped into a holding pond. Periodically, it was necessary to discharge water from this holding pond. To ensure compliance with provincial guidelines for discharged water, the pond water was analysed on an approximately weekly basis for gross alpha and gross beta activities, uranium and radium concentrations, and for other chemical constituents [12]. Water quality varied during the

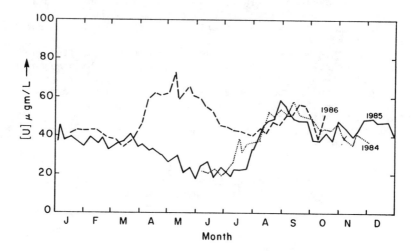

Figure 4.   Seasonal variation in uranium concentration in
            URL pond water.

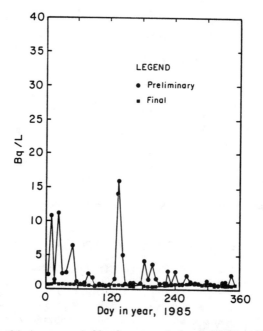

Figure 5.   Preliminary and final gross beta activity in holding
            pond water.
            Note: Gross alpha activity exhibited the same pattern
            as gross beta activity.

construction phase as the shaft passed through one major fracture zone and tapped water from a deeper, more saline zone. When shaft sinking was complete, the water quality stabilized.

From 1984 through 1986, however, it was observed that there was an elevation of the uranium concentration in the water during late summer and early fall. This seasonal elevation was subsequently observed in water taken directly from the major fracture zone and has been ascribed to a recharge phenomena. Wet weather during late spring and early summer leads to an inflow of oxygenated water into the fracture zone at the point where the zone is close to the surface and this results in a leaching of the uranium from the surficial sediments and rock along the fracture zone [13].

In 1986, a large elevation in uranium concentration was also observed in April through June (Figure 4). At present no explanation for this later elevation is evident.

Sporadic elevations of gross alpha and gross beta activities were also observed during pond water monitoring [12]. The elevated activities were short-lived, as may be seen from Figure 5; final activities (measured after 7-10 days) were considerably lower. The only possible source of this short-lived activity appears to be radon daughters. No pattern was evident in the elevations of gross alpha and beta activities. It would appear that pockets of radon-rich water exist within the fracture zones and these are gradually draining into the shaft. The radon is unsupported since no elevation in radium levels in the water has been observed.

(3) Atikokan Area Survey

A radioactivity survey, similar to that conducted at the URL, was conducted at a hydrogeological research area to the north-west of Atikokan, Ontario during the period July 22 to 28, 1984 [14]. The Atikokan site, like that of the URL, is located in a granitic pluton of Archean age.

Results of the analyses are shown in Table 6 [14]. In contrast with the results obtained in the Lac du Bonnet region, uranium and radium concentrations were found to be low and well within the Canadian federal drinking water guidelines (20 µg/L and 1 Bq/L respectively). The reason for the difference is not clear, but it was suggested that the greater enrichment of uranium in fractures in the bedrock in the Lac du Bonnet region may be a contributing factor. The high bicarbonate concentrations in Lac du Bonnet waters would also enhance uranium solubility by forming uranium carbonate complexes.

Table 6. Results of Uranium, Radium and Excess Radon
         Analyses Conducted in the Atikokan Area[1]

Site Description	Location	U (µg/L)	$\frac{^{234}U*}{^{238}U}$	^{226}Ra (Bq/L)	^{222}Rn (Bq/L)
Surface	Finlayson L	< 0.1	–	< 0.02	< 0.1
Waters	Eye L	< 0.1	–	< 0.02	0.1
	Forsberg L	< 0.1	–	0.7	< 0.1
	Dashwa L	< 0.1	–	< 0.02	< 0.1
	Eye L (-2 m)	0.1	–	< 0.02	< 0.1
	Eye L (-3 m)	< 0.1	–	< 0.02	0.2
	Dashwa L (-5 m)	< 0.1	–	< 0.02	0.2
	Stn. 14-1	0.1	–	< 0.02	< 0.1
	Stn. 95-1	0.1	–	< 0.02	< 0.1
	Boy Scout Camp	0.2	–	< 0.02	0.3
	Stn. 119-1	0.1	–	< 0.02	< 0.1
	Eye River trib.	< 0.1	–	< 0.02	0.2
Research	Eye L portage	< 0.1	–	< 0.02	15
Area Springs	Finlayson L.120-1	4.2	1.52±0.07	< 0.02	0.3
Shallow	FL-1-3, -6.1 m	0.6	–	< 0.02	117
drill	FL-2-3, -4.5 m	2.4	1.38±0.29	< 0.02	158
holes	FL-3-1, -10.2 m	< 0.1	–	< 0.02	7.3
(Forsberg	FL-4-1, -16.4 m	0.1	–	< 0.02	1.5
Lake Area)	FL-5-3, -4.7 m	< 0.1	–	< 0.02	82
	FL-6-1, -9.6 m	< 0.1	–	< 0.02	16
	FL-7-2, -10.3 m	3.4	1.44±0.26	< 0.02	23
	FL-9-2, -9.3 m	0.1	–	< 0.02	75
	FL-10-3, -9.8 m	10.2	1.31±0.09	0.04	162
	FL-11-1, -9.2 m	3.4	2.21±0.31	0.03	16
	FL-12-1, -14.1 m	3.2	1.27±0.24	< 0.02	193
Shallow	Site 1, -8.5 m	< 0.1	–	< 0.02	7.4
drill	Site 1, -19.4 m	1.9	4.87±0.95	< 0.02	28
holes	Site 2, -4.9 m	0.2	–	< 0.02	54
(Upper	Site 3,	< 0.1	–	< 0.02	15
Eye Basin)	Site 3, -13.8 m	5.7	6.76±0.42	< 0.02	485
Deep bore-	ATK2-2, -90 m	1.0	3.21±1.16	< 0.02	468
holes	ATK3-3, -90 m	1.9	2.39±0.63	< 0.02	15
(Forsberg	ATK4-4, -120 m	< 0.1	–	< 0.02	20
Lake Area)	ATK5-5, -90 m	< 0.1	–	< 0.02	29
Others	Hardy Dam Spring	< 0.1	–	< 0.02	11
	Powerline Spring	0.7	1.06±0.26	< 0.02	12
	Highland Park Well	0.1	–	< 0.02	4.9
	CN Well	0.8	1.65±0.28	< 0.02	0.7
	Airport Well	0.2	–	< 0.02	5.3
	Airport Spring	0.2	–	< 0.02	2.5
	Radisson Supply	< 0.1	–	< 0.02	< 0.1

*Activity ratios, not significant for uranium concentrations
 below ~ 1 µg/L.

[1]Table from reference 14.

Considerable variation in excess radon concentrations was observed in the groundwaters of the Atikokan area. No correlation was found with depth of sampling òr presence of uranium or radium in solution. There did, however, appear to be a geographical relationship. For example, in the Forsberg Lake area (see Table 6), the presence of significant excess radon correlated with proximity to fault zones in the granitic bedrock. In general, areas away from faults had radon concentrations less than 30 Bq/L, whereas sites close to faults have radon levels exceeding 50 Bq/L and reaching 485 Bq/L in one particular case.

SUMMARY AND CONCLUSIONS

High levels of uranium and radon were found in some individual wells in the Lac du Bonnet region of Manitoba. Considerable variation in concentrations was observed between individual wells located within a small geographic area. The cause of the individual high concentrations is thought to be a combination of localized enrichment in overburden and granitic bedrock and of the high bicarbonate oxygenated groundwater of the region. A similar survey was carried out in the Atikokan region of northwestern Ontario. Uranium concentrations were low, but high radon levels were observed in some drill holes. At the Atikokan site, the presence of significant excess radon correlated with proximity to fault zones in the granitic bedrock.

As a remedial measure, a uranium and radium removal system for individual household use was designed and tested and is now available commercially.

REFERENCES

1. McConnell, D.B. "The Canadian Nuclear Fuel Waste Management Program, 1984 Annual Report", AECL-8398, Atomic Energy of Canada Limited, April (1986).

2. Ministry of National Health and Welfare, Guidelines for Canadian Drinking Water Quality (1978).

3. Williamson, D.A. "Surface Water Quality Management Proposal", Water Standards and Studies Report N83-2 V1, Manitoba Department of the Environment and Workplace Safety and Health (1983).

4. Hess, C.T., J. Michel, T.R. Horton, H.M. Prichard and W.A. Coniglis. "The Occurrence of Radioactivity in Public Water Supplies in the United States", Health Physics 48, 553 (1985).

5. Lucas, H.F. "^{226}Ra and ^{228}Ra in Water Supplies", Journal AWWA, 57(1985).

6. Ivanovich, M. and R.S. Harmon eds. "Uranium Series Disequilibrium: Applications to Environmental Problems", Oxford University Press (1982).

7. Lathan, A.G. and H.P. Schwarcz. "Models of Uranium Etching and Leaching: Their Applicability to Estimating Rates of Natural Uranium Removal From Crystalline Igneous Rocks Using U-Series Disequilibrium Data from the Eye-Dashwa Lakes Pluton", TR-353, April (1986). Unrestricted, unpublished report, available from SDDO, Atomic Energy of Canada Limited Research Company, Chalk River, Ontario KOJ 1JO.

8. Betcher, R.N., D. Brown and M. Gascoyne. "Natural Uranium Concentrations in Groundwaters of Southeastern Manitoba, Canada", to be published.

9. Gascoyne, M. "A Highly Selective Method for Removing Natural Radioactivity from Drinking Water", paper presented at the Canadian Nuclear Society 7th Annual Conference, Toronto 1986 June 9-11.

10. Cross, F.T., N.H. Harley and W. Hofmann. "Health Effects and Risks from ^{222}Rn in Drinking Water", Health Physics 48, 649 (1985).

11. Dundulis, W., W. Bell, B. Keene and P. Dostio. "Radon-222 in the Gastrointestinal Tract: A Proposed Modification of the ICRP Publication 30 Model", Health Physics 47, 245, 1984.

12. Lemire, A.E. "Underground Research Laboratory Environmental Monitoring Program 1985", WNRE-570-4, Atomic Energy of Canada Ltd., 1986 October.

13. Gascoyne, M. "High Levels of Uranium and Radium in Groundwaters at Canada's Underground Research Laboratory", in preparation (1987).

14. Larocque, J.P.A. and M. Gascoyne. "A Survey of the Radioactivity of Surface Water and Groundwater in the Atikokan Area, Northwestern Ontario", TR-379, April (1986). Unrestricted, unpublished report, available from SDDO, Atomic Energy of Canada Limited Research Company, Chalk River, Ontario KOJ 1JO.

DETERMINATION OF BULK RADON EMANATION RATES BY HIGH RESOLUTION GAMMA-RAY SPECTROSCOPY

Nancy M. Davis, Rudolph Hon and Peter Dillon
Boston College, Boston, Massachusetts

INTRODUCTION

Naturally occurring radon consists of three different isotopes, each as an intermediate decay product of either of the U-238, U-235, or Th-232 decay series. Thoron (Rn-220) and actinon (Rn-219) are respective isotopes of radon in the thorium-232 and uranium-235 decay chains, and Rn-222 of the U-238 chain. Radon isotopes are short-lived chemically inert gases and are the only gaseous members of the decay schemes. The decay chain for U-238 is shown in Figure 1.

Being chemically inert gases, isotopes of radon are predictably more mobile than any other isotope (precursor or progeny) of either of the decay series. The greater mobility, particularly that of radon-222, is commonly evidenced by its frequently observed occurrences at unsupported, often anomalously high concentrations in air and in groundwaters associated with uranium enriched host rocks [1-3]. This suggests that radon gas, formed by a decay of Ra-226, migrates through these solids, and eventually escapes from these solids to a much greater extent than any of the other U-Th series radionuclides. The shorter half-lives of thoron and actinon (55 sec and 4 sec, respectively), as compared with the half-life of radon-222 (3.825 days), restrict the mobility of these two isotopes as they are more likely to undergo a decay before escaping from their host solids.

Measurements of radon in air and in natural waters provide information on the mechanisms of radon migration and release from host solids; about the physico-chemical controls on the release mechanisms; and on the nature of the uranium source in host solids. Radon anomalies and their higher than average occurrences in nature have recently become issues due to the suspected health threat posed by prolong exposures to radon and to several of its progeny [4-6]. There

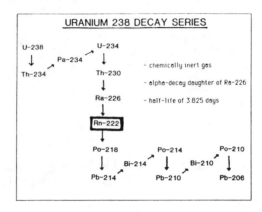

Figure 1. U-238 decay scheme highlighting the position of radon in relation to its precursors and progeny.

is, therefore, an immediate need for a better understanding of radon migration patterns and the development of predictive models of radon occurrences in nature.

Among instrumental and analytical techniques available for measurements of concentrations of uranium/thorium radionuclides in air, solids, and in groundwater samples are alpha-scintillation counting, gamma-ray spectroscopy, spectrophotometry, and atomic absorption. Although the determination of radionuclide concentrations with these methods generally require a complex and tedious preparation of samples prior to analyses, these methods have been proven in duplicate and replicate analyses to be quantitatively and statistically quite reliable and accurate [7-10].

In the present study, a new modified technique of gamma-ray spectroscopy was developed to quantitatively determine radon (Rn-222) emanation losses from bulk samples (size range of the order of 1 dm^3) of selected granitic rocks. Accurate determination of radon loss for each bulk sample is obtained by comparing apparent U-238 abundances calculated from activities of radon precursors (Th-234 and Ra-226) with those calculated from radon progeny (Pb-214 and Bi-214). Activities of these precursor and progeny nuclides were found from the gamma-ray spectra of the nuclides collected by a high resolution-high efficiency reverse electrode germanium (REGe) crystal gamma-ray detector equipped with a 8192 channel multichannel analyzer. This arrangement allowed precise and simultaneous acquisition of the desired gamma-ray spectra from bulk solids without the need for special preparation of samples.

The present study was undertaken for several reasons. One of our objectives was to accurately measure U and Th abundances in bulk samples by properly calibrating our equipment. Secondly, we hoped to develop an analytical procedure for determining bulk radon losses from solids, and finally, we sought to relate our data to predictive models for radon losses based on alpha-recoil and diffusion theories. These objectives were attempted on bulk solids using a wide range and a variety of solid samples.

EXPERIMENTAL

Selection of Solid Samples

The rocks chosen for the present study include: the Andover Granite (MA), the Cape Ann Granite (MA), the Chelmsford Granite (MA), the Concord Granite (NH), the Milford Granite (MA), the Peabody Granite (MA), the Rockville Granite (MN), and the Sharpner's Pond Diorite (MA). The Rockville Granite is often used as the exterior facing stone and sometimes as the interior stone in newly constructed buildings. Table 1 lists these rocks with brief petrographic descriptions.

Table 1. Petrographic Descriptions of the Study Samples.

rock name	age	petrographic description
Andover Granite	Sil. or Ord.	light gray, foliated, medium to coarse-grained muscovite-biotite granite
Cape Ann Granite	Ordovician	alkalic granite to quartz syenite containing ferro-hornblende
Chelmsford Granite	Devonian	light gray, even and medium grained muscovite-biotite granite, mildly foliated
Concord Granite	Carboniferous	fine grained, light gray to buff muscovite-biotite granite
Milford Granite	Proterozoic	light gray to pale orange-pink leucocratic biotite bearing granite
Peabody Granite	Devonian	alkalic granite containing ferrohornblende
Rockville Granite	Precambrian	hypersolvus, very coarse grained biotite granite
Sharpner's Pond Diorite	Silurian	medium grained, equigranular biotite-hornblende diorite

Reasons for Selection

Granites have been chosen for this study because some of the highest radon anomalies have been observed in and around granitic terrains [11-13], and because the amount of laboratory data which exists on radon emanation from crystalline rocks is still rather limited. The samples listed above were selected for the following reasons: availability of samples and/or petrographic data; mineralogical, petrographical, and textural variability among the samples; known or inferred uranium contents of the rocks; and because some of these rocks underlie areas where high radon anomalies have been identified.

Collection and Preparation of Samples

Bulk samples of fresh rock were collected and sized into various particle diameter fractions including: a < 0.5 cm fraction; a 2-4 cm fraction; a 5-10 cm fraction; and a 10-20 cm fraction. These fractions were prepared in our laboratory using standard rock pulverizing tools according to accepted methods. In addition to sizing by particle diameter, several of the rocks were cut into slabs and blocks of varying dimensions suitable to sit on the detector.

Radon loss was determined on volumes of the different particle sizes in order to relate radon loss to particle size (external surface area). Similarly, radon loss from the slabs was determined and related to slab dimensions. Sample weights (in grams) were recorded and used in analyses of uranium contents of the solids. The slab provides a more fundamental geometry for assessing depth of gamma-ray detection and depth from which radon escapes.

A potential problem with cutting the rocks into slabs on a water activated saw was the possibility that surface-bound uranium (radium) on the solid might be removed by the water in the process. As a check to this, several samples were cut on the kerosene saw, and another group was left uncut. $^{40}K/^{226}Ra$ activity ratios, as determined by gamma counting, were compared for uncut samples, water cut samples, and for kerosene cut samples. It was assumed that since potassium (^{40}K) is relatively im- mobile, any changes in this ratio resulting from cutting would reflect ^{226}Ra loss. Results of this analysis on three of the rocks showed that radium was not being removed from the solids during either of the cutting processes.

Another possible problem was the potential for radon loss during field collection and during the cutting and pulverizing processes. To ensure accuracy, most samples were allowed at least the minimally accepted 30 days storage after collection and/or preparation for radon ingrowth from and equilibration with its parent, radium (Ra-226). After 30 days or approximately 8 half-lives of radon has elapsed, radon should be in equilibrium with its parent radium. This means that radon is being produced at the same rate at which radium is decaying, and that radon escape rate is equal to its production/decay rate. Some samples, however, were counted on the detector within days of preparation while others had been in storage for periods up to four years.

THE DETECTOR

Detector Specifications

The Canberra brand reverse electrode Ge (REGe) detector was used for gamma-ray spectroscopy analysis. This detector is similar in geometry to other coaxial germanium detectors, but differs in having its electrode arrangement opposite from the conventional detectors. In the REGe, the p-type electrode (ion implanted boron) is on the outside and the n-type contact is on the inside. This electrode arrangement reduces window thickness which in turn extends energy response down to about 10 keV, and offers greater radiation damage resistance to the detector.

The germanium crystal used in this study has a relative efficiency of 16%, a resolution (full width-half maximum) of 1.80 keV at 1332 keV, and a peak to Compton ratio of 40:1 at 1332 keV. The crystal sits approximately 0.5 mm below the surface cover and is attached to the cryostat through a vertical dipstick mount. The detector was enclosed on all sides by 10 cm thick lead bricks to shield out any external radiation. On the interior of the detector cavity layers of copper and cadmium plating were inserted to minimize secondary and/or scattered radiation from the lead shield. The dewar, which holds the liquid nitrogen, was surrounded by styrofoam to further isolate the sensitive electronics to minimize electrical inter- ference caused by acoustical noise.

The signals were amplified with a Canberra amplifier (model 2021) and analyzed with a Nuclear Data ADC (model ND 581) and Nuclear Data ND 684, 8192-channel multichannel analyzer. The ADC has a fixed dead-time of 5 microseconds. The data were processed using a PDP11/23 computer under RSX11/M operating system. Our software includes standard industry and some custom written programs for optimal data reduction.

Calibration of the Detector

Absolute Detector Efficiency

The absolute efficiency of a detector is the ratio between the number of counts produced by the detector to the total number of gamma-rays produced by the source in all directions [14]. Absolute efficiency depends on three factors: 1) the source geometry, namely the shape of the source and its distance from the detector; 2) the intrinsic efficiency of the detector; and 3) gamma-ray energy.

Seven point-source standards (Isotope Products Laboratories, Burbank, CA), of known activities (traced to NBS certified standards) and covering an energy range from 80 keV to 1332 keV, were each counted in 12 specific geometries (vertical distances above and horizontally offset positions from the detector crystal). A charateristic value for efficiency (peak area : gamma activity) was obtained for every gamma-ray of each of the standards at all geometrical positions. The efficiencies were both hand calculated and calculated with the Nuclear Data Efficiency Calibration (EFF) Computer Program [15]. Agreement between the two methods was in all cases within 0.1%.

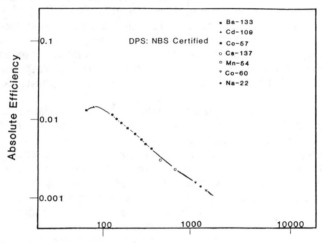

Figure 2. Absolute efficiency calibration curve obtained from NBS-traced calibration standards on CANBERRA REGe detector.
Position: on center, 6 cm from the detector head.

A log-log plot of efficiency vs. energy (keV) for each geometry gives a characteristic curve which describes the variation in detector efficiency with gamma-ray energy. The shape of the curve is more or less constant for a particular detector but will shift in position depending on sample geometry. Figure 2 shows the efficiency plots obtained for our seven standards counted at ~ 6 cm above the detector. Efficiency values are used as correction factors in equations to calculate absolute activities of gamma-emitting nuclides.

Acquisition of Gamma-Spectra from Bulk Samples

Counting Procedure

Bulk samples of the selected granites (2-20 cm diameter fractions of particles and slabs) were placed directly on the detector in fixed geometries for counting. The < 0.5 cm diameter fraction particles were placed in polyethylene sample containers which were then placed directly on the detector. Samples were counted on the detector for 12 to 24 hour periods of time in order to collect statistically accurate gamma spectra. Table 2 lists the nuclides of uranium-238 series which were of interest to this study as well as their associated gamma-ray energies. Note that radon gives off a gamma-ray at 510 keV, but this energy is too close to the 511 keV annihilation peak to be succesfully resolved by any method.

NUCLIDE	GAMMA-RAY ENERGIES
Th-234 $\longrightarrow$	63.28, 92.79 keV
Ra-226 $\longrightarrow$	186.18 keV
Rn-222 $\longrightarrow$	510 keV
Bi-214 $\longrightarrow$	295.21, 351.99 keV
Pb-214 $\longrightarrow$	609.31, 1120.28 keV

Table 2. Radon precursors and progeny of interest to this study and their associated gamma-rays.

DETERMINING BULK RADON LOSS

Analytical Method

Bulk radon loss from the various particle sizes of the selected granites has been determined. The analytical method involves first determining the absolute activities of radon precursors (Th-234 and Ra-226) and of radon progeny (Pb-214 and Bi-214) in each of the samples. The theory behind this analytical technique involves two principal considerations or assumptions: 1) secular equilibrium among the uranium-238 series nuclides in the solid has been established; and 2) a fraction of the radon atoms which form in the solid is able to and in fact does escape from the solid. A condition of secular equilibrium for the uranium-238 nuclides requires that the decay (production) rate of any daughter in the chain is equal to that of its parent. In a decay series consisting of a very long lived parent and a series of short-lived intermediate daughters (e.g. U-238), the condition of secular equilibrium is propagated through the entire series in such a manner that all intermediate nuclides have equal decay rates:

$$\lambda_1 N_1 = \lambda_2 N_2 = \lambda_3 N_3 = \ldots = \lambda_n N_n \qquad (1)$$

where λ^*N is the decay rate or activity of any nuclide in the decay chain. When secular equilibrium has been established, the number of atoms present of any of the nuclides in the chain may be calculated from the activity of any of the others:

$$N_x = (\lambda_y N_y)/\lambda_x \tag{2}$$

Secular equilibrium is assumed to be established in systems (in our case, granites) which are older than 8 times the longest half-life of any of the intermediate decay daughters (U-234 in this case with a half life of 248,000 years). Since our samples are all older than 330 million years (Table 1), we have used secular equilibrium principles by way of equation (2) to solve for uranium concentrations (N_x) from the known decay constant of uranium (λ_x) and from the activities ($\lambda_y N_y$) of Th-234, Ra-226, Pb-214 and Bi-214 as determined by gamma-ray spectroscopy.

Our second assumption was that radon is escaping from the host solids. Referring to Figure 1 and to equation 1, we suggest that if radon is escaping from the solid, then the activities of radon progeny Pb-214 and Bi-214 in the decay chain should be less than predicted by principles of secular equilibrium, and less than the activities of the precursor nuclides. The lower activities of radon progeny should then be reflected in the lower "apparent" concentrations of uranium based on these progeny nuclides. Comparison between precursor uranium values and progeny values then gives us a measure of radon loss from the solid. Uranium concentrations determined by this method using the precursor nuclides are equal to the actual uranium abundances because the requirement of secular equilibrium for these nuclides is satisfied and because there is no appreciable documented loss of any daughters between U-238 and Ra-226.

Calculation of Uranium Concentrations

Referring to Table 2, a uranium concentration has been determined from each of the gamma-rays associated with the nuclides shown. In all, seven uranium concentrations (actual and apparent) have been determined for each solid sample - three based on radon precursors (Th-234 and Ra-226) and four based on radon progeny (Pb-214 and Bi-214). A value for uranium concentration for each of the seven gamma-rays may be calculated from:

$$[\text{ppm U-238}] = \frac{1.0E06 \times 238.03}{\text{Sample wt. (gms)}} \times \frac{1}{\text{Avogadro \#}} \times \text{cps} \times$$

$$\frac{1}{\lambda(\text{U-238})} \times \frac{1}{\text{Eff.}} \times \frac{1}{\text{B.R.}} \times \frac{1}{0.9928} \tag{3}$$

where:
$\quad$238.08 = molecular weight of uranium-238 (grams per mole)
$\quad$cps = counts per second in the desired gamma-ray peak
$\quad\lambda$(U-238) = decay constant of U-238 (4.9159E-18 sec $^{-1}$)
$\quad$Eff. = coefficient of absolute bulk detector efficiency
$\quad$B.R. = branching ratio of the desired gamma-ray energy
$\quad$0.9928 = isotopic ratio of U-238 relative to total uranium

A check of units yields micrograms of U-238 per gram of host solid which is equal to ppm U-238 in the solid. For one sample of solid, the only values that change in the equation are the cps (obtained directly from the computer printout), the branching ratio (available in isotope tables), and the absolute bulk efficiency values. Among samples, these values also change as does the sample weight. The terms in equation 3 are all straight foward except for the coefficient of efficiency. Because our samples represent bulk solids, the efficiency values may not be obtained directly from the point source efficiency calibration curves shown in Figure 2. Instead, we have determined a bulk efficiency coefficient based on an efficiency calculation derived from the known potassium concentrations in each sample.

Since the K_2O content of a solid is given for each of the rock types, the contribution from ^{40}K may likewise be easily calculated. Gamma-ray analysis of the potassium peak in each sample compared with the calculated ^{40}K value then allows us to assign the bulk efficiency value to the potassium gamma peak (1460 keV). This value takes into account the absolute efficiency of the detector at this energy, and the geometry in which the sample was counted.

RESULTS AND DISCUSSION

Results

The results are shown plotted in bar graph forms in Figure 3. Each of the graphs represents the results for radon loss from a different particle size fraction. The solid panels in all cases show the actual uranium concentration (ppm) based on radon precursors or parents (Th-234 and Ra-226), and the light panels indicate the uranium apparent concetrations based on radon progeny or daughters (Pb-214 and Bi-214). Above each set of panels for a given rock is a percent value which indicates percent radon loss as determined from equation 4:

$$\%Rn \text{ loss} = [(\text{ppm U-238}_p - \text{ppm U-238}_d) / (\text{ppm U-238}_p)] \times 100 \qquad (4)$$

where the subscripts p and d indicate U-238 based on radon parents and daughters, repectively. Several points should be noted from these bar graph representations. Each set of bars for each rock is an average for the number of samples analyzed for that rock and for that particle size. We have tried to use at least three samples of each rock in a particular

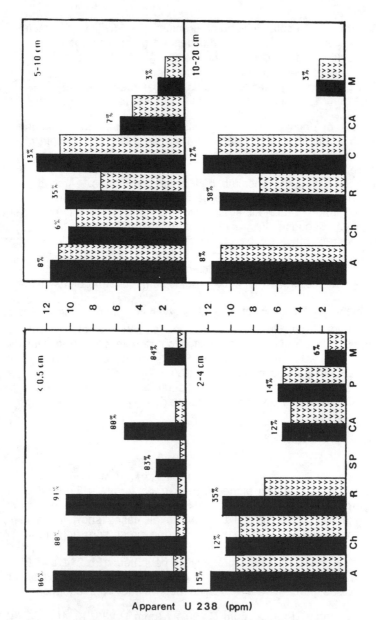

Apparent U 238 (ppm)

Figure 3.   Bulk % radon losses for different fraction sizes of selected granites.
In solid panels are shown the actual U-238 concentrations (based on Th-234
and Ra-226);   light panels (letters 'v') represent the apparent U-238
(based on Pb-214   and Bi-214).   Legend: A-Andover Granite;
CA-Cape Ann Granite; Ch-Chelmsford Granite; C-Concord Granite;
M-Milford Granite; P-Peabody Granite; R-Rockville Granite;
and SP-Sharpner's Pond Diorite.

size group. The height of the individual panels does not reflect radon loss, instead, radon loss for a particular rock sample is indicated by the relative difference between the solid and light panels (see equation 4). Finally, a point to note is that radon loss for each rock is incrasing with decreasing particle size, but at varying rates; the Rockville Granite shows the greatest percent radon loss over all particle sizes above the smallest particle fraction. Figure 4 shows these trends more clearly with all the data points plotted as % radon loss vs. particle diameter for all selected samples included in this study.

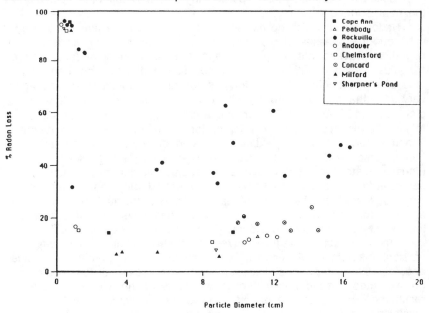

Figure 4.   Data from Figure 3, plotted as % radon loss vs. particle diameter (cm).

DISCUSSION

Mechanisms Affecting Radon Release from Solids

The fraction of radon atoms formed in a solid which escapes from the solid is known as the emanating power of that solid for radon [16]. Before analyzing these results, a discussion of the mechanisms known to affect radon release from solids is useful. Three different physical mechanisms have been shown to control radon release from solids: alpha-recoil, diffusion, and transportation. Contributions of each of these mechanisms toward releasing radon from the solids depends on certain physicochemical properties of radon and of the host solids.

Alpha-recoil derives its name from the process by which a radon atom recoils from a decaying parent radium atom. On decay, a radium atom (Ra-226) emits an alpha particle (He-4) to form radon (Rn-222).

Kinetic energy of this alpha-particle is, by principle of conservation of momentum, sufficient to cause the newly formed radon atom to be recoiled some distance in the opposite direction. This distance, known as the "alpha-recoil length" has been shown to vary between 10^{-6} and 10^{-3} cm depending on the medium through which the recoil occurs [17-18]. The recoil length increases with decreasing resistance of the host medium through which recoil is occurring. Despite the short recoil distances of radon, it is possible for a radon atom to escape from a host solid by this mechanism particularly if the recoil takes place in the near surface region.

Radon atoms which do not escape by recoil release may escape by subsequent processes of diffusion. Diffusion may be seen as a particle migration process with respect to a still ambient medium (solid, air or a liquid). As it is with the recoil process, the diffusion mechanism has also associated characteristic lengths for radon movement which again vary with the medium through which diffusion occurs. The diffusion distances (10^{-9} cm in solids, around 2 cm in liquids, and around 200 cm in air filled spaces) include a factor for radon half-life (3.825 days) and therefore represent maximum diffusion distances reached before a nearly complete decay of the radon has occurred [19-21]. These diffusion distances cited above also assume maximum diffusion coefficients for radon and linearly unobstructed diffusion. Since these conditions are not often met in real solids, factors must be included to account for non-linearity of the diffusive pathways; actual diffusion lengths should then be considerably shorter.

Transportation is particle movement by the action of a moving ambient medium (liquid or air). Because of the low primary porosities and low intrinsic permeabilities of crystalline rocks, air and water are not expected to readily flow through the bulks of these rocks. As a result then, transportation of radon by air or by water through crystalline rocks is not a mechanism that should exert significant control on radon loss. Alpha-recoil and diffusion are undoubtlessly the principal mechanisms by which radon escapes from granites and other crystalline rocks.

Controls on the Release Mechanisms

Andrews and Wood (1972) measured radon loss to water from a range of particle diameters of three sedimentary rock samples using alpha scintillation counting technique. Their samples included a sand (5-150 microns), a limestone (<149-25,000 microns), and a cemented sandstone (<149-25,000 microns). One of their objectives was to quantify radon loss from these solids as a function of particle diameter. The authors then related this functional relationship to the release mechanisms (alpha-recoil and diffusion) in terms of how the structure of the host solids might be influencing radon loss. A graphical representation of their results for percent radon loss vs. particle diameter is shown in Figure 5.

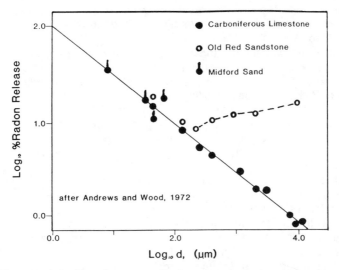

Figure 5.   Logarithm of percent radon loss to water vs. logarithm particle
diameter for three sedimentary rocks. After Andrews and Wood (1972).

The graph in Figure 5 shows a near linearly inverse relationship
between log particle diameter and log percent radon loss for the sand
and limestone samples examined by Andrews and Wood (1972). This
trend shows a decrease in percent radon loss from these solids with
increasing particle diameter.   Percent radon loss from the cemented
sandstone follows the same trend for particle dimensions up to that of
about 450 microns in diameter.   By contrast, for particle sizes above
approx. 450 microns of the cemented sandstone, however, radon loss
tends to, at first, increase with increasing particle sizes to about the
4,000 micron diameter fraction.   Radon loss beyond this size up to the
25,000 micron fraction remains nearly constant at levels of about 10-
15 %.

These authors interpreted their results by diferences in the
release mechanisms (alpha-recoil and diffusion) and by differences in
the host solid structures.   The specific surface area of a material
depends on its particle size, being proportional to 1/d where d is the
particle diameter.   Given this relationship, the authors expected to find
% radon release from these particles proportional to 1/d.   These
arguments were based on the assumption that the sand and limestone
samples are isotropic in radium distribution and structure.

The relationship between radon loss and particle size for the sand
and limestone samples as shown in Figure 5 indicates that radon
release is instead proportional to $1/d^{-1/2}$.   Andrews and Wood (1972)
concluded that radon release from these two samples was occurring
primarily by the recoil mechanism with some contribution from
diffusion along grain contacts.   They also noted that release by these
two mechanisms should depend on "the extent to which grain

boundaries or imperfections intersect the particle surface.   This probability of grain boundary intersection with a surface is proportional to $1/d^{-1/2}$.

In case of the cemented sandstone the same radon release mechanism is postulated to prevail for particle sizes up to 450 microns.   The authors suggested that radon release from sandstone particles larger than 450 microns in diameter "must be due to the rapid diffusion of radon through the cementing material from deep within the rock fragments, release into the cementing material having taken place either by recoil or diffusion from the mineral fragments embedded within the cement or by release from grains of the cementing material itself."

From the discussion above and from the results of the Andrews and Wood (1972) study, we suggest that several factors will tend to maximize radon release from uranium enriched host solids:
- concentration of uranium (radium) near an escape surface
- fine particle size of the host solid (large surface areas)
- microporosity in the host solid
- interconnectedness of this porosity with surfaces
- concentration of uranium (radium) along microporous channels.

These and additional factors have been outlined and examined in other studies [22-25].   Figure 6 is a graph of our results as shown in Figure 4 but superimposed over the line trends of Andrews and Wood (1972) (Figure 5).

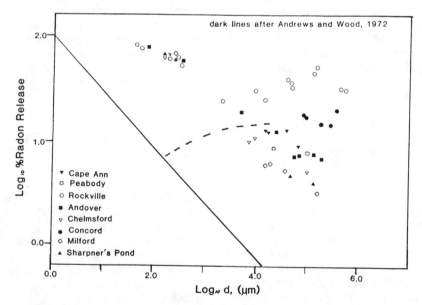

Figure 6.   Log-log plot of % radon release vs. particle diameter for samples of granites included in the present study. Our data same as Figure 4; line trends from Figure 5 after Andrews and Wood (1972).

SUMMARY AND CONCLUSIONS

Our results show a general trend of decreasing radon loss with increasing particle diameter. From Figure 6, we note that the rates of decrease vary within and among rock types in an unpredictable way. We have noted that, in addition to scatter, percentages of radon loss from selected granites included in this study are somewhat higher than those determined by Andrews and Wood (1972). Our linearly inverse trends lie somewhat higher than but still parallel to the trend observed by Andrews and Wood (1972). Given our results in comparison with those obtained by Andrews and Wood (1972), we would like to suggest that radon release from our granites is occurring likewise by processes of alpha-recoil and diffusion.

The results on the Rockville Granite are of particular interest to us. This rock shows the greatest percentage radon loss across the whole range of particle fractions. In fact, radon loss from these rocks is on average 2-3 times higher than for equivalent sized particles of the other granites. We believe that radon loss from the Rockville Granite is real, does in fact deviate from a linearly inverse trend, and does follow more clearly the trend found for the cemented sandstone used by Andrews and Wood (1972).

Such excessive radon losses from the Rockville Granite are unexpected for two reasons: 1) preliminary analysis of the rock does not indicate that the bulk of the uranium (radium) is concentrated on nor at surfaces of these particles; and 2) the average diameter of grains composing the Rockville Granite is ~1 cm, which is relatively large and should result in small overall surface area of the rock. One criterion for maximizing radon release from solids is to have small particle (and grain) diameters which in turn increase surface area for radon emanation. On this basis, we would have expected lower radon losses from the Rockville and Cape Ann (Peabody) Granites (coarse-grained rocks) and higher radon losses from the finer-grained samples such as the Milford, Chelmsford, Andover, and/or Concord Granite samples.

We suggest that some of the scatter (deviation from linearity) observed for larger particle diameters of our finer-grained granites might be due to radon diffusion through their larger internal surface areas. For the Rockville Granite, we hypothesize that there are imperfections such as inter- or intragranular microcracks in this solid which may be serving as conduits for rapid diffusion of radon from depths to escape surfaces. The ease with which we were able to break the Rockville Granite suggests that this rock may indeed be extensively permeated with microcracks. The presence of such microcracks could possibly offset the negative effects of small emanation surface areas on radon escape. Radon escape from the Rockville Granite by diffusion along such hypothetical microcracks would, of course, be enhanced if the radon were to be forming along these microcracks, i.e., if the microcracks were partially healed with uranium (radium) bearing minerals to provide a direct, near-surface supply of radon.

FUTURE WORK

We plan to continue to use and improve this method of gamma-ray spectrocsopy to examine radon loss from these and other bulk solids. A number of other granites may be included in the study. Our primary interest is in further investigating the possibility of enhanced radon loss due to microporosity in the solids. We plan to analyze our samples for microcracks and for radiation damage induced microporosity. Both types of porosity have been suggested as enhancers of radon release from solids [26-29].

ACKNOWLEDGEMENT

We would like to express our thanks to numerous colleagues and friends who kindly made available to us the various samples of granites included in this study. This paper is part of a M.Sc. dissertation thesis of Nancy Davis at the Department of Geology and Geophysics, Boston College.

References

1. Asikainen, Matti and H. Kahlos. "Anomalously high concentrations of uranium, radium and radon in water from drilled wells in the Helsinki region," Geochim. Cosmochim. Acta. 41:1681-1686 (1979).

2. Cowart, J.B. and J.K. Osmond. "Uranium isotopes in groundwater: Their use in prospecting for sandstone type uranium deposits," J. Geochem. Explor. 8:365-379 (1977).

3. King, P.T., J. Michel and W.S. Moore. "Groundwater geochemistry of Ra-228, Ra-226 and Rn-222," Geochim. Cosmochim. Acta. 46:1173-1182 (1982).

4. Hess, C.T., R.E. Casparius, S.A. Norton, and W.F. Brutsaert. "Investigations of natural levels of radon-222 in groundwater in Maine for assessment of related health effects," in: The Natural Radiation Environment, III, USDOE CONF-780422, 1:529-546 (1978).

5. Hess, C.T., C.V. Weiffenbach, and S.A. Norton. "Environmental radon and cancer correlations in Maine," Health Physics. 45:339-348 (1983).

6. Lyman, Gary H., Carolyn G. Lyman, and Wallace Johnson. "Association of leukemia with radium groundwater

contamination," The Journal of the American Medical Association.   254:621-626 (1985).

7.   Prichard, H. and T. Gesell. "Rapid measurements of Rn-222 concentrations in water with a commercial liquid scintillation counter," Health Physics. 33:577-581 (1977).

8.   Michel, J., W.S. Moore, and P.T. King. "Gamma-ray spectrometry for determination of radium-228 and radium-226 in natural waters," Analy. Chem. 53:1885-1889 (1981).

9.   Sarmiento, J.L., D.E. Hammond, and W.S. Bruecker. "The calculation of the statistical error for Rn-222 scintillation counting," Earth Plan. Sci. Lett. 32:351-356 (1976).

10.   Fleischer, R.L. and A.C. Delany. "Determination of suspended and dissolved uranium in water," Analy. Chem. 48:642-645 (1976).

11.   Hall, F.R., P.M. Donahue, and A.L. Eldridge. "Radon gas in ground water in New Hampshire," in Proceedings of the Second Annual Eastern Regional Ground Water Conference (Worthington, Ohio: National Water Well Association, 1985), pp. 86-100.

12.   Asikainen, Matti. "State of disequilibrium between U-238, U-234, Ra-226 and Rn-222 in groundwater from bedrock," Geochim. Cosmochim. Acta. 45: 201-206 (1981).

13.   Andrews, J.N., I.S. Giles, R.L. Kay, D.J. Lee, J.K. Osmond, J.B. Cowart, P. Fritz, J.F. Barker, and J. Gale. "Radioelements, radiogenic helium and age relationships for groundwaters from granites at Stripa, Sweden," Geochim. Cosmochim. Acta. 46:1533-1543 (1982).

14.   Canberra Products Catalogue, Canberra Industries, Inc., 1981-1982.

15.   Nuclear Data RSX-11M Application Software Operator's Manual, Nuclear Data, Inc., 1985.

16.   Giletti, B.J. and J.L. Kulp. "Radon leakage from radioactive minerals," American Mineralogist. 40:481-496 (1955).

17.   Tanner, A.B. "Radon migration in the ground: A review," in: The Natural Radiation Environment, J.A.S. Adams and W.M. Lowder Eds. (Chicago: Chicago University Press, 1964). pp.161-190.

18.   Andrews, J.N. and D.F. Wood. "Mechanisms of radon release in rock matricies and entry into groundwaters," Institution of

Mining Metallurgy: London Transactions Sec. B81 (Applied Earth Science). 197-209 (1972).

19.   Ibid. 16

20.   Ibid. 17

21.   Ibid. 18

22.   Ibid. 17

23.   Lanctot, E.M., A.L. Tolman, and M. Loiselle. "Hydrogeochemistry of radon in ground water," in Proceedings of the Second Annual Eastern Regional Ground Water Conference (National Water Well Association, 1985).

24.   Brutsaert, W.F., S.A. Norton, C.T. Hess, and J.S. Williams. "Geologic and hydrologic factors controlling Rn-222 in groundwater in Maine," Groundwater. 19(4), 407-417 (1981).

25.   LeGrand, Harry E. "Radon and radium emanations from fractured crystalline rocks- a conceptual hydrological model," Groundwater. Vol :59-69 (1986).

26.   Fleischer, R.L. "Alpha-recoil damage and solution effects in minerals: uranium isotopic disequilibrium and radon release," Geochim. Cosmochim. Acta. 46:2191-2201 (1982b).

27.   Rama and W.S. Moore. "Mechanism of transport of U-Th series radioisotopes from solids into groundwater," Geochim. Cosmochim. Acta. 48:395-399 (1983).

28.   Simmons, G. and L. Caruso. "Uranium migration and microcracks in Sherman Granite, Wyoming," Preprint. (1985).

29.   Barreto, P.M.S., R.B. Clark, and J.A.S. Adams. "Physical characteristics of radon-222 emanation from rocks, soils and minerals: Its relation to temperature and alpha-dose," in: The Natural Radiation Environment II, USERDA Conf-720805-P2 (1972).

NANCY DAVIS, received a Bachelor of Arts degree with a major in geology in 1982 from Smith College, Northampton, Massachusetts. At the present time, she is a graduate student completing a Master of Science degree in the Department of Geology and Geophysics at Boston College, Chestnut Hill, Massachusetts. Her area of interest is environmental geochemistry with specific research emphasis on the mechanisms of radon emanation from crystalline rocks. She is also employed as a hydrologic field assistant by th U.S. Geological Survey-Water Resources Division, Boston, Massachusetts. There, she is involved with a water-table mapping project using ground penetrating radar on Cape Cod, Massachusetts.

RUDOLPH HON, is Professor of Geochemistry and Petrology in the Department of Geology and Geophysics, Boston College, where he teaches and conducts research in the fields of geochemistry, environmetal geochemistry, theoretical petrology, distribution of trace and minor elements in rocks and the behavior of elements during geologic cycles. He received a Ph.D. from the Massachusetts Institute of Technology in 1976 and spent three years at the University of Oregon and as a visiting researcher also at the Arizona State University, Tempe. One current research interest is a study of various naturally occurring radioactive isotopes, particularly their distribution in the upper crustal regions of the earth. Results of this and other studies have been presented at national and international meetings and published in scientific journals.

PETER DILLON, received a Bachelor of Science degree in geology in 1983 from Boston College, Chestnut Hill, Massachusetts. He is currently enrolled in the Masters of Science program in the Department of Geology and Geophysics at Boston College. His area of study is igneous petrology and trace element geochemistry. He is also employed as an intern at the Department of Environmental Quality Engineering (DEQE) where he is developing a program to monitor radon in groundwater throughout Massachusetts.

Mailing Address: Department of Geology and Geophysics
Devlin Hall
Boston College
Chestnut Hill, MA 02167

RADON IN GROUNDWATER OF THE
LONG VALLEY CALDERA, CALIFORNIA

Steve Flexser, Harold A. Wollenberg, and Alan R. Smith,
Lawrence Berkeley Laboratory, Berkeley, California

Abstract: In the Long Valley caldera, an area of recently (~ 550 y) active volcanism and current seismic activity, ^{222}Rn concentrations in hot, warm, and cold spring waters have been measured since 1982. Rn contents of the waters correlate inversely with temperature and specific conductance, with high concentrations (1500 to 2500 pCi/l) occurring in dilute cold springs on the margins of the caldera, and low concentrations (12 to 25 pCi/l) in hot to boiling springs. Rn correlates only slightly with the uranium contents of the rocks which host the hydrological system feeding the springs, which encompass a wide range of rock types.

Anomalous changes in groundwater Rn contents may accompany or precede earthquake activity, and a continuous Rn monitoring system was installed in 1983 to monitor short-term variations. A gamma detector is submerged in a natural pond fed by ~ 11°C spring waters with ~ 700 pCi/l Rn, and measured gamma activity is due almost entirely to ^{222}Rn in the water. The gamma record, which is integrated hourly, shows a consistent, pronounced diurnal variation (~ 30% of mean count rate), and weaker higher frequency variations. This pattern correlates well with small variations (< 1°C) in water temperature at the Rn monitoring point, and is strongly influenced by precipitation and by patterns of water flow in the pond. It does not adhere closely to a tidal pattern.

These environmental effects on the radon record may mask responses to small or distant seismic events. To date, anomalous changes in waterborne Rn have been observed in connection with at least one earthquake, which occurred close to the monitoring site. This continuing study points out that an understanding of the geological setting, its associated hydrological system, and environmental influences is necessary to properly evaluate concentrations and changes in groundwater radioactivity.

INTRODUCTION

The Long Valley caldera is an area of active volcanism and ongoing seismic activity. Because of the potential seismic and volcanic hazards, as well as the intrinsic geological importance of the area, various geophysical and geochemical parameters have been and continue to be monitored at sites within and adjacent to the caldera. Among these prameters are hydrogen, helium, and radon in soil gas, and chemical constituents of groundwaters. Long Valley is therefore an excellent site for close comparison between variations in groundwater radon and other geochemical and geophysical variations.

Accordingly, since 1982 we have been analysing radon contents of water from hot, warm, and cold springs. Concurrently, rocks encompassing the hydrologic systems feeding the springs were analyzed for their radioelement contents, in an effort to better understand the source term of the ^{222}Rn in the water. Finally, for observation of short-term variations, a continuous monitoring system for measuring water-borne radon at a spring has been in place since 1983. Early results and interpretations of these studies were described by Wollenberg et al. [1]. This report incorporates more recent data, with particular emphasis upon the continuous monitoring system.

HYDROGEOLOGIC SETTING

The hydrologic system of Long Valley has most recently been described by Sorey et al. [2]. The Long Valley caldera, whose location is shown in Fig. 1, is situated on the eastern front of the Sierra Nevada in eastern California. Volcanism in the Long Valley area has continued inter-

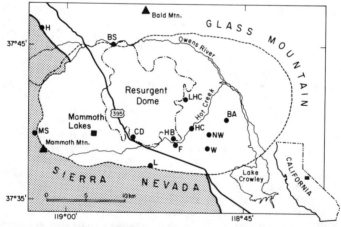

Figure 1.   Outline map of the Long Valley caldera (after Sorey et al. [2]), showing locations of the springs sampled. Dashed line: border of caldera; dot-dashed line: outline of resurgent dome. Letter symbols keyed to Table 2.

mittently over approximately 3 million years, the most recent activity occurring 500 to 600 years ago. The major event that formed the caldera occurred about 0.7 million years ago, when a very large volume of predominantly rhyolitic material (the Bishop Tuff) erupted, followed by collapse of the roof of the magma chamber. The present configuration of the caldera is that of steepsided bowl demarcated by ring fractures. It is filled predominantly by the Bishop Tuff, but also includes more recent volcanics and fluvial, lake-bed, and glacial deposits. The western part of the caldera is occupied by a resurgent rhyolitic dome which is bounded on the north, south, and west by a moat partly filled with rhyolitic and basaltic volcanic rocks.

Sorey et al. [2] point out that two principal groundwater subsystems, one shallow and at near-ambient temperature and one relatively deep and considerably hotter, make up the hydrologic system of Long Valley. Water in both of these subsystems is derived from meteoric sources, primarily runoff from the Sierra Nevada. The post-caldera volcanic rocks and sediments are relatively permeable, and a stratification of hot and cold aquifers in these rocks is evident from geothermal well logs. By contrast, the underlying Bishop Tuff, predominantly densely welded, is relatively impermeable, and fluid movement in that unit probably is confined to fault zones and fractures.

Chemical geothermometry by Mariner and Willey [3] indicates that water at depth in the geothermal systems at Casa Diablo Hot Springs and Hot Creek gorge attains temperatures well in excess of 200°C. A generalized concept of the hydrothermal system then envisions cold-water recharge from the Sierra Nevada on the west and south sides of the caldera, water moving eastward through caldera fill and becoming heated and at the same time mixing with locally recharged groundwater. The warm and hot water then occasianally 'daylights' at springs whose locations are controlled mainly by faults and fronts of volcanic flows.

RADIOELEMENT CONTENTS OF ROCKS IN THE LONG VALLEY AREA

The uranium and thorium contents of rocks through which groundwater percolates control the initial abundances of their daughter products in the groundwater. With regard to this we measured radioelement contents of rocks in the Long Valley region by field and laboratory gamma spectrometry. Results of these analyses are listed in Table 1. Low concentrations of uranium and thorium are evident in basalt of the north and south moats and in carbonate rocks that border the caldera in the Sierra Nevada to the south. Metamorphosed volcanic and clastic (sedimentary) rocks bordering the caldera have intermediate radioelement abundances, while the bordering Sierran granitic rocks and rhyolites within the caldera are of relatively high radioactivity. The influence of the rocks' radioelement concentrations on the radon content of the groundwater is discussed in a following section.

The radioelement content of spring deposit material in Long Valley has not yet been fully investigated. Samples have been collected and analyzed from three spring areas: Hot Creek, Casa Diablo, and north of Whitmore. These are predominantly siliceous sinter deposits with low radioelement contents, in keeping with observations of spring deposits in

Table 1. Radioelement Concentrations of Long Valley Rocks

	U(ppm)	Th(ppm)	K(%)
Caldera fill			
Bishop Tuff[a]			
'Early'	6.5	22	4.0
'Late'	2.6	12	4.6
Rhyolite			
Tuffs	4.9	11.2	1.6
Flows and domes	5.6	15.7	4.1
Obsidian, Inyo domes	6.5	20.4	3.8
Rhyolite of moat	5.0	18.8	3.0
Mean:	5.2(± 1.4)	16.6(± 4.5)	3.5(± 1.2)
Basalt			
South moat	0.8	2.0	0.8
North moat	1.3	4.2	1.3
Mean:	1.0	3.1	1.1
Precaldera rocks			
Granodiorite			
Northwest border	6.6	26.1	3.6
South border	4.6	16.3	3.2
Mean:	5.6	21.2	3.4
Paleozoic metamorphics[b]			
Carbonates	1.2	1.7	0.3
Clastics	4.4	7.3	1.8
Mean:	2.8	4.9	1.1
Mesozoic metavolcanics	2.9	8.9	2.4

[a]Data from Hildreth [4].

[b]Data for Paleozoic carbonate and clastic rocks from Wollenberg and Smith [5].

northern Nevada (Wollenberg [6]), where siliceous sinter is of low radioactivity. Further investigation is required to determine whether some of the Long Valley deposits are of calcium carbonate, shown in the Nevada studies to contain appreciable radium, a near-surface source of springwater radon.

DISCRETE RADON MEASUREMENTS OF WATERS

Water samples were collected from several springs in the Long Valley region to provide a baseline of radon and water-chemistry data, for comparison with future resamplings, and to choose sites for long-term continuous monitoring of radon concentrations.

The location of the water sources sampled is shown in Fig. 1. The sources include many of those sampled and analyzed over the past decade by members of the U.S. Geological Survey (Mariner and Willey [3]). Sampling was done at orifices where boiling occurs as well as at those of intermediate and cold temperature. Temperature, pH, specific conductance, and radioactivity were measured at the time of sampling. Samples for laboratory gamma-counting for ^{222}Rn were obtained by filling two 500 ml thick-walled Nalgene bottles, which were then capped and taped tightly. Samples were also collected for analyses of major and trace elements by the filtering of water into two polyethylene bottles at each site, one acidified to preserve the dissolved silica contents in solution. Unfiltered samples were also taken in small glass bottles for oxygen and hydrogen isotope analyses (only the radon data are reported and discussed here).

At the laboratory the total gamma radioactivity of the two 500 ml samples was measured with a NaI(T1) detector in a low-background counting facility; the total gamma radioactivity of the water was observed to be due entirely to the presence of radon and its daughter products in the water. Following a procedure described by Smith et al. [7], the initial counting rates were corrected for the 3.8-day half-life decay of ^{222}Rn and for a slight leakage of radon through the walls of the bottles. The corrected counting rates were then converted to radon concentrations (in picocuries per liter) with the calibration constant of 1 count/min = 0.84 pCi l^{-1}.

Field-measurement data and results of laboratory gamma spectrometric measurements of the waters' ^{222}Rn concentrations are listed in Table 2. An examination of the table indicates a wide range of radon concentrations, generally corresponding inversely to the temperature and specific conductances measured at the springs. These are illustrated in Fig. 2. The highest radon concentrations are in cold springs on the western and southern margins of Long Valley (Hartley, Laurel, and Minaret Summit), while the waters sampled at several of the hottest springs (Casa Diablo, Hot Creek, and Hot Bubbling Pool) are very low in radon content. Springs of intermediate temperature in Long Valley proper--Big Alkali Lake, Whitmore, and north of Whitmore--have easily measured radon concentrations, as do the cool springs at the Fish Hatchery and Big Spring.

The rough inverse correlation between radon concentration and surface temperature of the springwater is attributed to the diminished solubility of radon with increasing temperature (Wilhelm et al. [9]). A similar inverse correlation between radon and specific conductance can be attributed to the strong correlation between the temperature and the specific conductance of water. There is a poorer correlation between ^{222}Rn concentrations and expected temperatures of unmixed thermal water, calculated by Fournier et al. [8] with chemical geothermometers. The very low contents of ^{222}Rn observed in the water of the hottest springs, Casa Diablo and Hot Creek, and of Hot Bubbling Pool, are a attributed to flushing of radon from near-surface water due to boiling or bubbling of exsolving gas.

As shown in Fig. 2, there is only a slight apparent correlation between ^{222}Rn concentrations in water and uranium in the rocks from which the springs flow. The highest concentrations are associated with the dilute cold springs at Hartley, Laurel, and Minaret in the granitic and metamorphic rocks, while the concentrations of the cool springs flowing from the basalt of relatively low radioactivity at the Fish Hatchery and Big Spring are significantly lower. The concentrations of the basalt springs, however, are higher than those of intermediate-temperature and hot springs, whose sys-

Table 2. Radon Concentrations and Other Parameters of Springwater

Name	Rock type[a]	(pCi l⁻¹)	No. of analyses	Spec. conductance (μmho-cm²)	pH	Temp. (°C)	Na-K-Ca geother- mometer[b] (°C)
BA Big Alkali Lake	al	225	1	1700-1750	6.4	59	200
BS Big Spring	b	601 ± 67[c]	4	175-180	6.6	10	83
CD Casa Diablo	r	23	1	1400-1425	7.7	94	238
F Fish Hatchery	b						
'CD' pool		789± 170[c]	4	210-220	6.7	14	50
'H-2 & 3' pool		667± 39[c]	9	130-160	6.7	11	
H Hartley Spring	g	2490	1	40-60	6.6	2	
HB Hot Bubbling Pool	r	43	1	1800-1850	7.6	63	189
HC Hot Creek Gorge	r	16	2	1700-1750	8.2	90	192
L Laurel Spring	g	2467± 214[c]	45	90-110	8.6	11	11
LHC Little Hot Creek	r,l	319	2	4700	7.4	80	172
MS Minaret Summit	m	1430	1	210-220	6.7	3	
NW North of Whitmore	r,l	200	1	1400-1450	7.5	53	
W Whitmore	r,al	395	1	625-650	7.1	31	

[a]Rock types encompassing spring systems: al = alluvium, b = basalt, g = granitic rock,
l = lake-bed deposits, m = metavolcanic rocks, r = rhyolite.

[b]From Fournier et al. [8].

[c]Values are means ± standard deviation.

tems primarily flow through the more radioactive rhyolites, lake deposits, and alluvium. Thus, at Long Valley the ^{222}Rn concentrations of springs are most strongly determined by the temperature of the shallow portions of the hydrothermal system and, where present, by boiling or bubbling of gases from solution. They are determined to a much lesser degree by the radioelement contents of the rock through which the water is percolating.

Repeated sampling has been conducted at one of the most radioactive springs, Laurel, flowing from granite on the south border of the caldera. A time record of the ^{222}Rn concentration of this spring is shown in Fig. 3. Sharp variations over relatively short periods are evident at several points. These exceed the range of standard deviation (10%) but cannot be ascribed to significant seismic or crustal deformation events, partly because of the wide and variable spacing between sampling. Samples were also collected at other springs in the caldera, though less frequently than at Laurel Spring.

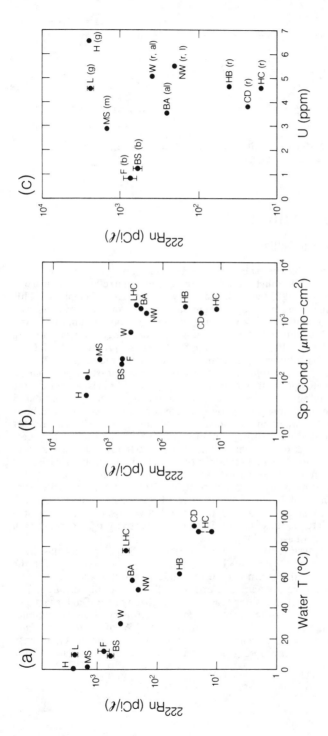

Figure 2.   Variation of radon with temperature (A) and specific conductance (B) of springwaters, and with uranium concentrations of rocks that encompass the spring systems (C). Letter symbols keyed to Table 2.

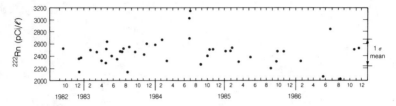

Figure 3.   Variation with time of radon concentration in water of
               Laurel Spring.

CONTINUOUS MONITORING

Instrumentation and Setting

In order to observe short-to-intermediate-term groundwater radon
fluctuations, a continuously monitoring gamma detection system was
installed in August, 1983, at Fish Hatchery spring "H-2,3" (table 2). This
site was chosen on the basis of discrete sampling of the springwaters, the
availability of a source of power and shelter for instrumentation, and
because a pond is situated at the outflow of the spring. The monitoring
system is similar to that described by Smith et al. [7, 10]. A NaI(Tl)
gamma detector is submerged in a pool so that it is surrounded on all sides
by at least 1 m of water. The data are acquired by a gain-stabilized gamma
spectrometer, whose output is organized into several broad gamma-energy
windows. The integrals in these windows are recorded hourly as the radon
data. The one-hour integration time is appropriate to the several-hour
residence time for water in the pond, and to the 35-40 min half-life for
ingrowth of the gamma-emitting daughters of Rn^{222}. The gamma-ray spectrum from this location, shown in Fig. 4, demonstrates that variations in
radioactivity measured in this manner are due almost entirely to variations
in the radon content of the water. As shown in the Figure, the recorded
interval in the gamma-ray spectrum contains contributions primarily from
the Bi^{214} daughter of Rn^{222}. The absence of appreciable peaks from K^{40}
and Tl^{208} in this spectrum attests to the effectiveness of the water surrounding the detector in shielding it from radioactivity of the rock, soil,
and concrete dam that line the pond.

The spring system that feeds the Fish Hatchery springs has been
described by Sorey [11] and by Farrar et al. [12]. The recharge for the
system is in the southwestern part of the caldera, and the water flows eastward, with a small component of northward flow, primarily through fractured basalt. Some of the water may also flow through the Quaternary glacial debris that underlies the basalt. From the chemistry of the springwater
Sorey [11] estimated that a small amount (1% to 3%) of geothermal water
mixes with the eastward-flowing water. An estimate of the effective range
of radon carried by the springwater may be obtained by means of the

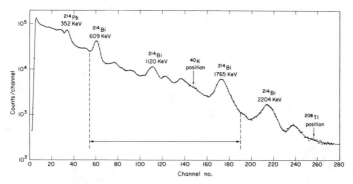

Figure 4.   Gamma-ray spectrum measured by a NaI(Tl) detector
suspended in water of the Fish Hatchery spring. Arrows:
interval measured for radon content.

model of Stoker and Kruger [13]. Using measured emanation values for
the basalt, rock porosities between 5% and 15%, and considering radial and
linear flow models, radon is being contributed to the springs from a maximum of between 0.3 and 1.0 km.

The pond in which the gamma-detector is located is shown in Fig. 5a.
It is fed by a line of springs along the southern margin of the pond, at the
base of a basalt flow. The temperature of the spring averages about 11°C,
and although constant over the short term it shows a yearly cyclic variation
of about 0.5°C. The average flow rate is about 140 l/sec, also varying seasonally, from high flows in early summer to low flows in late fall and
winter. The residence time of water in the pond--given a surface area of
7800 ft^2 and an average depth of between 5 and 8 ft--is of the order of 2-3
hours. Outflow from the pond takes place over a small concrete dam, to
Hatchery fish ponds below. In order to discourage the growth of algae and
interference with flow to the fish ponds, Hatchery personnel stretched a
tarp over the surface of the pond in March, 1984. The gamma detector
was removed before the tarp was emplaced, then re-inserted through a
hole to its original position about 6 ft out from the dam, and 10 ft to the
side of the outflow. The position of the detector beneath the tarp is shown
in Fig. 5b.

Character of the Radon Record

Although the original intent in installing the continuous system was to
monitor changes in groundwater radon concentration associated with
seismic events, it soon became apparent that environmental factors were
strongly affecting radon concentration and could mask geophysical effects.
In order to gain a better understanding of the environmental effects on
radon concentration, we have examined continuous records of water temperature and specific conductance measured adjacent to the gamma detector by Chris Farrar of the U.S. Geological Survey. These have been moni-

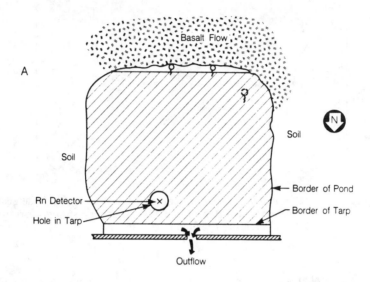

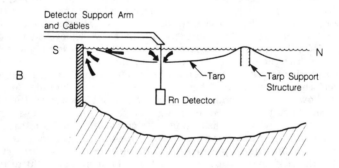

Figure 5.   (A): Plan view of the main features of the Fish Hatchery pond, including tarp covering surface, and position of gamma detector. (B): North-south cross-section through detector showing gaps in tarp at dam and above detector. Heavy arrows: flow of water toward outflow (out of plane of section), and of cold water below tarp. Not to scale.

tored since February 1985, using a power source independent of that of the radon system. In addition, daily precipitation records, maximum and minimum air temperatures, and selected barometric pressure readings were provided by the U.S. Forest Service and Mammoth Lakes Airport for several locations within the caldera.

Comparison of these data with radon concentration shows pronounced correlations between radon, water temperature at the detector, and precipitation. Specific conductance shows a lesser, but distinct correlation with radon, but this may be due mainly to the dependence of conductance upon temperature. Changes in barometric pressure often correlate with disruptions in the daily radon pattern, but the effects of barometric changes usually cannot be separated from those of associated precipitation. Continuous barometric records have not been available, so that effects from short-term barometric changes would probably not be detected.

Shown in Fig. 6 are portions of the radon record along with the corresponding precipitation and water temperature records. The radon cycle typically shows a daily variation of 25% to 35% about a mean of about 6×10^4 counts/hr, with a nighttime peak. It shows consistent seasonal characteristics, among which are higher amplitude daily cycles and earlier peak positions in colder weather. The record of water temperature at the detector usually has a well-defined daily cycle like radon, with a daily variation of 0.5°C or less. Compared with radon, temperature has broader peaks that are often truncated at the top. But it has the same tendency for peaks to decrease in amplitude and to shift toward later hours, with increasing air temperature. It also exhibits a phase shift with respect to the radon pattern that remains despite the seasonal variations; peaks in water temperature fall off several hours before the fall-off of the radon peak. Both radon and water temperature show a clear response to precipitation. Larger amounts of precipitation generally suppress or eliminate the daily radon and water temperature peaks, and most precipitation events appear as perturbations in the radon and temperature patterns that often parallel one another closely (e.g., Fig. 6d).

Fig. 7 shows Fourier transforms of the water temperature and radon records for the time periods displayed in Fig. 6. (The time intervals used in Fig. 7 differ somewhat, in several cases, from those in Fig. 6, as the former were chosen to avoid precipitation events.) The seasonal differences in the radon record are pointed up in Fig. 7a and 7b: The strong 24-hour periodicity typical of winter and fall is seen in 7b, while 7a is typical of the warm weather pattern in which 12-hour and higher-frequencies are more prominent. In the water temperature record, there is usually a stronger 24-hour periodicity than for corresponding radon patterns (corresponding pairs in Fig. 7 are a-e and b-d), and this persists through most warm weather periods. However, during periods in which minimum air temperatures remain above the temperature of incoming springwater (~ 11°C), the water temperature record can be nearly constant, with greatly reduced periodic components (Figs. 6c, 7f). (It should be noted that the Fourier transforms tend to be somewhat distorted, particularly for the temperature patterns, due to the presence of angular features such as truncated peaks. The program attempts to fit these features with higher-frequency harmonics of the 24-hour peak.)

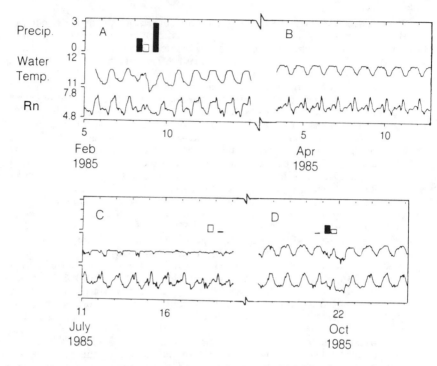

Figure 6.   Four segments of radon record at the Fish Hatchery spring, with corresponding records of water temperature at the gamma detector, and daily precipitation totals. Radon expressed in (gamma counts/hr)x10^4, temperature in °C, precipitation in inches. Solid bar: precipitation at Mammoth Lakes, elev. 7800 ft; open bar: precipitation at Little Hot Creek, elev. 6960 ft (see fig. 1 for locations). Elevation of Hatchery is 7200 ft.

Causes of the Radon Pattern

The close association of radon concentration and water temperature suggests that much of the variation in the radon pattern is due to dilution by descending, cold, radon-depleted surface water. This effect would be more pronounced and occur earlier in the day at colder air temperatures, and this is consistent with the higher amplitude radon and water temperature patterns in winter.

The paths by which surface water can descend toward the gamma detector are limited by the tarp covering the pond. Some water may descend through the gap along the northern margin of the pond, but the main route is probably the hole above the detector (Fig. 5). The hole is near the bottom of a sagging section of tarp, and although in effect an annulus only 2 to 3 inches wide, it may serve as a funnel for cold surface water or precipitation to descend from a much larger area (Fig. 5b). This could account for the magnitude of the precipitation effect on radon and water

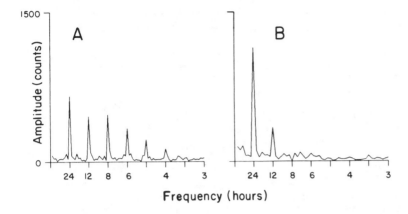

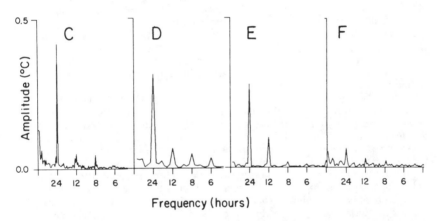

Figure 7.   Fourier transforms of radon (A, B) and water temperature (C, D, E, F) patterns for approximately the time periods shown in fig. 6. Time periods, and corresponding plots in fig. 6, are (A): Apr 3-13 (fig. 6B); (B): Oct 23-Nov 8 (fig. 6D); (C): Feb 10-Mar 7 (fig. 6A); (D): Oct 23-Nov 2 (fig. 6D); (E): Apr 3-13 (fig. 6B); (F): Jul 8-25 (fig. 6C). Narrow peaks in (C) due to longer time period.

temperature, which is much larger, given the flow rate of the spring, than the average dilution of pond water by precipitation. It would also explain the consistently lower amplitude of the radon pattern prior to the covering of the pond in March, 1984. This is seen in Fig. 8, which compares January radon records for 1984 and 1985. The daily variation is much reduced in the 1984 record, but peaks still occur at the same part of the day as in the later record. Thus the descent of surface water was probably a factor in radon concentration prior to covering the pond, but the effect was much less pronounced than in the present configuration.

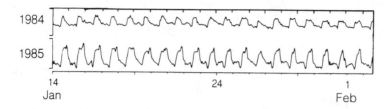

Figure 8.   Portions of the radon record for same time periods in 1984 and 1985, showing effect of pond cover on amplitude of 1985 pattern. Limits on radon axes are as in Fig. 6.

As described above, during periods in summer with high minimun air temperatures, water temperature at the detector can be nearly constant, as well as unaffected by the infrequent rainfall (e.g., Fig. 6c). However, the radon record retains its daily variation during these periods. This indicates that desending cold surface water does not entirely account for the radon variation pattern. There must be other factors responsible for that pattern, either environmental factors operating on water in the pond (such as circulation of radon-depleted water by some mechanism other than density-driven descent from the surface), or factors influencing the radon concentration of incoming springwater.

Earth tidal forces could be a mechanism for introducing regular diurnal and semi-diurnal variations in groundwater radon. In Fig. 9, several segments of the radon record are shown along with corresponding plots of theoretical earth tides. The theoretical tides have been checked against recording gravimeter measurements at several different times, and they follow the measured gravity variations closely, with a phase difference of 1.5 hours. From Fig. 9 it can be seen that the radon and tidal patterns are not in close agreement. This cannot be attributed to the effects of precipitation, which was negligible (with the exception of October 20 and 21, 1985; Figs. 9a and 6d) during the periods shown. Changes in amplitude of the tidal envelope are not reflected in the radon pattern (e.g., Fig. 9c), nor are significant variations in the relative amplitudes of the tidal pattern (e.g., the first and second halves of Fig. 9b, and portions of Fig. 9a).

The differences in relative amplitudes between earth tide and radon are brought out by comparison of their Fourier transforms. Fig. 10a shows the frequency distribution for the main portion of the tidal pattern in Fig. 9a, and Fig. 10b shows the same for a smaller section of that pattern. The strong 24-hour periodicity shown particularly in Fig. 10b contrasts with the frequency distribution of the corresponding radon pattern (Fig. 7b). Fig. 10c plots the Fourier transform of a long segment of the tidal pattern, showing that the prominence of the 12-hour component (as well as a splitting of the 12- and 24-hour components) is a typical feature of the earth tides. The difficulty in reconciling this with the typical dominance of the 24-hour component in the radon record is certainly due in part to the influence on radon of descending surface water, with its strong 24-hour periodicity. However, as seen in Figs. 9b and 9c, the concordance of the radon and tidal records does not appear to increase during warm weather, when the 24-hour component of the radon record is weakest (as seen in Fig. 7a, corresponding to Fig. 9b). These considerations do not rule out a

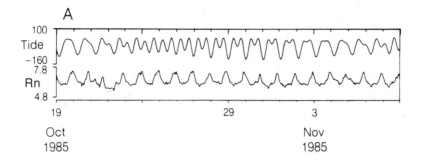

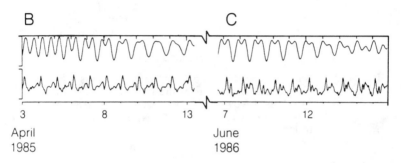

Figure 9.   Three segments of radon record, with plots of theoretical
              earth tides for same time periods. Earth tide expressed in
              μgal, radon as in Fig. 6.

relation between earth tides and radon concentration, but they show that
other factors which could be masking that relation must first be understood or eliminated before it can be discerned.

Association of Radon and Seismic Events

A major goal of the continuous monitoring program since its inception
has been to observe changes in the radon record assiciated with seismic
events. This has not, however, been the most fruitful aspect of the program. To date, there has been one significant perturbation in the radon
record that appears to correlate clearly with earthquake data. This occurred
on November 14, 1983, soon after the inception of the monitoring program. The radon record for that time is shown in Fig. 11, along with the
occurrence of caldera earthquakes of magnitude greater than 1.5. A distinct spike is observed in radon concentration that is approximately coincident with several small earthquakes just preceding a stronger earthquake
of magnitude 3.4 (epicenter approximately 5 km southwest of the Hatchery
springs).

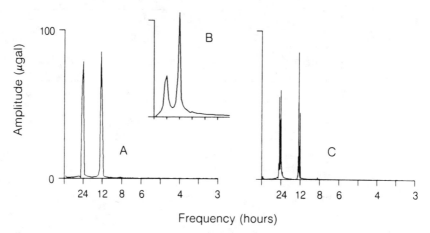

Figure 10.   Fourier transforms of earth tide patterns. (A) and (B) correspond to portions of the tidal pattern shown in Fig. 9A: (A) Oct 23-Nov 8; (B) Oct 23-30. (C) corresponds to tidal pattern for first half of 1985. Narrow peaks in (C) are due to longer time period.

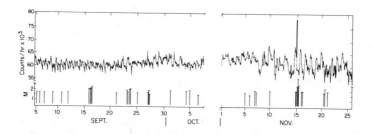

Figure 11.   Initial portion of continuous radon record at the Fish Hatchery spring, September-November, 1983. Earthquakes of magnitude > 1.5 in the caldera shown by vertical lines below radon record. Earthquake data provided by R. Cockerham (private communication, 1984).

The initial portion of the continuous radon record, beginning in September, 1983, is also shown in Fig. 11. It will be seen that the character of this early record is significantly different from subsequent portions, shown in Figs. 6, 8, and 9, and this is probably due in large part to the lack of a cover on the pond at that time. One reason that other correlations between radon anomalies and seismic events have not been observed may be that such correlations are now more easily masked by the larger "normal" variation of the radon pattern related to the covering of the pond. But another reason is the sparseness of the seismic record. Only one other event of magnitude equal to or greater than that of the November, 1983 earthquake occurred within the caldera during operation of the monitoring system.

CONCLUSIONS

The principal conclusions of the investigation to date are as follows:

1. The range of radon concentrations of springs in the Long Valley caldera covers several orders of magnitude. The concentrations are controlled primarily by water temperature and near-surface exsolution of gases, and to a lesser degree by the radioelement concentrations of the rocks that encompass the hydrologic systems. The cold springs on or near the western, northern, and southern margins of the caldera, which may represent water that is recharging the caldera's hydrothermal system, have high radon concentrations relative to those of the cool, warm, and hot waters of the caldera.

2. Radon in groundwater is readily monitored by an in-situ gamma-radiometric system. The radioactivity of the water is due almost entirely to its ^{222}Rn content.

3. Radon concentration of springwater monitored in a pond exhibits strong diurnal, and weaker semi-diurnal variations. It is strongly influenced by precipitation, and by temperature-sensitive flow patterns of radon-depleted surface water in the pond. It does not appear to be strongly influenced by earth tides.

4. In the current configuration of the monitoring system, radon variations associated with small or distant geophysical changes probably stand little chance of being detected. We will need to understand better the nature of the "normal" variations in the radon record in order to recognize correlations between radon anomalies and seismic or other geophysical events.

Operation of the present system continues, and we plan several modifications in the near future. One is the completion of the covering over all portions of the pond, in order to close off all routes to descending surface water. We also plan to improve the monitoring system to reduce the frequency of breaks in the radon record, and possibly to introduce continuous monitoring of meteorological parameters. These measures could aid greatly in deciphering causes of the radon variation that are not yet understood, in observing longer term periodicities and trends in the record, and in detecting influences, like that of earth tides, which may at present be masked or distorted. Deployment and concurrent operation of several systems at springs and, with some modification, wells in the caldera would permit us to observe better the relationship betwen radon and geophysical changes.

ACKNOWLEDGEMENTS

We appreciate the cooperation of Robert Iselin, Jim Eichman, and their colleagues at the Hot Creek Hatchery, California Department of Fish and Game, who provided the site, and have helped in data collection, for the continuous monitoring system. Chris Farrar of the U.S. Geological Survey provided water temperature and other data, Mark Clark of the U.S. Forest Service helped in sampling of springwaters, and Rob Cockerham of the U.S. Geological Survey provided information on the occurrence of earthquakes in the caldera. Our thanks to Dwayne Mosier for guidance with instrumentation, and to Harry Bowman and Richard Rachiele for valuable discussions. This work was supported by the U.S. Department of Energy's Office of Basic Energy Sciences through U.S.D.O.E. contract No. DE-AC03-76SF-00098.

REFERENCES

1. Wollenberg, H. A., A. R. Smith, D. F. Mosier, S. Flexser, and M. Clark. "Radon-222 in Groundwater of the Long Valley Caldera, California," Pure and Applied Geophysics 122, 327-339 (1985).

2. Sorey, M. L., R. E. Lewis, and F. H. Olmsted, "The Hydrothermal System of Long Valley Caldera, California," U.S. Geol. Survey Prof. Paper 1044-A (1978).

3. Mariner, R. H., and L. M. Willey, "Geochemistry of Thermal Waters in Long Valley, Mono County, California," J. Geophys. Research 81(5), 792-800 (1976).

4. Hildreth, W. "The Bishop Tuff: Evidence for the Origin of Compositional Zoning in Silicic Magma Chambers," Spec. Paper, Geol. Soc. Am. 180, 43-75 (1979).

5. Wollenberg, H. A., and A. R. Smith, "Radiogenic Heat Production in Prebatholithic Rocks of the Central Sierra Nevada," J. Geophys. Res. 75(2), 431-438 (1970).

6. Wollenberg, H. A. "Radioactivity of Nevada Hot Spring Systems," Geophys. Res. Lett. 1 (8), 359-362 (1974).

7. Smith, A. R., H. R. Bowman, D. F. Mosier, F. Asaro, H. A. Wollenberg, and C.-Y King, "Investigation of Radon-222 in Subsurface Waters as an Earthquake Predictor," IEEE Trans. Nuclear Sci., NS-23 (1), 694-698 (1976).

8. Fournier, R. O., M. L. Sorey, R. H. Mariner, and A. H. Truesdell, "Geochemical Prediction of Aquifer Temperatures in the Geothermal System at Long Valley, California." U.S. Geol. Survey Open File Rept. 76-469 (1976).

9. Wilhelm, E., R. Battino, and R. J. Wilcox, "Low-pressure Solubility of Gases in Liquid Water," Chem. Revs. 77(2), 219 (1977).

10. Smith, A. R., H. A. Wollenberg, and D. F. Mosier, "Roles of Radon-222 and Other Natural Radionuclides in Earthquake Prediction." In Natural Radiation Environment III (eds. T. F. Gesell and W. M. Lowder) U.S. Dept. Energy CONF-780422 1, 154-174 (1980).

11. Sorey, M. L. "Potential Effects of Geothermal Development on Springs at the Hot Creek Fish Hatchery in Long Valley, Mono County, California," U.S. Geol. Survey Open File Rept. 75-637 (1976).

12. Farrar, C. D., M. L. Sorey, S. A. Rojstaczer, C. J. Janik, R. H. Mariner, and T. L. Winnett. "Hydrologic and Geochemical Monitoring in Long Valley Caldera, Mono County, California, 1982-1984," U.S. Geological Survey Water-Resources Investigations Report 85-4183 (1985).

13. Stoker, A. K., and P. Kruger, "Radon in Geothermal Reservoirs," in Proceedings of the 2nd United National Symposium on the Development and Use of Geothermal Resources, San Francisco (1975), pp. 1797-1803.

Session III : Mining Impacts on the
Occurrence of Radon,
Radium and Other
Radioactivity in
Ground Water

Moderator: DeWayne Cecil,
U.S. Geological Survey

RADIOACTIVITY IN HOCKING RIVER BASIN

Moid U. Ahmad and Roger W. Finlay, Ohio University,
Athens, Ohio

INTRODUCTION

Numerous studies have examined the environmental
impacts and health hazards associated with man-made
radioactive pollution. Within the last decade,
however, there has been a growing awareness of the
natural radiation environment. Very little is known
about the way the natural radiation environment is
affected when subjected to artificial stresses imposed
by man's activities. The prolonged exposure from low
level radioactivity may significantly affect our
environment by altering the distribution of naturally
occurring radionuclides through such processes as
mining, water resources management or land reclamation.
It is the intent of this study to examine more closely
the way in which this radiological burden is altered by
coal mining activities in an Appalachian watershed.

PREVIOUS WORK

The uranium boom of the 1950's resulted in the
widespread exploration for commercial deposits. Coal
and organic black shale formations were among the
various types of deposits and lithologies studied.
Presumably, the Chatanooga Shale and its stratigraphic
equivalents constituted the single largest low grade
deposit of uranium. However, the extremely low concentrations of uranium cannot be economically extracted
from the formation [1-2] One of the earliest reports to
determine potential uranium prospects for Ohio Coals
was performed by Snider [3]. Five of the fifty samples
analyzed by Snider [3] showed uranium concentrations

ranging from 0.001% U_3O_8 to 0.003% U_3O_8. Other Appalachian coals analyzed in Pennsylvania have uranium concentrations ranging from 10 to 100 ppm [4]. Analysis of major coal seams located in north-northeast Ohio (Brookville No. 4 through the Waynesburg No. 11) indicate that thorium and uranium concentrations are less than 10 ppm [5].

Regional studies performed concluded that approximately 2 million pounds of uranium are carried in solutions from conterminous U.S. streams. [6] Caldwell [4] demonstrated that the leaching of abandoned strip mines along the Kiskminetas River in Pennsylvania contributed to the radioactivity of the stream. He showed that total activity per day was a stronger function of flow rate rather than concentration. Alpha activity concentrations for the Kiskminetas River ranged from 1 pCi/L to more than 100 pCi/L. The maximum alpha activity for potable water supplies is 15 pCi/L (National Interim Primary Drinking Water Regulations (1976).

High resolution gamma-ray spectrometry is a widely utilized technique in the study of radionuclides. The U.S. Geological Survey has utilized Ge(Li) detectors to establish guidelines and accuracy limits for other non-destructive multi-element analysis techniques such as the recent development of epitherman neutron-activation analysis [7]. GE(Li) detectors and sondes, because of their superior resolution facilitated the study of geochemical mobilities and disequilibrium associated with the leaching and migration of uranium and thorium [8-11].

PURPOSE

The purpose of this study is to examine the natural radiation environment of the Hocking River Basin. The erosion of unreclaimed strip mines may accelerate the leaching of radionuclides from exposed coal seams and spoil piles. The migration and subsequent redeposition of radionuclides at favorable locations downstream may create point sources of low level radioactive contamination to the environment. An analysis of the host lithologies, stream sediments and stream water is performed in order to estimate the environmental impact resulting from the migration of radionuclides through the watershed.

The objective of this study is to identify specific nuclides and their respective concentration in various mediums. The concentrations of ^{40}K, ^{238}U, ^{235}U, ^{232}Th and their respective daughter products are determined for the major mineable coal seams, their associated roof shales and other miscellaneous lithologies located in the Hocking River Basin. In addition, the concentrations of these nuclides are also determined for stream sediments and water samples from the Hocking River and its major tributaries.

STUDY LOCATION

The Hocking River Basin constitutes an ideal study location for several reasons. The geology and hydrogeology of the basin is typical of other watersheds in the Appalachian Plateau. The small size of the Hocking River is more readily sampled than adjacent larger river basins. Pennsylvanian and Permian age formations containing most of the major mineable coal seams crop out or underlie a portion of the basin. A variety of active and abandoned mines exist within the basin for each major coal seam. With the exception of Clear Creek, each major creek basin has at some time experienced extensive strip mining. The majority of the mines have not been reclaimed and are strongly susceptible to erosion.

The Hocking River Basin is not contaminated by the direct discharge of fission or fusion products released from atomic weaponry, nuclear reactors or nuclear waste disposal sites since these facilities are not located within the basin. The bulk of the radioactivity observed results from the decay of naturally occurring nuclides.

SAMPLE LOCATIONS

The location for a majority of coal and shale samples lies within Athens and eastern Hocking Counties. A significant portion of these samples were obtained from the Monday Creek basin. Each sample location (Quadrangle location) is described in detail in Appendix A [12]. At least one major coal seam and overlying roof shale was sampled for each major creek basin as well as the main Hocking River Valley. The exception to this is Margaret and Clear Creeks in which no formations were sampled.

Stream sediment samples were obtained from at least two locations in each creek basin. Stream sediment samples from the Hocking River were obtained from six locations between Lancaster and Coolville, Ohio. In most cases, a water sample was obtained from the same stream sediment sample location. When possible, water samples were collected during relatively high stages of stream discharge.

THE NATURAL RADIATION ENVIRONMENT

Deposition of Uranium in Coal

Several uranium minerals have been identified in lignites and coals. Denson [13] in the study of Lignites in the Williston Basin found that meta-antimite, metanyaminite and abernathyite were the most common. Nekrasova [14] in the study of Jurassic age coals identified pitchblende, uraninite; with minor amounts of sulfates, phosphates carbonates, sulfo-carbonates and vanadates also occurring in the oxidized zone. Uraninite and coffinite were the primary types of uranium minerals identified by Breger [15] in his analysis of coalified logs of Triassic and Jurassic age.

The major leading hypotheses on the origin of uranium mineraliation in coals points to epigenetic deposition [13, 14, 16, 17, 18]. Mineralization is generally confined to coal beds which overlie or under-lie coarse, permeable sandstones [14]. Coals having the highest concentrations are usually overlain by a coarse permeable sandstone. Massive and sooty pitch-blende mineralized into small irregular stockworks of veinlets are associated with epigenetic pyrite. Ore mineralized occurs in the crest and limbs of the anti-clines and in zones of intense fracturing where there is an increase in permeability. In general, the ana-lyses of Illinois coals, the elemental concentrations are more commonly observed in the top and/or bottom of the coal seam. He further stated that most elements occur in significantly higher concentrations in the fine-grained sedimentary rocks associated with the coal (roof shales, underclays, and shale partings) [19].

Another restriction in the mineralization of ura-nium in coals and rocks is the amount of organic matter or hydrogen sulfides present. Either of these two materials are capable of reducing the uranium and pre-cipitating it out of solution [13-16]. Masursky [20] noted that the accumulation of molybdium, thorium, and rare earths were associated with uranium enriched coals.

Moore [21] found that peat, lignite and subbitu-minous coal irreversable removed 98% of uranium from a solution containing 196 ppm U. Subbituminous coal was the most absorbing material of uranium from an aqueous solution. Moore [21] concluded that the uranium probably precipitated out as a metallo-organic compound. Vine [18] proposed that the environment of coal, similar to poorly drained peat bogs, received most of the uranium in streams. However, these streams were heavily charged with soluble brown organic matter so that the uranium ions would tend to remain fixed by the organic matter in solution and be carried out to sea [18].

Deposition of Uranium in Marine Black Shales

More than 200 formations in the United States containing marine black shales, ranging from Precambrian to Tertiary in age, were examined during the period 1944- 1957 for commercial deposits of uranium. An estimated 8 to 10 thousand samples were fluoromerically analyzed for their uranium content [1].

The term "black shale" is very vague, however, it is at best defined as that large class of sedimentary rocks composed chiefly of mineral grains of clay and silt size, and containing sufficient organics, iron sulfide, or manganese oxide to give the rock an overall dark gray to black color. Terms such as "humic shale", "oil shale", "graptolitic shale", "alum shale", "exunic shale", "sapropelite shale", and othes may also be thought of as black shales [1].

The Chattanooga shale and its stratigraphic equivalent or partial equivalents (New Albany, Ohio, Antrim, Dunkirk, Mountain Glen, Woodford, Lodgepole) have been studied extensively by several authors [1, 2, 17, 22]. The Gassaway Member of the Chattanooga shale showed an average content of .006% uranium. Marine black shales in cyclothems of Pennsylvanian age in Illinois, Kansas, and Oklahoma showed a range of .004 to .010% uranium several authors [1, 22] have developed and identified the criteria between the uranium content and black shales.

Nuclides in Streams

The uptake and transport of radionuclides by streams and stream sediments is an extremely complex process. In general, the uranium content for streams within the United States ranges from 0.05 to 10.0 ppb with a median value of 1.5 ppb [23]. Exceptionally high amounts, however, have been reported in such areas like the Colorado Plateau. Radionuclides may enter a stream either as a soluble or relatively insoluble material and can be transported downstream during which time either dilution or possible concentration occurs. Once in the stream, radionuclides may experience changes in state, remain in solution or interact with various components of the aquatic system.

The criteria for identifying hazardous streams involve several factors. In order to establish the potential liability for a particular stream, the uptake and concentration of significant quantities of radionuclides by stream sediments must be analyzed. In addition, stream geometry and sediment conditions condusive to extensive local deposition of contaminated sediments and the rate of accumulation of radioactivity in deposits which exceeds the natural rate of decay are other parameters which influences the potential liability of any given stream [24]. White and Gloyna [25] give the important factors that influence the concentration of radionuclides.

Pickering [26] among others have intensively studied the Clinch River in Eastern Tennessee due to the release of low level radioactive wastes by the Oak Ridge National Laboratory. The more pertinant conclusions of these studies are:

1. the radioactive sediments are found along the sides of the stream channel (where the velocity is reduced) near the point of discharge.
2. the radioactive sediment extends clear across the stream channel and attains its greatest thickness further downstream.
3. the same general pattern of variation of gross gamma radioactivity with depth correlated with the pattern of annual releases of radionuclides from the Oak Ridge National Laboratory.

EXPERIMENTAL METHOD

Samples of coal, shale and sediments were studied for gamma ray activity using standard techniques. The gamma rays were detected with a Ge(Li) spectrometer that was enclosed in a thick lead house to reduce background radiation. Since the sample activity was generally quite low, the effectiveness of the lead shielding was crucial, and improvements to the shielding were made throughout the course of this work as materials became available. Early measurements were performed with a 79 cm^3 ORTEC Ge(Li) detector. That detector was damaged through failure of the liquid nitrogen dewar so measurements were completed with a 39 cm^3 PGT detector with an accompanying loss of sensitivity.

Samples were contained in 250 cm^3 stainless steel cans that were placed inside the Pb shield. Counting times between 2700 and 7200 seconds were used for all samples. Spectra were accumulated in a multi-channel analyzer system and transferred to computer disk for subsequent analysis. The energy scale was calibrated with a standard ^{228}Th rod source. The intensities of the peaks of interest were obtained by Gaussian peak fitting and were corrected for background through analysis of the sample-out spectra.

Since the sample cans were placed directly against the Ge(Li) detector, a correction for the counting geometry was necessary. Three separate approaches were used to obtain this correction. First, a small, well-characterized, sealed sample of U_3O_8 was measured in various geometries to determine the effective efficiency of the detector as a function of gamma ray energy and the solid angle correction factor. Second, "standard" samples of $Th(N)_3)_4 \cdot 4H_2O$ and of KCl were dissolved in water and measured in the same geometry as the samples under investigation. This procedure suffers from possible disequilibrium effects and both procedures suffer from energy dependent self-absorption in the sample volume. Finally, the effective solid angle was calculated by numerical integration. Agreement between the various procedures was moderately good, but a systemmatic error of +20% for Th and ^{40}K and $\pm11\%$ for U in not excluded.

Errors due to counting statistics varied from a few percent for the more active U and Th samples to about 30% for the weaker samples. For ^{40}K, statistical errors were typically smaller by a factor of about 3. The sensitivity limit for this work was dominated by background counting rates and was found to be a few parts in 10^{-2} ug/g for U and Th and a few parts in 10^{-3} g/g for ^{40}K.

RESULTS AND DISCUSSION

Introduction

A possible route of exposure from radionuclides occurs from uranium-bearing coal formations through acid mine drainage. Oxygenated water circulating through the coal seam and spoil banks converts pyritic materials in the coal and shale to sulfuric acid. Because uranium is capable of forming soluble sulfate complexes in the presence of sulfuric acid, the leaching of uranium from coal is a potential problem [27].

Disequilibrium Considerations

It is necessary to assume that the daughter products are in secular equilibrium with their parent products for the coal and shale samples. This validity of this assumption is supported by studies performed by authors [15, 28]. Examination of the data in Appendix A, [12] also supports this assumption. Within the bulk of the data, intermediate daughter products (^{226}Ra, ^{235}U) characterized by a 185 KeV gamma-ray energy appears to be in equilibrium with stable end daughter products such as bismuth and lead.

Secular equilibrium probably does not exist between the ^{238}U parent and its respective daughter products for the stream sediment samples or the water samples. ^{234}U is preferentially mobilized over ^{238}U, especially in aqueous environments   Analysis of the waters activity ratio also indicates disequilibrium between ^{234}U and ^{238}U.   It is also assumed that the ratio of $^{235}U/^{238}U$ is 1:137.5 [29].

Identification of Isotopes

The major isotopes identified from the gamma-ray spectrum displayed energies ranging from 185 to 2614 KeV.   Within this energy region, the daughter products (radiogenic lead, bismuth, thallium, radium, and actinium) can be identified. ^{40}K is identified by a single energy which is emitted at 1460 KeV.   Major isotopes are used to calculate the apparent thorium content as depicted in Table 4, [12].

Concentration of thorium, uranium potassium-40 in roof shale and coal

The apparent concentrations of thorium uranium, ^{40}K for each coal seam and their respective roof shale are depicted in Fig. 1 and Fig. 2.   Thorium concentrations remain fairly constant within the same coal formations as indicated by the Lower Freeport Coal, Upper Freeport coal and Middle Kittanning coal.   Of the 21 coal samples the mean concentration of thorium is 1.95 ug/g. The greatest concentration of thorium is association with the Redstone No. 8A coal (7.32 ug/g) which is followed by the Lower Freeport No. 6A coal (2 samples, 5.00 ug/g and 5.78 ug/g), the Quakertown coal (4.02 ug/g) and the Upper Kittanning coal (3.58 ug/g).   The remaining coal samples have thorium concentrations of less than 2.00 ug/g.

Thorium concentrations for the overlying roof shale samples are generally greater than the thorium concentrations of the coal samples.   The average thorium concentration for 21 samples is 4.35 ug/g.   It is also evident that the thorium content is somewhat variable within the same formations.   For example, the Harlem shale displays the greatest variability ranging from 7.19 ug/g to 1.69 ug/g.

The average uranium concentration for the 19 samples is 1.70 ug/g.   The two greatest uranium concentrations are associated with the Redstone No. 8A coal (14.97 ug/g) and the Pittsburgh No. 8 coal (5.55 ug/g). The remaining samples generally have concentrations of less than 1.00 ug/g.   The low thorium and uranium content may indicate rapid burial which would decrease the potential for precipitating these elements out of solution [30].

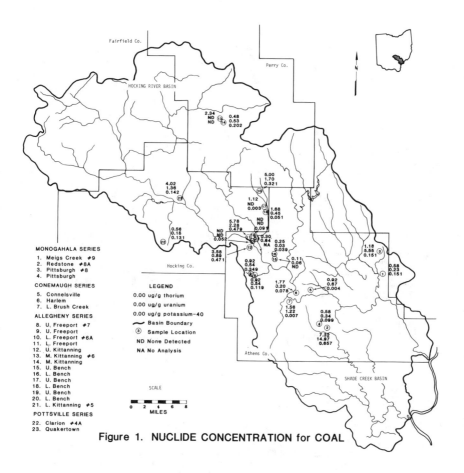

Figure 1. NUCLIDE CONCENTRATION for COAL

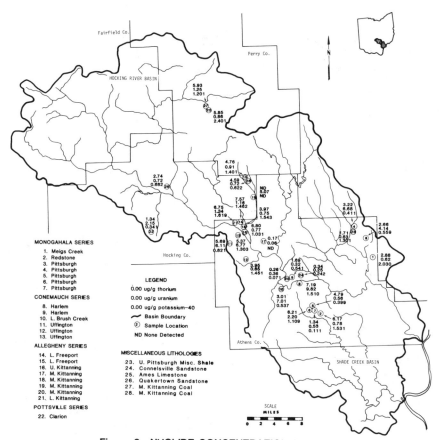

Figure 2. NUCLIDE CONCENTRATION for SHALE

The ^{40}K content of the coal samples range from 0.003 ug/g to 0.479 ug/g with an average value of 0.180 ug/g. In contrast the ^{40}K content of the shale samples ranges from 0.111 ug/g to 2.401 ug/g with an average value of 1.106 ug/g. The average values of ^{40}K are less than the average values for uranium and thorium.

Stream Sediment Samples

The distribution and concentration of nuclides for stream sediment samples are depicted in Figure 3. Three tributaries, Federal, Sunday and Monday Creeks, show an increase in thorium and uranium content downstream. However, this trend is not apparent in the other stream basins. Rush and Clear creeks show a significant decrease of thorium and uranium downstream. The stream sediments collected along the Hocking River between Lancaster and Nelsonville show a decrease in thorium and uranium content. However, stream sediment samples collected at Athens and Coolville show an increase for thorium and uranium. In most cases the thorium and uranium content of the sediment samples is considerably below that of the coal and shale samples. Thorium concentrations range from 2.89 to 0.07 ug/g and uranium concentrations range from 1.12 to 0.02 ug/g for the stream sediment samples. The concentrations of ^{40}K, however, is approximately of the same range for the coal and shale samples analyzed.

Several gob piles that were analyzed showed concentrations of thorium and uranium that were within the same range for concentrations computed for the stream sediments.

Water Samples

U.S. E.P.A. Analysis 1977-78. Nineteen water samples were collected using one gallon containers and preserved with 15 ml of concentrated Nitric acid. An analysis of the samples failed to produce a gamma spectrum that could be differentiated from background. As such, a new set of samples was collected and sent to the U.S. EPA, Las Vegas, Nevada District Office for alpha-ray pulse height analysis[1]. The alpha activity (pCi/L) for ^{235}U ^{238}U, ^{234}U, ^{230}Th and ^{232}Th is depicted in Figure 4.

[1] Analysis of the water samples was performed by Mr. Mullins, U.S. EPA, Las Vegas, Nevada. The analytical procedure used by U.S. EPA is described in the 13th Edition in which the sensitivity is $\pm$ 2 pCi/L for total alpha and $\pm$ 5 pCi/L for total beta.

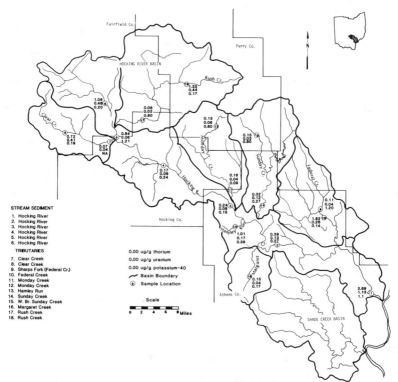

Figure 3. NUCLIDE CONCENTRATION for STREAM SEDIMENTS

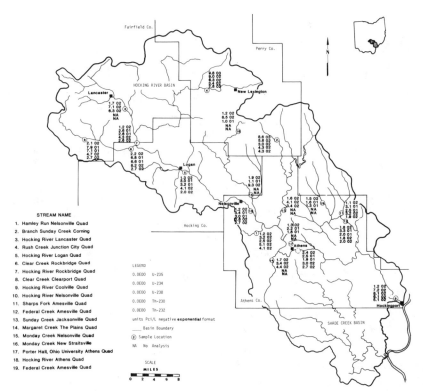

Figure 4. ALPHA ACTIVITY ANALYSES, USEPA, 12/77 to 3/78

The alpha activity of the water samples is considerably below the levels set by the National Interium Primary Drinking Water Standards (NIPDWS). The gross alpha activity of potable water supplies cannot exceed 15 pCi/L according to NIPDWS (1976). Table 1 lists the average activity of the water samples that were analyzed for this report.

Table 1. Average and Range of Alpha Activity; US EPA Analyses (pCi/L).

Radionuclide	Minimum	Maximum	Average
235_U	0.0086	0.024	0.015 ± .004
234_U	0.0032	0.790	0.258 ± .235
238_U	0.0025	0.710	0.228 ± .216
230_{Th}	0.0180	0.082	0.042 ± .016
232_{Th}	0.0190	0.051	0.029 ± .0009

A water sample obtained from O.U. Porter Hall, Athens Athens City (treated water) was also included in the river water samples. The city obtains the water from the wells near the river Hocking. The Porter Hall water sample yielded:

235_U	0.013 pCi/L
234_U	0.220 pCi/L
238_U	0.180 pCi/L

which is slightly below the average alpha activity of the river water samples.

CONCLUSIONS

The maximum and minimum uranium concentrations for coal are 14.97 ug/g (#8A Redstone Coal) and 0.03 ug/g (#6 Middle Kittanning Coal) respectively.
The maximum and minimum uranium concentrations for shale are 9.82 ug/g (Harlem shale) and 0.06 ug/g (Middle Kittanning roof shale) respectively.
The average uranium concentration for shale (2.97) ug/g) is greater than the average uranium concentration for coal (1.70 ug/g).
The maximum and minimum thorium concentrations for coal are 7.32 ug/g (#8A Redstone Coal) and 0.11 ug/g (#6 Middle Kittanning Coal) respectively.
The maximum and minimum thorium concentrations for

shale are 7.57 ug/g (Lower Freeport roof shale) and 0.17 ug/g (Middle Kittanning roof shale) respectively.

The average thorium concentration for shale (4.35 ug/g) is greater than the average concentration for coal (1.95 ug/g).

The average concentration of the Hocking River sediments for uranium (0.33 ug/g), thorium (0.90 ug/g) and ^{40}K (0.529 ug/g) is greater than the average concentration of uranium (0.14 ug/g), thorium (0.50 ug/g) and ^{40}K (0.457 ug/g) for its tributary sediments.

The alpha activity of a treated water sample from the Athens municipal supply is

a.   0.013   pci/l ^{235}U

b.   0.220   pci/l ^{234}U

c.   0.180   pci/l ^{238}U

which is slightly below the average alpha activity of the river water samples.

The alpha activity of the river water samples discussed within this study is considerable below the 15 pci/l gross alpha activity maximum contaminant level (MCL) set by the USEPA for potable drinking water supply standards.

The similarity of alpha activities between river water samples and that of the Athens municipal water supply suggests that other contaminants within the river water are also infiltrating into the ground water which serves as the primary source of potable water for many communities and individuals within the basin.   It also shows that present conventional treatment methods do not effectively remove these trace elements at these concentrations.

A review of USGS files indicated no record of radioactivity analysis for the Hocking River at the time of this study (1980).   The older Ohio EPA measurements were conducted using very much less sensitive instruments.   Therefore, it is not possible to draw any conclusion about the nature of increase of radioactivity.   Therefore, these measurements may be considered as a baseline and may be used for comparison with the recent observations.

ACKNOWLEDGEMENTS

The authors have condensed the thesis-The Alteration of the Natural Radiation of the Hocking River Basin by John T. Massey-Norton who obtained his M.S. in Hydrogeology at O.U. in 1980 under the supervision of the authors.   We appreciate very much the work performed by John.

REFERENCES

1. Swanson, V. E. "Geology and Geochemistry of uranium in marine black shales," a review, Geological Survey Professional Paper 356-C, 1961, pp. 67-112.

2. Brown, A. "Uranium in the Chattanooga shale of Eastern Tennessee," Geological Survey Professional Paper 300, (1956), pp. 457-462.

3. Snider, J. L. "Radioactivity of some coal and shale of Pennsylvanian Age in Ohio," (USGS-USAEC TE1-404, 1954, p. 22.

4. Caldwell, R.D., R.F. Crosby, and M.P. Lockard. Radioactivity in coal mine drainage: Environmental surveillance in the vicinity of nuclear facilities, Reinig, William C., Ed. (Charles C. Thomas-Publisher, Springfield, Illinois, 1970), pp. 438-445.

5. Botoman, G., and D. A. Stith. "Analysis of Ohio Coals," Ohio Geol. Circ. No. 47, 1978, 148p.

6. Mallory, Jr.; E.C., J.O. Johnson and R.C. Scott. "Water Load of Uranium, Radium, and Beta Activity at Selected Gaging stations," Water Year 1960-61, U.S. Geological Survey Prof. Pap. 1535-0, 1969,) 31p.

7. Rowe, J.J., and E. Steinnes. Determination of 22 minor and Trace Elements in 8 new USGS standard rocks by Instrumental Activation Analysis with Epithermal Neutrons, (U.S. Geological Survey Tour. of vol. 5, no. 3, 1977), pp. 397-402.

8. Moxham, R.M., and A.B. Tanner. High-Resolution Gamma-Ray Spectrometry in Uranium Exploration, U.S. Geological Survey, Jour. of Research, vol. 5, no. 6, (1977), pp. 783-795.

9. Tanner, A.B. R.M Moxham and F.E. Senftle. "Assay for uranium and determination of disequilibrium by means of in situ high resolution gamma-ray spectrometry," (USGS open file rept. 77-571, 1977), 22p.

10. Senftle, F.E., R.M. Moxham, A.B. Tanner, G.R. Boynton, P.W. Philbin and J.A. Baicker. Intrinsic Germanium Detector used in Borehole Sonde for Uranium Exploration, Nuclear Instruments and Methods, Vol. 138, North-Holland Publ. Co., 1976. pp. 371-381.

11. Phelps, P.L, L.R. Anspaugh, S.J. Roth, G.W. Huckabay and D.L. Sawyer. D.L. Ge(Li) Low Level In-Situ Gamma-Ray Spectrometer Applications IEEE Transactions on Nuclear SCI. Vol. ns-21 no. 1, 2/1974.

12. Norton, J.T.M. "The Alteration of the Natural Radiation of the Hocking River Basin", MS Thesis, Ohio University, Athens, Ohio (1980).

13. Denson, N.M., and J.R. Gill. "Uranium Bearing Lignite and Carbonaceous shale in southwestern part of the Williston Basin - A regional study, "Geological Survey Professional Paper 463, 1965.

14. Nekrasova, Z.A. The Origin of Uranium mineralization in coal, Soviet Journal of Atomic Energy, Atom Naya Energiya (supplement number 6) Atomic Press, Moscow, translated by Voprosy urana (1958) published by Chapman & Hall, LTD., London pp. 29-42, 1957.

15. Breger, I.A. "The Role of Organic matter in the Accumulation of Uranium: The organic geochemistry of coal-uranium association" Formations of Uranium Ore Deposits, Proceeding Symposium-Athens, Greece International Atomic Energy, 1974.

16. McKelvey, V.E., D.L. Everhart, and R.M. Garrels. "Summary of Hypotheses of Genesis of Uranium Deposits," Geological Survey Professional Paper 300, 1956, pp. 41-54.

17. Vine, J.D. "Uranium bearing coal in the United States," Geological Survey Professional Paper 300, 1956, pp. 387-404.

18. Vine, J.D. "Geology of uranium in coaly carbonaceous rocks," Geological Survey Professional Paper 356-D, 1962, pp. 113-170.

19. Gluskoter, H.J., R.R. Ruch, W.G. Miller, R.A. Cahill, G.B. Dreher and D.K. Kuhn. "Trace Elements in coal: occurrence and distribution," Illinois State Geological Survey Circular 499, 1977, 154p.

20. Masursky, H. "Trace Elements in Coal in the Red Desert, Wyoming," Geological Survey Professional Paper, 300, 1956, pp. 439-444.

21. Moore, G.W. "Extraction of Uranium from aqueous solution by coal and some other materials," Economic Geology Volume 49 number 6, 1954, pp. 652-658.

22. Conant, L.C., and V.E. Swanson. "Chattanooga shale and related   rocks of Central Tennessee and nearby areas," <u>Geological Survey</u> Professional paper 357, 1961, 91p.

23. Davis, S.N., and R.J.M. Dewiest. "Radionuclides in Ground Water," <u>Hydrogeology,</u> John Wiley & Sons, Inc., New York, 1966, pp. 129-155.

24. Sayre, W.W., H.P. Guy, and A.R. Chamberlain. "Uptake and Transport of Radionuclides by Stream Sediments," <u>Geological Survey Professional Paper</u> 433-A, 1963, p. 35.

25. White, A., and E.F. Gloyna. "Radioactivity transport in water-mathematical simulation," Tech. Rept. 19 (EHE-70-04) (CRWR-52) (ORO-19) 1970.

26. Pickering, R.J. "Distribution of Radionuclides in bottom   sediments of the Clinch River Eastern Tennessee," Geological Survey Professional Paper 433-H, 1969, p. 25.

27. Lee, H., T.O. Peyton, R.V. Steele and R.K. White. "Potential Radioactive Pollutants Resulting from Expanded Energy Programs," EPA-600/7-77-082, August 1977.

28. Bates, T.F. "An investigation of the mineralogy and petrology of uranium bearing shales: Atomic Energy Com." NYO 9605, 1961.

29. Osmond, J.K. and J.B. Cowart. "The Theory and uses of Natural Uranium Isotopic variations in Hydrology: Atomic Energy Review", vol. 14, no. 14, International Atomic Energy Agency, Vienna, 1976, pp. 621- 679.

30. Horne, J.C.; Ferm, J.C.; Caruccio, F.T. and Baganz, B.P., Depositional Models in Coal Exploration and Mine Planning in Appalachian Region. AAPG vol. 62, no. 12, 1978. pp. 2379-2411.

FACTORS CONTROLLING URANIUM AND RADIUM ISOTOPIC DISTRIBUTIONS IN GROUNDWATERS OF THE WEST-CENTRAL FLORIDA PHOSPHATE DISTRICT

Cynthia L. Humphreys
Seaburn and Robertson
A Division of Law Environmental, Inc.

Phosphorite sediments are enriched in uranium-238 and its decay daughters including uranium-234 and radium-226. It has been suggested that the extent to which these radioisotopes interact with and are distributed throughout the hydrosphere is enhanced by phosphate mining operations. The results of this investigation indicate that the hydrogeologic framework and related hydrogeochemical environment were found to override the impacts of the mining process with regard to the distribution and concentration of uranium and radium-226 in groundwaters underlying the phosphate district and adjacent areas in west central Florida.

Groundwater and aquifer materials from mined, mineralized but unmined and from economically unmineralized areas were analyzed to determine uranium and radium-226 content and 234U/238U and 226Ra/238U activity ratios. Samples were obtained from all three aquifers that underlie the area: the surficial, secondary artesian and Floridan aquifers. The data were then correlated with prevailing hydrogeochemical conditions to determine to what degree mining operations alter natural conditions with respect to the distribution of uranium-series isotopes.

It appears that mining disrupts uranium isotopic distributions and mobility in the intermediate and Floridan aquifers to some degree, and may slightly increase uranium concentrations in the surficial aquifer. However, the oxidation-reduction potential and the ionic strength of the groundwater, directly related to its residence time within the aquifer, were identified as the major factors controlling uranium and radium-226 concentrations and activity ratios in groundwaters of the study area.

INTRODUCTION

The processes involved in mining of uranium-rich phosphate ore are suspect in the liberation of uranium and its decay products to groundwater. Carbonate fluorapatite, the principal-phosphate bearing mineral in phosphorites of the study area, commonly contains 50 to 200 ppm of uranium [1,2]. Present along with this uranium are decay products of the uranium-238 series including long-lived daughters uranium-234 and radium-226 (Figure 1).

Of concern in this investigation were the concentrations and relative distributions of uranium and radium-226 in groundwaters and sediments of the land-pebble phosphate district of west- central Florida.   Concentrations and activity ratios of uranium-238, uranium-234 and radium-226 in soils and natural waters of mined and unmined terrains were compared within the phosphate area.   In addition, groundwaters and sediments from an adjacent unmineralized area, in which radium-226 concentrations in groundwater were known to be elevated, were also studied in an attempt to better understand the hydrogeologic and geochemical factors affecting the distribution of these radionuclides.

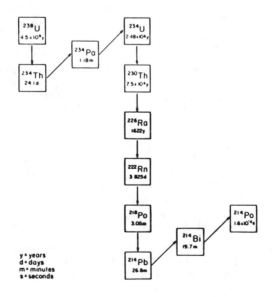

Figure 1. - Uranium decay series.

AREA

Mining

The study area, including the land-pebble phosphate district and adjacent unmineralized areas, is shown in Figure 2.   The west-central Florida phosphate district is one of the leading producing areas of phosphate ore in the world.   Phosphate rock occurs at shallow depths within the Bone Valley and Hawthorn Formations.   These deposits are currently exploited by surface mining in the northern and central portions of the district.

Strip mining has resulted in the alteration of the near-surface hydrogeologic system by excavation of water-bearing phosphatic strata and overlying sediments. Additional disturbances of the system occur during dewatering of the surficial aquifer in preparation for mining.   Dewatering is accomplished by the installation of shallow injection or recharge wells which drain groundwater from the surficial aquifer, through a confining bed, into the underlying intermediate aquifer.

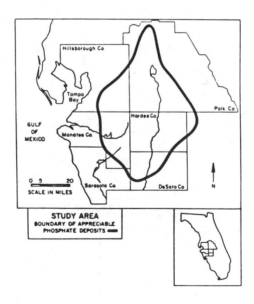

Figure 2. - Location of study area.

Hydrogeology

There are three aquifer systems present throughout
the study area and a fourth system occurs locally in
the central portion of the phosphate district (Figure
3). The three major aquifers include the unconfined
surficial aquifer, the secondary artesian aquifer and
the Floridan aquifer. The boundaries and degree of
interconnection of these aquifers are highly variable.

The water table or surficial aquifer is comprised
of undifferentiated sands and clays. These deposits
are underlain by the Pliocene-age Bone Valley Formation
in the northern and central parts of the study area
[3]. In the northern part of the study area, the
upper portion of the Bone Valley Formation is composed
of weathered phosphate, sand, and clay. This unit
functions hydraulically as part of the surficial
aquifer. All of the surficial sediments are removed as
overburden in preparation for mining.

In the central portion of the phosphate district,
the uppermost artesian aquifer, a localized water
bearing zone lies beneath the surficial aquifer. This
unit is composed entirely of the phosphate pebble
conglomerate of the Bone Valley Formation which is the
primary ore material extracted through strip mining.

The secondary artesian aquifer is a confined
aquifer, composed of limestone and dolostone of the
lower Hawthorn Formation. This aquifer receives

AGE	FORMATION	AQUIFER		ASSOCIATION with PHOSPHATE	MINING TERMINOLOGY
Pleistocene	terrace deposits	surficial		none	overburden
Pliocene	Bone Valley Formation	upper	surficial	usually depleted in phosphate due to weathering	overburden
		lower	uppermost artesian (Polk Co.)	phosphate pebble cgl.; enrichment due to reworking of Hawthorn Fm.	matrix
Early and Middle Miocene	Hawthorn Formation	upper	confining unit	source of Bone Valley phosphate; may contain minable ore	possible matrix
		lower	secondary artesian	parent material for reworked Bone Valley Fm.	bedrock
Early Miocene	Tampa Formation	SECONDARY ARTESIAN / CONFINING UNIT ; CONFINING UNIT / FLORIDAN			
Late Oligocene	Suwannee Limestone	Floridan			
Late Eocene	Ocala Limestone	Floridan			
Middle Eocene	Avon Park Limestone	Floridan		from: Cathcart (1966) and Kaufman and Bliss (1977)	

Figure 3. - Hydrogeology of west-central Florida.

artificial recharge from the surficial and uppermost artesian aquifers when they are dewatered in preparation for mining.

The Floridan aquifer is the primary artesian aquifer in the area. It is separated from the secondary artesian aquifer by clays that are thin to absent in the northern part of the study area and thicken toward the south. The Floridan aquifer is comprised of a thick sequence of limestones and dolostones. The transmissivity of the aquifer is generally higher in the northern portion of the study area due to the existence of well-developed cavernous zones [3,4].

Groundwater Recharge, Residence Time and Quality

The primary recharge area for the Floridan aquifer is located in the extreme northern portion of the study area. In this area the Bone Valley and Hawthorn Formations are absent and the aquifer is unconfined. Indirect recharge by leakage through confining beds also occurs primarily in the northern portions of the study area through thin semi-confining beds or those breached by sinkholes. The regional direction of groundwater flow within the Floridan aquifer is to the southwest.

Groundwater quality of the Floridan aquifer is primarily influenced by the solubility and composition of aquifer materials, and the residence time of groundwater in the system. Residence time of groundwater within an aquifer is a function of flow velocity and the distance between the point of recharge to point of discharge [5].

As groundwater moves away from its point of recharge, it dissolves aquifer materials and the concentration of dissolved mineral constituents, or total dissolved solids (TDS), increases. The oxygen content of groundwater is usually highest at the time of recharge, since in its previous form as meteoric or surface water it was in intimate contact with atmospheric oxygen. As residence time increases, groundwater commonly loses much of its dissolved oxygen through chemically and biologically mediated reactions.

Field measurements conducted during this investigation indicated that recharge waters entering the Floridan aquifer in the northern portion of the study area have a lower TDS content and a much higher oxidation-reduction potential (Eh) than groundwater tested for the same parameters in the southern portion of the study area.

Groundwater quality in the surficial aquifer varies due to the open nature of the system and highly variable sediment composition. Generally, TDS concentrations are less than 220 mg/l in the northern portion of the area and in excess of 200 mg/l in the southern portion. The Eh values are quite variable depending on the organic content of the soil.

Groundwater within the secondary artesian aquifer shows an increase in TDS from northeast to southwest with concentrations ranging from 100-300 mg/l to 300-1,500 mg/l, respectively. The Eh measurements taken during this study indicate oxidizing conditions exist in this aquifer in the northeastern part of the study area and reducing conditions are prevalent in the southwest. In the southwest portion of the study area, 90 percent of the wells sampled exhibited redox values of less than -200 millivolts.

URANIUM AND RADIUM GEOCHEMISTRY

The occurrence and abundance of uranium and radium in natural waters are controlled by their chemical properties, and the prevailing hydrogeochemical conditions and mechanisms of radioactive decay.

Uranium exists in several oxidation states and its solubility varies depending on its valence. The hexavalent (U^{6+}) and tetravalent (U^{4+}) forms are the most commonly occurring oxidation states. The hexavalent state has the greatest solubility and is found in oxidizing waters in the form of the uranyl ion, UO_2^{2+}. Upon reduction to U^{4+}, uranium becomes insoluble and is immobilized by adsorption onto aquifer materials. The U^{4+} ion has an ionic radius close to that of Ca^{2+} and, in this form, may substitute for calcium in the apatite structure.

Redox conditions control not only the amount of uranium in solution but are also a major factor influencing isotopic distributions. Of the three naturally occurring isotopes of uranium, two are significant in natural waters. These two isotopes, ^{238}U and ^{234}U, are part of the same decay series. Uranium-238, with a half-life of 4.5×10^9 years is the long-lived parent of ^{234}U. Uranium-234 has a half-life of 2.48×10^5 years (Figure 1).

The fractionation of ^{234}U and ^{238}U was first documented by Cherdyntsev, et al in 1955 [6]. Fractionation of uranium-series isotopes is most pronounced in groundwater. In an aquifer system where groundwater and aquifer materials are in intimate contact, leaching, precipitation and adsorption play major roles in the fractionation process.

Because the chemical properties of ^{234}U and ^{238}U are identical, the fractionation of ^{234}U and ^{238}U is a function of mechanisms of radioactive decay. The relative abundance of ^{234}U to ^{238}U is measured in terms of alpha activity ratio, $^{234}U/^{238}U$. At secular equilibrium the activity ratios is unity. Excess ^{234}U is denoted by a ratio of greater than one, a deficiency by a ratio of less than one. The decay process may result in direct ejection of ^{234}Th (the intervening decay product) into solution. This dislocation into less stable lattice position [6], may facilite the preferential leaching of ^{234}U into solution or the stripping of electrons resulting in the oxidation of ^{234}U [7].

There are four naturally occurring radium isotopes found in groundwater. Only ^{226}Ra of the ^{238}U decay series was studied in this investigation. Radium-223 and ^{224}Ra were considered geologically insignificant due to their extremely short half-lives. Radium-228 of the ^{232}Th decay series was not included because of the very low concentrations of thorium in sedimentary phosphorites [1]. The abundance of ^{226}Ra in groundwater is a function of the ^{238}U and ^{234}U content of the aquifer materials and groundwater, the residence time of the groundwater and chemical interaction between radium and the hydrosphere.

High radium concentrations tend to indicate a primary enrichment of parent uranium within the aquifer matrix or the presence of uranium as a secondary accumulation. The half-life of ^{226}Ra is 1620 years, so that in a highly transmissive aquifer radium may be transported a considerable distance from its source. Radium is introduced into solution by alpha recoil or leaching. The rate that any radioisotope is introduced into the aqueous phase by alpha recoil mechanisms is dependent on its half-life. The chemistry of the environment will then determine if it will remain in solution.

In a slow moving system, groundwater has a longer residence time within the aquifer and will develop higher concentrations of radium than in a highly transmissive aquifer with matrix materials of comparable uranium enrichment. Slower moving water has an increased contact time with the radium source.

The susceptibility of radium to leaching and its retention in groundwater is enhanced by a solution with high ionic strength such as groundwater containing high total dissolved solids. In low ionic strength solutions, radium can readily be adsorbed on grain surfaces [8]. In mineralized groundwater, many other ions are available in concentrations many orders of magnitude higher and these ions will be more readily adsorbed leaving radium in solution (i.e. the common ion effect).

Radium exists only as a divalent cation and redox conditions do not influence its solubility. If concentrations of radium are sufficiently high, it can complex with sulfate, forming highly insoluble $RaSO_4$. Maximum concentrations of radium in groundwater, however, are much too low to reach the solubility product (K_{sp}) for $RaSO_4$ of $10^{-10.4}$. High sulfate concentrations do not seem to affect radium solubility unless large amounts of barium are available for co-precipitation of the radium with $BaSO_4$ [8].

SAMPLES

A suite of groundwater samples was collected from 120 locations. Groundwater samples were obtained from all three aquifers present in the study area. Samples were taken from domestic, industrial, and municipal wells in mineralized regions (both mined and unmined) as well as unmineralized areas. Phosphate production and recharge wells were also sampled in mined areas. Sampling locations are shown in Figure 4.

Sediment and rock samples were obtained from mining sites in the northern portion of the study area and from cores taken in the unmineralized southern portion of the study area. Samples from active mines were taken from a mine face and consisted of unconsolidated sands (overburden), upper Bone Valley Formation (leached sandstone), middle and lower Bone Valley Formation (phosphate ore), and the upper Hawthorn Formation (both a leached zone and phosphatic zone). Core samples from unmineralized Sarasota County, in the southwest portion of the study area, show the presence of moderately-to-well indurated carbonates of the Bone Valley and Hawthorn Formations. A sand-sized phosphate fraction was present in the samples but never exceeded 5 percent of the sample.

METHODS

Each suite of groundwater samples consisted of 17.5 liter and 1 liter samples for uranium and radium analyses, respectively. Field measurements of pH, Eh, specific conductivity, and temperature were performed

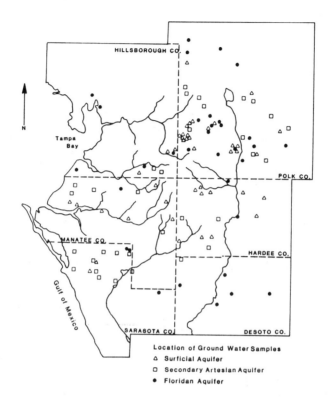

Figure 4. - Locations of groundwater samples obtained
during the investigation.

at each sampling location. Wells were pumped for a
sufficient period of time to assure representative
measurements and samples.

Uranium samples were acidified with nitric acid to
a pH of one at the end of each day. In addition, a
ferric nitrate carrier and an artificial uranium tracer
were added to each sample. The recovery process for
uranium consisted of ferric hydroxide co-precipitation,
organic solvent extraction and ion exchange. After
purification, uranium was electro-deposited on a
stainless steel planchet and analyzed using a high
resolution alpha spectrometer [7].

Radium was concentrated and scavenged by coprecipitation with barium sulfate. This precipitate
was dissolved using an EDTA solution. The solution was
stored in a sealed glass bubbler for a minimum of 96
hours to allow in-growth of ^{222}Rn. Samples were de-

emanated, using the Rn emanation method [9,10], into
an alpha scintillation cell [11] and counted using a
photomultiplier tube.

RESULTS

Hydrogeochemical conditions within the various
aquifers of the study area, coupled with the magnitude
of uranium enrichment of aquifer materials are the
primary controlling factors with respect to uranium and
radium concentrations and distributions. The phosphate
mining process appears to slightly disrupt natural
radioisotopic patterns in groundwaters of all three
aquifers.

Groundwater Data

Uranium concentrations were found to be moderate
to low in all groundwater samples generally ranging
from 0.01 to 0.5 ug/l. Highest concentrations occurred
in mined areas within the surficial aquifer (Table 1).
T-tests comparing average uranium concentrations in
mined, unmined, and unmineralized areas indicate that
these higher uranium concentrations in proximity to
mining operations are significant These "high"
concentrations were less than .80 pCi/l (2.36 mg/l).

Uranium concentrations and activity ratios varied
widely over the study area and it was difficult to
identify or separate aquifers on the basis of uranium
data. However, the greatest degree of variation was
found to occur in data from mined areas. F-tests
indicated this variation was significant in samples
obtained from the intermediate and Floridan aquifers in
mined terrains.

Radium concentrations on the whole were several
orders of magnitude higher than uranium concentrations.
Radium concentrations in groundwaters from all three
aquifers of the study area increase to the southwest,
downgradient of the phosphate district. Highest
concentrations were found in the unmineralized,
southern portion of the study area. Lowest
concentrations were found to exist in mined areas of
the phosphate district (Table 1). The average ^{226}Ra
concentration in groundwaters of all aquifers in the
southern part of the study area exceeded the state
and federal drinking water standard of 5pCi/l for
combined activities of radium isotopes.

The majority of samples analyzed displayed
^{226}Ra/^{238}U activity ratios much greater than unity.

Table 1. - Average Values for Radiosotopic Data.

Aquifer	Terrain	U Conc. ug/l	234U/238U	226 Ra/Conc. pCi/l	226Ra/238U
Surficial	Mined	.725 ±106.4%	1.70±58.5%	2.11±155.6%	45.86±145.0%
	Unmined	.150 ± 90.3%	2.35±72.8%	3.46±127.9%	130.19±135.9%
	Unmineralized	.010 ± 98.7%	4.26±44.8%	10.4 ± 23.6%	5373.8 ± 80.6%
Secondary Artesian	Mined	.217 ±101.8%	1.56±63.9%	3.49± 77.4%	376.16±145.5%
	Unmined	.197 ± 94.8%	1.75±19.2%	3.60± 99.7%	549.69± 67.9%
	Unmineralized	.025 ± 93.7%	3.24±45.7%	7.82± 87.5%	1595.2 ± 84.4%
Floridan	Mined	.497 ±159.4%	1.72±36.8%	2.13± 78.3%	69.42±129.0%
	Unmined	.0359± 97.2%	2.47±61.5%	4.07± 61.2%	415.04±110.4%
	Unmineralized	.0316± 31.6%	2.94±23.8%	5.77± 33.2%	1864.5 ± 5.84%

Activity ratios greater than 1.00 are common in groundwater because radioactive daughters are usually more mobile in solution than the parent isotope. $^{226}Ra/^{238}U$ ratios were found to increase markedly from the northeastern mined and unmined areas to the unmineralized areas in the southwest part of the study area in all three aquifers.

Sediment Data

Sediments analyzed from the phosphate-rich Bone Valley and Hawthorn Formations obtained from a mine face of an active mining site in southern Polk County contained an average of 130 ppm uranium and 102 uranium equivalents of ^{226}Ra (Tables 2 and 3). These high concentrations are directly related to the phosphatic nature of the sediments. These samples exhibited approximate secular equilibrium between $^{226}Ra/^{238}U$ and $^{234}U/^{238}U$ with activity ratios averaging 1.04+25.5 percent and .99+5.13 percent, respectively.

Table 2. - Rock and Sediment Data from a Stratigraphic Section
at USS Rockland Mine, Polk County.

Stratigraphic Position	U Conc. ppm	234U/238U	226Ra Conc. dpm/g	226Ra/238U
Upper Bone Valley Fm. Leached Sandstone	4.76± .20	1.04±.172	3.722± .293	1.05 ±.18
Pleistocene Sands (overburden)	2.29± .34	.96±.11	5.243± .032	1.09 ±.15
Middle Bone Valley Fm.	126.64± .53	1.07±.08	124.10 ± .984	1.43 ±.20
Lower Bone Valley Fm.	189.39±2.04	.94±.05	196.96 ±1.54	1.04 ±.09
Upper Hawthorn Fm. (leached)	261.54±2.97	.96±.04	117.15 ±1.33	.604±.122
Upper Hawthorn Fm.	65.51±1.23	1.02±.27	49.22 ± .720	1.04 ±.31

Table 3. - Rock and Sediment Data from Two Cores from Sarasota County.

Core Number and Formation	U Conc. ppm	234U/238U	Ra Conc. dpm/g	226Ra/238U	Depth ft.
W-11908 Bone Valley Fm.	378.38±3.62	1.02 +.06	97.96± .55	.349±.104	18
W-11908 Hawthorn Fm.	73.27±1.24	.885±.047	20.11±	.370±.197	25
W-11908 Hawthorn Fm.	139.95±1.69	.988±.069	100.12±1.13	.964±.240	328
W-12983 Hawthorn Fm.	16.28± .197	.972±.037	11.83± .38	.844±.163	112
W-12983 Hawthorn Fm.	19.21± .126	.997±.022	10.19± .34	.715±.154	150
W-12983 Hawthorn Fm.	33.56± .109	1.01 +.019	10.38± .02	.417±.119	173

In contrast to the above data, two cores taken from central Sarasota County, in the southwest part of the study area (Figure 4) exhibit extreme deficiencies with respect to ^{226}Ra (Table 2). Although radium deficient, ^{234}U/^{238}U ratios in both cores were close to, but slightly less than unity. These samples consisted of moderately-to-well indurated carbonates with sand-sized phosphate component of less than 5 percent. Uranium concentrations greatly exceeded concentrations normally found in carbonate sediments.

DISCUSSION

Regional Trends in the Floridan Aquifer

In a simple aquifer system where neither excessive leaching or secondary accumulation of uranium has occurred, the rule of high uranium concentrations in combination with low ^{234}U/^{238}U activity ratios in groundwaters near recharge areas and low uranium concentrations coupled with high activity ratios in downgradient reducing areas is found to apply [12, 13]. In aquifers where flow systems have been modified by sea level changes and thus water table fluctuations resulting in periods of both leaching and accumulation, a clear isotopic pattern is not generally evident. This latter situation seems to be the case in the Floridan aquifer of west-central Florida where only an overall pattern is identifiable. This overall trend is depicted in Figure 5 where ^{234}U/^{238}U is plotted against reciprocal concentration for samples obtained from the Floridan aquifer. Each line represents a flow system in which ground water evolves from higher uranium concentration, lower activity ratio water in the northern, upgradient oxidized zones to lower concentration, higher activity ratio water to the south in the downgradient reduced portions of the aquifer.

Uranium Distributions and Phosphate Mining

Uranium concentrations were found to be slightly higher, in proximity to mining operations in all three aquifers sampled. Statistical analysis of data indicate that these higher concentrations were significant in waters of the surficial and Floridan aquifers. Figure 6 illustrates, in a generalized sense, the uranium isotopic evolution of groundwater in both mined and unmined areas as compared with patterns found to be representative of isotopic evolution in high and low transmissivity systems [8, 14].

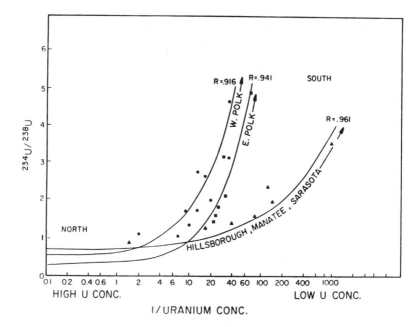

Figure 5. – $^{234}U/^{238}U$ versus 1/uranium concentrations in the Floridan aquifer.

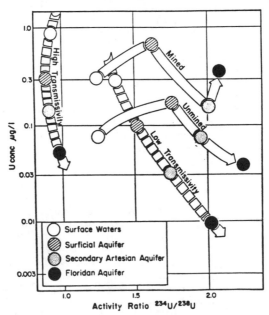

Figure 6. – Evaluation of groundwater with depth in mined and unmined areas.

The study area generally follows the low transmissivity curve indicating that the mining process has not had a profound effect on the uranium mobilization system except with regard to the Floridan aquifer in proximity to mined areas. In these areas, uranium solubility appears enhanced, both with respect to the low transmissivity curve and with respect to the concentration of uranium in the secondary artesian aquifer. This might be attributed to an enhancement of lateral groundwater flow within the Floridan aquifer and coincident leaching of aquifer materials due to drawdown of the aquifer in this area from large groundwater withdrawals by the phosphate industry.

Radium Distributions

Two factors appear responsible for the elevated ^{226}Ra concentrations and high $^{226}Ra/^{234}U$ ratios observed in waters of the unmineralized, southwestern part of the study area: (1) a secondary accumulation of uranium within aquifer materials in this area; and (2) higher total dissolved solids in groundwaters of all three aquifers.

A uranium source must be present to generate high ^{226}Ra concentrations in groundwaters of the unmineralized area. Rock and sediment samples from two cores located in this portion of the study area indicate that the carbonate sediments present there have an anomolously high uranium content. Due to the very small phosphate fraction in these materials, this enrichment was inferred to be of a secondary nature. This may have occurred as uranium, mobile in groundwaters flowing southwest from the primary recharge area, was immobilized upon encountering reducing conditions existing in aquifers of this region.

The very low $^{226}Ra/^{238}U$ ratios, coupled with high uranium concentrations in the sediments, suggest that ^{226}Ra generated by this secondary accumulation of uranium, is preferentially held in solution. The ionic strength (TDS content) of groundwaters in the southwestern part of the area is sufficiently high to retain radium in solution once it enters groundwater by mechanisms of radioactive decay.

The TDS content of groundwaters within the phosphate district is not high enough to retain ^{226}Ra in solution. Sediments there display an overall balance in ^{226}Ra and ^{238}U activities. Although some leached zones are deficient in ^{226}Ra, mobilization is short-lived.

Groundwater Residence Time and Radioisotopic Distributions

In a general way, distance from a recharge area can be used as a parameter indicative of the chemical evolution of groundwater. In Figure 7, the average sampling distances from the primary recharge area were plotted against each of the radioisotopic variables of interest. These plots indicate a similar trend of increasing $^{234}U/^{238}U$, ^{226}Ra concentration, and $^{226}Ra/^{238}U$ as the distance from the recharge area increases.

Increasing $^{234}U/^{239}U$ ratios in the downgradient direction can be attributed to the reducing conditions encountered within aquifers of the southwest part of the study area. Increasing ^{226}Ra concentrations and $^{226}Ra/^{238}U$ activity ratios are a result of the higher TDS content and low redox potential of groundwater in downgradient waters. Decreasing uranium concentrations with distance from the recharge area are due to immobilization of uranium as it encounters the diffuse reducing barrier in the southwest part of the study area.

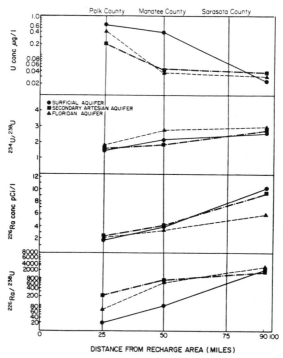

Figure 7. - Radioisotopic distributions versus distance from areas of recharge.

CONCLUSIONS

Impacts of the phosphate mining process appear to be overridden by natural conditions with respect to radioisotopic distributions of uranium and radium in groundwaters of the study area. These natural phenomena include:

1. the presence of a uranium source, either a primary enrichment or a secondary accumulation; and

2. the residence time of ground water within the flow system which directly influences oxidation-reduction potential and total dissolved solid content of ground water.

Data indicate that mining operations appear to slightly alter natural uranium distributions. This is evidenced by slightly elevated uranium concentrations observed in groundwater samples from the surficial and Floridan aquifer. Uranium concentrations did not exceed .80 pCi/l (2.36 ug/l) in these samples. These slightly increased concentrations may possibly be attributed to alterations in flow patterns due to dewatering of the surficial aquifer and high pumpage rates by the phosphate industry.

Radium concentrations and distributions were not affected by mining operations. Highest concentrations of ^{226}Ra and excessively high $^{226}Ra/^{238}U$ activity ratios were found to occur in aquifers of the downgradient, unmineralized portion of the study area. Average concentrations of ^{226}Ra exceeded the U.S.EPA drinking water standard of 5.0 pCi/l for combined activities of radium isotopes in this portion of the study area.

Groundwater quality in the unmineralized downgradient area, specifically, high total dissolved solids and low oxidation-reduction potential, are a product of a longer residence time. Radium solubility is enhanced by the high total dissolved solid content of groundwaters in the downgradient area. The radium source in that area was determined to be a secondary accumulation of uranium deposited upon the aquifer matrix as oxidizing, uranium-bearing waters from the phosphate district encountered reducing conditions prevalent in this area.

REFERENCES

1. Altschuler, Z.S., Clarke, R.S. and E.J. Young, 1958, "Geochemistry of Uranium in apatite and phosphorite," U.S. Geol. Surv. Prof. Paper 314-D.

2. Cathcart, J.B., 1956, "Distribution and Occurrence of Uranium in the Calcium Phosphate zone of the Land-Pebble Phosphate District of Florida," U.S. Geol. Surv. Prof. Paper 300, p. 489-494.

3. Stewart, H.G., 1966, "Groundwater Resources of Polk County," Florida Geol. Survey, R.I. 44.

4. Wilson, W.E., 1977, "Groundwater Resources of DeSoto and Hardee Counties," Florida Bur. of Geology, R.I. 83.

5. Drever, J.I., The Geochemistry of Natural Waters, (Prentice-Hall, Inc., 1982) p. 292-294.

6. V.V. Cherdynster, P.I. Chalov and G.Z. Khaidarov, in: Transactions, 3rd Session of the Committee for Determination of Absolute Ages of Geologic Formations, Izv. Akad. Nauk SSSR (1955) p. 175.

7. Osmond, J.K., and J.B. Cowart, "The Theory and Uses of National Uranium Isotopic Variations in Hydrology," Atomic Energy Review, v. 14, no. 4, p. 621-636 (1976).

8. Gilkeson, R.H., Cartwright, K., Cowart, J.B. and R.B. Holtzman, "Hydrogeologic and Geochemical Studies of Selected Natural Radioisotopes and Barium in Groundwater, Bureau of Reclamation, U.S. Dept. of Interior, Wash., D.C., UILU-WRC-83-0180, (1983).

9. Broecker, W.S., 1965, in Symposium on Diffusion in Oceans and Fresh Water, T. Ichiye, Ed., Lamont Geological Observatory, Palisades, N.Y.

10. Mathieu, G.G., 1977, ^{222}Rn-^{220}Ra Technique of Analysis in Annual Tech. Rep. Coo-2185-0 to ERDA, Lamont-Doherty Geol. Obs., Palisades, N.Y.

11. Lucas, H.F., 1957, Improved Low-Level Alph-Scintillation Counter for Radon, Rev. of Scientific Inst. v. 28, no. 9, p. 680-683.

12.   Cowart, J.B., and J.K. Osmond, 1974, ^{234}U and ^{238}U in the Carizzo Sandstone Aquifer of South Texas, International Atomic Energy Agency, Isotope Techniques in Groundwater Hydrology, v. 2, p. 131-149.

13.   Osmond, J.K. and J.B. Cowart, 1977, Uranium Series Isotopic Anomalies in Thermal Groundwaters from Southwest Florida, in the Geothermal Natural of the Florida Plateau, Fl. Bur. of Geology, S.P. 21, p. 131-147.

14.   Osmond, J.K., verbal comm., 1983.

15.   Cowart, J.B., Kaufman, M.I. and J.K. Osmond, 1978, Uranium Isotope Variations in Groundwaters of the Floridan Aquifer and Boulder Zone of South Florida, Jour. of Hydrol. 36, pp. 161-172.

Session IV: Sampling and Analysis of Radon, Radium and Other Radioactivity in Ground Water

Moderators: DeWayne Cecil, U.S. Geological Survey

Michael Weber, U.S. Nuclear Regulatory Commission

SAMPLING AND ANALYSIS OF DISSOLVED RADON-222 IN SURFACE AND GROUND WATER

In Che Yang,
U.S. Geological Survey, Denver, Colorado

INTRODUCTION

Radon-222 is a radioactive gas known to occur in ground water. It diffuses throughout a ground-water aquifer, and its concentration can be many orders of magnitude larger than the concentration of radium or uranium. Radon-222 is a daughter product of ^{226}Ra and has a rather short half-life (3.8 days); however, long-lived components of the decay products (those of environmental concern) are ^{210}Pb (half-life of 22 years and a weak beta emitter), and ^{210}Po (half-life of 138.4 days and an alpha emitter). The way that ^{222}Rn enters the body tissue of a person is not well-understood, possibly involving the stomach wall. However, inhalation of ^{222}Rn, as experienced by miners working in uranium mines, can cause lung cancers from the daughter products of ^{222}Rn that remained at the bifurcations of the lung. A maximum contamination level (MCL) for ^{222}Rn in drinking water is under consideration by the U.S. Environmental Protection Agency [1].

Sampling and measurement of ^{222}Rn are complicated by the volatility of the gaseous element. For example, ^{222}Rn in water appears to be removed effectively by aeration. Therefore, onsite techniques for sampling water for ^{222}Rn analysis need to minimize aeration of the water, and avoid transfer of samples through the open air. Losses of ^{222}Rn during transportation and storage may occur, if sampling vessels are permeable to gas, or if they have leaky stopcocks and caps.

Analytical methods that have been used in other laboratories for the determination of ^{222}Rn are briefly described hereafter, followed by detailed descriptions of the methods (both sampling and analysis) used in this study at the U.S. Geological Survey's laboratory in Denver, Colo.

Gamma Spectroscopy

In the gamma-spectroscopy method, gamma rays from the ^{222}Rn daughters, ^{214}Pb and ^{214}Bi, are measured with a NaI(Tl) crystal or Ge(Li) semiconductor detector. This method requires negligible sample preparation; however, the detection limit is about 10 pCi/L for a 1-L sample [2, 3].

Direct Liquid-Scintillation Counting

In the direct liquid-scintillation counting method, a 10-mL water sample is mixed with 5 mL of toluene-based liquid-scintillation fluid, followed by measurement of radioactivity in a liquid-scintillation counter. The method lacks specificity. Any alpha or hard-beta emitters, other than ^{222}Rn and its daughters, in a form soluble in toluene, also will contribute to the count rate. The detection limit, if no other alpha or beta contaminants are present, is about 10 pCi/L for a 10-mL water sample [4].

Extraction Concentration and Liquid-Scintillation Counting

In the extraction concentration and liquid-scintillation counting method, a large-volume water sample, about 500 mL, is extracted with 20 mL of toluene in a closed system, and toluene is separated from the water for liquid-scintillation counting. The same contamination problems may exist as in the direct liquid-scintillation counting method. The detection limit is about 0.2 pCi/L for a 500 mL sample [5].

EXPERIMENTAL METHODS

Water Sampling

An all-glass deemanation bubbler (Figure 1) is used for collecting water samples for analysis of ^{222}Rn. Before sampling, evacuate the bubbler (assembled as shown in Figure 1) to less than 10 millitorr in the laboratory or laboratory trailer through stopcock 1

INLET PORT

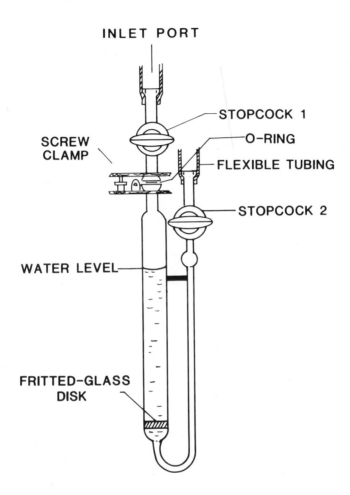

STOPCOCK 1

O-RING

SCREW
CLAMP

FLEXIBLE TUBING

STOPCOCK 2

WATER LEVEL

FRITTED-GLASS
DISK

Figure 1.   All-glass deemanation bubbler used for
collecting water samples for analysis
of radon-222.

(stopcock 2 is closed). Then, close stopcock 1. Gently
place the evacuated sample bubbler into the surface
water until the space between stopcock 1 and the inlet
port is filled with water. Open stopcock 1 to draw
sample water directly into the main chamber of the
bubbler, until the chamber is 60% full. Close stopcock 1. Label bubbler with: (1) Sample number; (2)
place of collection; and (3) collection time (day, hour,
and minute). Use masking tape to secure the stopcocks 1
and 2 in the closed position. Place bubbler inside a

foam-packed cardboard box and send by airmail to the laboratory as soon as possible. For ground-water samples, use a peristaltic pump or a submersible pump to supply an uncontaminated, full-column flow of sample water through a short, flexible tubing to the stopcock-1 end of the evacuated bubbler. When the main chamber is about 60% full (as shown in Figure 1), close stopcock 1. Label and transport samples to the laboratory in the same manner described for the surface-water samples.

Transportation and Storage

Three replicates of similar bubblers were used to collect two sets of source waters (one set from a low-activity source and the other set from a high-activity source) for this investigation. One replicate was analyzed soon after sampling; a second replicate was sent to University of Texas Health Science Center at Houston, Tex., for return immediately after receipt (about 1 week for a round trip), and a third was kept in our laboratory, until the shipped replicate was returned from Houston. Samples that were sent to Houston and returned to Denver, and samples that were stored in Denver were analyzed at the same time.

Analytical Method

Radon-222 in water occurs from two sources: one source is the in-situ decay of ^{226}Ra already in water; the other source is external, such as fallout from air, or inflow of other water containing mainly dissolved ^{222}Rn with little parent ^{226}Ra (commonly called excess or nonsupported ^{222}Rn). To measure the excess or non-supported ^{222}Rn, the quantity of ^{222}Rn in equilibrium with the ^{226}Ra has to be determined after the total ^{222}Rn measurement [6]. The relation among various components of ^{222}Rn can be expressed in the following equation:

$$^{222}\text{Rn (total)} - {}^{222}\text{Rn (supported, in equilibrium with } {}^{226}\text{Ra)}$$
$$= {}^{222}\text{Rn (nonsupported)}$$

Here the supported ^{222}Rn is assumed to be in equilibrium with the ^{226}Ra; the assumption is reasonable because of the short half-life of ^{222}Rn.

Deemanation

When ^{222}Rn samples are received in the laboratory, they need to be analyzed immediately. Use clamps to attach the bubbler to the deemanation system (Figure 2).

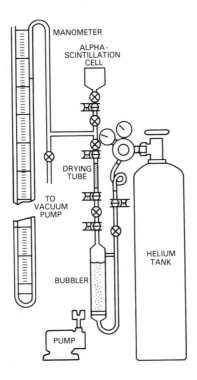

Figure 2. Radon-deemanation system.

This deemanation system consists of the following parts:
(1) Bubbler; (2) a glass drying tube packed with
drierite (anhydrous magnesium perchlorate) to remove
moisture, and ascarite (NaOH on asbestos plus soda-
lime) to remove CO_2; (3) a manometer to indicate the
pressure inside the system during transfer of ^{222}Rn from
the bubbler to the alpha-scintillation cell, and also to
detect any leak in the system; (4) a helium tank to
purge dissolved ^{222}Rn into the cell; (5) an alpha-scintillation cell; and (6) a vacuum pump. With the bubbler
stopcocks closed, evacuate the system and the alpha-
scintillation cell to a vacuum of 10 millitorr. Close
stopcock to the vacuum pump and allow the system to
equilibrate for about 2 min. If the manometer menicus
remains stable, connect the helium line (3-5 lb/in^2) to
the other side of the bubbler. Purge trapped air in the
line between the helium tank and bubbler with helium gas
by momentarily opening the O-ring connection on the
helium-line inlet a few times. Record the time (day,
hour, and minute); then carefully open the bubbler-

outlet stopcock to purge ^{222}Rn directly into the alpha-scintillation cell. Allow the vacuum to equilibrate slowly; otherwise, an excessive risk exists of blowing the liquid sample from the bubbler into the drying tube. Continue the ^{222}Rn purging by slowly opening the bubbler-inlet stopcock, checking the fritted-glass disk carefully for rising bubbles. Allow the pressure to increase slowly, control the manometer decrease rate, and complete purging in 10 to 15 min. Fill the cell to 4 mm of mercury less than the atmospheric pressure, and close the cell stopcock. Close all other stopcocks in the deemanation system. Record the time (day, hour, minute). Quickly remove the bubbler from the assembly, and open the outlet stopcock momentarily to release helium pressure. Turn off the room light; remove the alpha-scintillation cell for ^{222}Rn-activity measurement. Measure the water volume inside the bubbler using a graduated cylinder.

Radon-222-Activity Counting System

The ^{222}Rn-activity counting system consists of: (1) An alpha-scintillation cell internally coated with silver-activated ZnS, with a 2-in.-diameter quartz window; (2) a 3-in.-diameter scintillation-photomulti-plier tube; (3) a preamplifier connected to the scintil-lation-photomultiplier base; (4) an amplifier; (5) a scaler and counter/timer; and (6) a high-voltage power supply.

Before counting, turn off the room lights; remove the light shield from the alpha-scintillation cell, and place the cell in the light-proof counting chamber (Figure 3). Allow about 20 min for the ^{222}Rn daughters to ingrow; then count the sample until 1,000 counts have been obtained. Flush out the ^{222}Rn from the cell after obtaining the data. The detection limit is about 0.2 pCi/L for a 60-mL sample.

Calibration of Alpha-Scintillation Cell and Equipment

The counting efficiency of each alpha-scintillation cell varies between cells and between counting instru-ments. Consequently, counting efficiency of each alpha-scintillation cell was determined in each instru-ment in which it was used. This was done by counting ^{222}Rn transferred from a bubbler containing a measured volume of ^{226}Ra standard solution. A minimum of four or five bubblers containing same standard, enables four or five cells to be calibrated simultaneously. Use the following procedures: (1) Pipet 10 mL of 10 pCi/mL

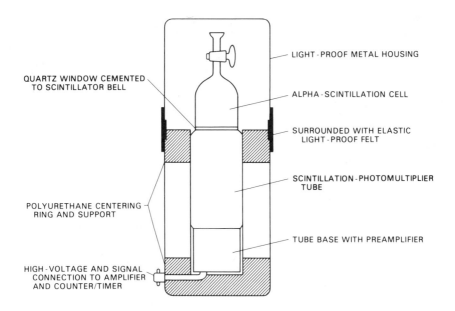

Figure 3.   Alpha-scintillation cell and light-proof housing.

^{226}Ra standard solution directly into each of four to five bubbler tubes and dilute with distilled water to 60% full; (2) deemanate the bubbler to determine the zero-in-growth time for ^{222}Rn; after 1 month, redeemanate the standard to determine instrument and cell efficiency.   Barring any spillage or breakage, the standards will last indefinitely, and can be deemanated to provide ^{222}Rn calibration.   The counting efficiency of the system used at the U.S. Geological Survey laboratory was in the range of 4.9 to 5.6 counts/min per 1 pCi of ^{222}Rn, with a background-count rate of 0.03 to 0.09 count/min.

RESULTS AND DISCUSSION

The method is specific for ^{222}Rn; no direct interferences occur.   Radon-220 and ^{219}Rn, if present, will have decayed completely by the time of the analysis, because of their short half-lives of about 4 s for ^{219}Rn and 56 s for ^{220}Rn.   Normally, these constituents will not be supported by any substantial concentration of the parent ^{224}Ra and ^{223}Ra in the sample; thus, they usually are insignificant.

Alpha-scintillation-cell contamination can be a major problem. Ground water may contain large concentrations of dissolved ^{222}Rn, which, if deemanated into the alpha-scintillation cells and allowed to remain in them for several hours, will result in considerably increased background count rates. Increased background-count rates of more than several tenths of 1 count/min preclude the use of contaminated cells for the analysis of most environmental samples. Therefore, ^{222}Rn in the cells needs to be counted and flushed out immediately after obtaining the data. The contaminated cells can be disassembled, cleaned, and refabricated, by using fresh phosphor and a new quartz window if necessary.

Loss of Radon-222

Analytical results for loss of ^{222}Rn during transportation and storage for low- and high-activity samples are listed in Table 1.

Table 1.   Transportation and Storage Loss of Radon-222.

| Sample type | Date | | | ^{222}Rn[a] (pCi/L) | ^{226}Ra[a] (pCi/L) |
	Collected and shipped or stored	Returned	Analyzed		
Orig-inal	1-10-79	--	1-10-79	47.0±2.6	<1
	1-11-79	--	1-11-79	8,180±78	5.41±0.21
Trans-ported	1-10-79	1-15-79	1-16-79	54.5±1.1	<1
	1-11-79	1-17-79	1-18-79	7,850±77	5.41±0.21
Stored	1-10-79	--	1-16-79	52.0±1.1	<1
	1-11-79	--	1-18-79	8,450±84	5.41±0.21

[a]Counting error is two sigmas, based on counting statistic.

The loss of ^{222}Rn during transportation and storage, after 1 week, is insignificant for a low-activity sample (within three sigmas of counting errors); the loss of ^{222}Rn during transportation and storage, for a high-activity sample after 1 week, indicates a loss of 4% during transportation and no loss during storage. The ^{226}Ra concentrations in both high- and low-activity samples also were measured and determined to be less than 1 pCi/L for a low-activity sample and about 5 pCi/L for a high-activity sample. Therefore, contribution of ^{222}Rn from a ^{226}Ra source can be neglected. The data presented here are limited; general conclusions are limited to the same degree.

Precision and Interlaboratory Comparison

Precision of the method may be expressed in terms of the relative deviation (sigma) at different concentrations, as listed in Table 2.

Table 2.   Precision of Radon-222 in Water Analysis Determinations.

Sample[a]	Number of runs	Measured mean concentration (pCi/L)	Relative deviation (%)
1	2	3.12	19
2	3	48.0	5
3	2	105.0	4
4	3	756.0	3
5	3	4600.0	2
6	3	8160.0	1.5
7	2	24,690.0	1.0

[a]Samples were prepared by precipitating ^{226}Ra with barium sulfate twice, with minimal stirring, from ^{226}Ra stock solutions stored for 2 months after preparation. The clear solution was decanted as ^{222}Rn water samples.

Interlaboratory comparisons of the ^{222}Rn determinations in water analyses (from the same source of water) by the U.S. Geological Survey (USGS) laboratory and other laboratories are listed in Table 3.

Four different analytical methods were used in these analyses. Concentrations in low-activity samples ranged from 650 to 930 pCi/L; concentrations in high-activity samples ranged from 4,050 to 5,500 pCi/L. Our concentrations for low-activity samples incidently match the average concentration; our concentrations for high-activity samples are slightly larger than the average concentration.

SUMMARY

Sample collection for determining the concentration of dissolved ^{222}Rn in surface water is performed best by placing one end of an evacuated bubbler into the water, and then drawing a water sample directly into the main chamber of the bubbler until the bubbler is about 60% full. For collection of ground-water samples, a peristaltic or submersible pump is needed to supply an unaerated water sample through a short flexible tubing connected to one end of the evacuated bubbler, until the

Table 3.   Interlaboratory Comparison of Radon-222
Determinations in Low- and High-Activity
Water Sources.

Sample type	Collection (date and time)	Deemanation (date and time)	^{222}Rn (pCi/L) USGS laboratory[a]	^{222}Rn (pCi/L) Average of six laboratories[b]	^{226}Ra (pCi/L) USGS laboratory[a]
Low activity	1-26-79 1:00 p.m.	2-1-79 8:42 a.m.	800±25	800	2.15±0.11
High activity	1-26-79 11:00 a.m.	2-1-79 9:00 a.m.	5,120±100	4,900	3.90±0.15

[a]All errors are two-sigma values based on counting statistics; USGS, U.S. Geological Survey.

[b]Different methods of ^{222}Rn analyses were used in the six laboratories: one laboratory used gamma spectroscopy; one laboratory used deemanation and alpha-scintillation cell counting; two laboratories used direct liquid-scintillation counting; two laboratories used extraction concentration and liquid-scintillation counting.

main chamber is about 60% full. Loss of ^{222}Rn during transportation from the sampling site to the laboratory (within 1 week) was negligible for low-activity samples and was 2 to 3% for high-activity samples.

Current methods for the determination of dissolved ^{222}Rn in water include: (1) Gamma spectrometry; (2) direct liquid-scintillation counting; (3) extraction concentration and liquid-scintillation counting; and (4) direct deemanation and alpha-scintillation counting. The direct deemanation and alpha-scintillation counting method described in this report is more sensitive (0.2 pCi/L for a 60-mL sample) and has better precision compared to other methods for most environmental samples.

REFERENCES

[1.]   U.S. Environmental Protection Agency "Radionuclides in the Interim Regulation, and Other Radionuclides Under Consideration," Federal Register, Vol. 48, No. 194, pp. 45520 (1983).

[2.]   Lucas, H. F., Jr. The Natural Radiation Environment, Adams, J. A. S., and Lowder, W. M., Eds. (University of Chicago Press 1964), pp. 315.

[3.] Irfam, M., and J. H. Read "Low-Level Activity of ^{226}Ra in Water Measured with a 3x3 inch NaI(Tl)," Nuclear Instrument and Methods, 179:389-392 (1981).

[4.] Prichard, H. M., and T. F. Gesell "Rapid Measurements of ^{222}Rn Concentrations in Water with a Commercial Liquid Scintillation Counter," Health Physics 33:577-581 (1977).

[5.] Homma, Y., and Y. Murakami "The Study on the Applicability of the Integral Counting Method for the Determination of ^{226}Ra and Various Sample Forms Using a Liquid Scintillation Counter," J.of Radioanalytical Chemistry, 36:173-184 (1977).

[6.] Yang, I. C. "Improved Method for the Determination of Dissolved ^{226}Ra," in Proceedings of the 5th International Conference on Nuclear Methods in Environmental and Energy Research, Puerto Rico (ONF-840408, J.R. Vogt, Ed., U.S. Department of Energy, 1984) pp. 191-201.

RADON-222 CONCENTRATION OF GROUNDWATER FROM
A TEST ZONE OF A SHALLOW ALLUVIAL AQUIFER
IN THE SANTA CLARA VALLEY, CALIFORNIA

Lewis Semprini Department of Civil Engineering,
Stanford University, Stanford, California

INTRODUCTION

Current interest in groundwater contamination problems and
studies on fate and transport of contaminants in groundwater have
resulted in a need for new techniques to investigate hydrogeologic
characteristics of aquifers. Interactions of contaminants with
the solid matrices of aquifers play an important role in the fate
and transport of contaminants in groundwater. Examination of the
radon concentration of groundwater is a useful method of studying
the interactions of the groundwater with the solid matrix. Radon
studies can also help determine aquifer characteristics, such as
porosity.
Several factors make radon an interesting geochemical tool for
subsurface investigations. Its source is the decay of radium-226
within the solid matrix of the aquifer. Radium, with a half-life
of 1600 years, supplies a continuous source of radon to the pore
fluid. Radon is radioactive, with a half-life of 3.86 days.
The half-life of radon is relatively short compared to the long
time scales associated with groundwater transport. The constant
source of radon to the pore fluid, and its own decay, result in a
rapid approach of the pore fluid radon concentration to a steady-
state value, in equilibrium with emanation rate from the aquifer
solids. The equilibrium concentration is achieved when the rate
of radon emanation into the pore fluid is equal to its rate of
decay in the fluid. Because it is a noble gas, radon does not
interact strongly with the aquifer's solid matrix once the gas is
in the pore fluid: thus, radon is advectively transported with
the groundwater. The short half-life of radon, its rapid
equilibration in the pore fluid, and its inability to react with
the solid matrix make the radon concentration a good indicator of

changes in the emanation source caused by changes in local geologic conditions. These changes should be reflected in spatial variation in radon concentration of the groundwater.

In this study, the radon concentration in groundwater from a experimental test zone in a shallow alluvial aquifer in the Santa Clara Valley, California, was measured to study its relationship to the hydrogeologic characteristics of the test zone. The test zone is currently being used to demonstrate in-situ biodegradation methodologies for the restoration of aquifers contaminated with halogenated hydrocarbons. The hydrogeologic characteristics of the test zone have been fairly well-characterized, based on aquifer core samples, pump tests and tracer tests.

The objectives of the radon study were: 1) to measure the spatial variability of the groundwater's radon concentration over small distances to determine the concentration's relationship to the local hydrogeology, 2) to study the emanation rate of radon from aquifer core samples as a function of particle size, 3) to estimate the porosity of the aquifer based on emanation values and groundwater radon concentration, and 4) to assess whether emanation characteristics of the aquifer solids are consistent with the hydrogeologic relationships.

BACKGROUND

Recent measurements of radon concentrations in groundwater have been performed in order to: 1) evaluate health hazards due to the presence of radon in potable water supplies [1]; 2) study nuclear waste migration [2]; 3) explore for uranium deposits [3]; 4) predict earthquakes [4]; and 5) estimate fluid age based on He[4] and radon content [5].

Table 1 summarizes reported radon concentrations of groundwater for different sedimentary aquifers [5-14].

Table 1.   Radon Concentration of Groundwater from Sedimentary and Alluvial Aquifers

STUDY	LOCATION	AQUIFER DESCRIPTION	NO. WELLS	RADON RANGE (pCi/kg)	RADON MEAN (pCi/kg)	CV %
Tanner (1964b)	Utah	Alluvial Sands	11	223-810	342	51
Andrews & Wood (1972)	England	Milford Sands	9	481-757	593	14
		Red Sandstone	4	153-643	488	69
		Limestone	15	34-418	169	57
Andrews & Lee (1979)	England	Sandstone	24	46-608	251	49
Gorgoni et al. (1982)	Emilia	Alluvial Deposits	43	100-400	188	-
	Italy: Lombardy	Sand-gravels	21	260-1100	392	-
	Verona	Thermal 12-26°C	7	80-250	166	-
			9	250-1130	ND	-
Mitsch (1982)	North Carolina (Phosphate Mining Area)	Water Table	26	8-5733	210	-
		Yorktown	12	8-575	250	-
		Castle Hayne	4	7-181	44	-
Gilkeson et al. (1983)	Illinois	Sandstone	80	40-670	ND	-
Prichard & Gesell (1983)	Midwest USA 8 States	Sedimentary	209	-	117-228	-
King et al. (1982)	South Carolina	Sedimentary	104	5.1-1700	158	-
Heaton (1984)	South Africa	Sandstone	14	330-3230	1204	68
Hussain & Krishnasami (1980)	India	Sandstone	12	102-409	312	-

An order of magnitude variation in the mean concentrations is shown among the different formations studied. Within a given geologic formation, concentrations are shown to vary by an order of magnitude.

The variations in concentration among and within aquifers may result from geologic changes, as discussed by [5]. A change in rock type can result in a change in the parent radium-226 concentration and/or a change in emanation power, which represents the fraction of radon produced in the solid matrix that escapes to the pore fluid. The emanation rate to the pore fluid is the product of these two factors. Concentrations are also dependent on the water/rock ratio which, in a sedimentary or an alluvial aquifer, is the porosity of the aquifer.

The equation that describes the equilibrium radon concentration in a pore fluid is given by:

$$C_F = [C_{RA} E_P p_B] \frac{1}{\theta} \frac{}{p_F} \qquad (1)$$

where,

C_F = radon mass concentration in the pore fluid (nCi/kg)
C_{RA} = rock radium concentration (nCi/kg)
E_P = emanation power
p_B = rock bulk density (kg/m^3)
p_F = fluid density (kg/m^3)
θ = porosity (m^3/m^3)

Forms of Eq. 1 are given by [7,15,16]. The term in brackets represents the volumetric emanation coefficient, E_V (nCi/m^3), which is dependent on properties of the solid matrix of the aquifer, as discussed above. The emanation power is also a function of temperature and moisture content. However, these parameters are relatively constant in most groundwater aquifers. The fluid density term in Eq. 1 does not vary widely in groundwater. Thus, the properties of the rock matrix are likely to be responsible for most of the variability in groundwater radon concentration.

In groundwater systems, the assumption of equilibrium conditions is generally valid, due to the long contact time of the fluid with the solid aquifer matrix. Based on Eq. 1, the radon concentration of groundwater can be used as a geochemical tool to estimate properties of the aquifer. For instance, in order to estimate aquifer porosity, the solid matrix emanation coefficient must be determined, as well as the radon concentration of the groundwater.

Several investigators have evaluated both radon groundwater concentration and emanation coefficients from aquifer solid samples. [7,13,17] found reasonable agreement between predicted values based on Eq. 1 and measured values for sedimentary aquifers. Fractured aquifers and aquifers with secondary porosity gave poorer agreement.

In this study, measurements of the radon concentration of groundwater and radon emanation coefficients from aquifer core samples were performed. These measurements were used to study the relationship of radon concentration to hydrogeologic characteristics of the aquifer and to estimate aquifer porosity.

EXPERIMENTAL METHODS

Groundwater samples from wells at the test site were obtained by pumping at the surface with a peristaltic pump. Two methods of sample collection and analysis were used for radon measurements. The first method involved the collection of a groundwater sample in an evacuated vessel. The analytical method used was that described by [18]. This method involved the laboratory sorption of radon onto activated carbon using a vacuum extraction line. Radon was then transferred to a ZnS alpha-scintillation flask (Lucas cell) using helium as a carrier gas. Finally, alpha disintegrations of Rn-222 and its short-lived daughters, Po-218 and Po-214 were counted by a photomultiplier tube and a single-channel pulse-height spectrometer. The system was calibrated using an 80 pCi NBS radium solution.

The second method of analysis developed for field use eliminates the sample processing time required in the vacuum line method. A groundwater sample was collected in a 100 ml Spectrum™, gas-tight syringe that was fitted with a glass stopcock valve at the entry port. The empty and filled syringe was weighed to determine the sample size. Approximately 70 grams of groundwater are normally collected. The radon in the sample was transferred to an evacuated Lucas cell that was fitted with a rubber septum. The transfer procedure entailed four successive extractions of the groundwater sample using 30 ml of helium. Helium was introduced into the syringe, and the syringe was vigorously shaken to partition the radon into the helium effectively. The transfer efficiency was determined by measuring the radon remaining in the sample after four transfers. Because fifteen percent of the radon remained in the groundwater sample, a transfer efficiency of 85% was included as a calibration factor in calculating radon concentration using the field method.

A cross-comparison test of the laboratory and the field methods was performed using replicate samples from the test zone. The results of the cross-comparison test are presented in Table 2. The laboratory and field measurements are shown to be in good agreement for both multiple and single samples. Based on a series of replicate measurements, the coefficient of variation for the field method is 7 to 12%, which is greater than the 5% obtained by the laboratory method. The field method, however, has adequate precision for most field investigations.

Laboratory emanation coefficients for aquifer solids as a function of particle size were measured. An unconsolidated core sample of the aquifer material was dried and sieved using standard sieve analysis. A sample from a specific particle size range was placed in a leak-tight vessel and saturated with aquifer water. A small headspace of air was maintained under partial vacuum conditions to check for leaks. After an emanation period of 30 days, the radon in the vessel was removed for analysis by bubbling helium through the water. The laboratory method of analysis described above was used to determine the quantity of radon that emanated into the pore fluid.

Table 2.   A Cross Comparison of the Laboratory
and Field Method of Radon Analysis

Well	Laboratory Method (nCi/kg)	Field Method (nCi/kg)
	0.224	0.224
S1	0.220	0.255
	NA	0.269
S2	0.205	0.215
S3	0.260	0.249
	NA	0.226
E2	0.363	0.331

NA = Not available

DESCRIPTION OF THE TEST SITE

The test site is located in the Santa Clara Valley, California, one of the large alluvial filled intermontane valleys of the Coast Range Province. The site, located at the Moffett Naval Air Station in Mountain View lies on the lower part of the Stevens Creek alluvial fan, about 5 km south of the southwest extremity of San Francisco Bay. The test zone is located in the shallowest of several aquifers which are separated by clay and silt layers. The test zone is confined above and below by a clay aquitard, with a piezometric surface approximately 10 ft above the base of the upper clay layer.

The aquifer solids are composed of sands and gravels. The thickness of the aquifer in the area of the test zone is approximately 5 ft, at a depth ranging from 14 to 19 ft. The particle distribution of core samples from the test zone is shown in Figure 1. Core samples from wells 4 and 6 taken at a depth of 18 ft and 17.5-18 ft, respectively, have similar particle size distributions. A large fraction of these solids are coarse to medium sands and gravel. The core from well 5, taken at a depth of 13.5-14 ft, shows a greater fraction of fine sand and silt. The wide range of particle sizes in the cores indicates the heterogeneous nature of the aquifer solids. Petrographic analysis indicates that the aquifer solids consist of rock fragments, even at the smallest particle size studied (0.075-0.150 mm), corresponding to a sieve mesh size of 100-200. The rock fragments consist of sandstone and volcanic rocks, shales, and cherts.

Pump tests indicate that the aquifer is leaky, with a transmissivity of 10,000 gpd/ft. Based on an aquifer thickness of 5 ft and the transmissivity value, the hydraulic conductivity of the test zone is approximately 100 m/d. This value is in the range expected for an aquifer composed of sands and gravels [19].

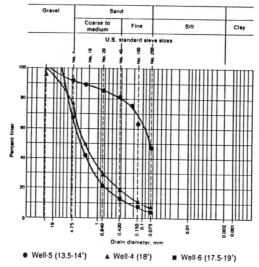

● Well-5 (13.5-14')   ▲ Well-4 (18')   ■ Well-6 (17.5-19')

Figure 1.   The particle size distribution from test zone cores.

Figure 2 presents a planar view of the wellfield studied, which covers an areal extent of approximately 50 m². The wells sampled under natural gradient conditions include observation wells (S1 S2 S3), (E1 E2 E3), and (N1 N2 N3). The observation wells are partially penetrating sandpoints, slotted over a 2 ft interval. The wells were hand-driven to minimize disturbances to the aquifer. The slotted intervals of the wells were centered at a depth of approximately 16.5 to 18.5 ft below the surface which, based on fully penetrating well logs, represents the sand and gravel zone.

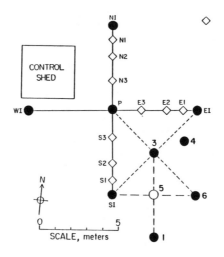

Figure 2.   Locations of the wells at the field test site.

RESULTS OF THE FIELD EXPERIMENTS

A series of groundwater radon measurements was made at the field site from December, 1985, through July, 1986, under natural gradient flow conditions. Table 3 summarizes the results of the measurements for the nine observation wells. The average concentrations show some variability among wells, with values ranging from 0.218 to 0.384 nCi/kg. Groundwater radon concentrations of individual wells did not vary greatly, as indicated by the range of minimum and maximum values and the coefficients of variation, which were less than 12% for all wells except E2.

Table 3.   Summary of radon concentrations the groundwater under natural gradient flow conditions

Well	Number of Observations	Minimum (nCi/kg)	Maximum (nCi/kg)	Average (nCi/kg)	Coefficient of Variation
S1	4	0.197	0.249	0.223	0.087
S2	3	0.214	0.247	0.225	0.068
S3	3	0.215	0.262	0.237	0.081
N1	4	0.206	0.240	0.220	0.057
N2	2	0.196	0.239	0.218	0.098
N3	2	0.239	0.264	0.252	0.049
E1	7	0.283	0.336	0.324	0.100
E2	6	0.283	0.692	0.384	0.362
E3	5	0.219	0.292	0.253	0.118

Along a south-north transect of wells (S1 S2 S3) and (N1 N2 N3), concentrations fell within a narrow range, with average values ranging from 0.218 to 0.264 nCi/kg. Two wells along the east-west transect (E1 E2) have significantly higher average radon concentrations of 0.324 and 0.384 nCi/kg, respectively. Well E3, which is nearer to the south-north transect, has an average concentration of 0.253 nCi/kg, similar to that of the north-south wells.

The radium-226 concentration of the groundwater was determined for a sample from well N1, using the method described by (18). The radium concentration was 0.0013 nCi/kg. Based on the average radon groundwater concentration from well N1 of 0.22 nCi/kg, the radium concentration in the groundwater accounts for less than 1% of the radon present. Thus, emanation from aquifer solids is the main source of radon concentration in the groundwater.

The constant lower radon concentrations along the south-north transect and the higher concentrations along the east-west transect suggest that hydrogeologic features of the aquifer in the

area of the test zone may be affecting the radon concentration.
Piezometric data from regional surveys and from wells spaced
approximately 20 meters apart at the test site indicate a
hydraulic gradient of 0.003 from south to north at the site.
The velocity of the groundwater, based on the gradient, the
hydraulic conductivity, and an aquifer porosity of 0.33 has been
calculated at approximately 1 meter/day. Both the velocity and
the direction of groundwater flow were confirmed by a natural
gradient tracer experiment performed at the site. The spatially
constant and lower radon concentrations are aligned with the
south-to-north direction of groundwater flow. The radon concentrations suggest that the aquifer's solid matrix provides a fairly
uniform source of radon to the groundwater in the direction of
flow. Also, since the groundwater velocities are high, the radon
concentrations should reflect a source strength integrated over a
scale of meters upgradient of the point of sampling. Thus, small
scale aquifer heterogeneities, which might affect the radon concentration if groundwater velocities were lower, are not observed
over the 10 meter flow length studied in the direction of
groundwater flow.

Higher radon concentrations were observed along the east-west
transect, perpendicular to the groundwater flow direction.
The higher concentrations may be associated with lower porosity
and/or a greater fraction of higher emanation solids in the area
of the aquifer, affecting the observed radon concentrations. Well
logs obtained during drilling indicate a greater change in the
aquifer characteristics in an east-west direction compared to the
north-south direction. A fence diagram constructed from well log
information is shown in Figure 3a. An east-west cross-section of
wells SI,5,6 shows the aquifer around well 5 being composed mainly
of sand, while the aquifer around wells SI and 6 is composed

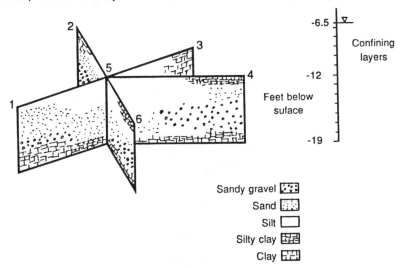

Figure 3a.   A fence diagram of the aquifer based on logs from wells
(1,2(SI),3,4,5,6).

mainly of sandy gravel.   The different composition of the aquifer
solids in the area of well 5 is supported by the particle size
distributions shown in Figure 1.   The distribution of the sample
from the well 5 core contains a much greater fraction of smaller
size particles including fine sands and silts, compared to that of
samples from wells 4 and 6 which contain medium to coarse sand and
gravel.   It should be noted that the depths from which the
particle size determinations were made are not the same for
well 5 and wells 4 and 6.   Additional sieve analyses of core
samples as a function of depth are required to confirm the fence
diagram constructed from well logs.

The aquifer along the south-north transect of wells (SI P NI)
consisted mainly of sandy gravel.   The fence diagram in Figure 3b
constructed from fully penetrating well logs shows a sandy gravel
layer of fairly constant thickness in the north-south direction.
The constant lower concentrations in this direction indicate a
similar composition of aquifer solids in the direction of
groundwater flow.

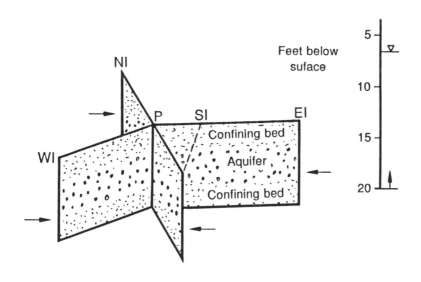

Figure 3b   A fence diagram based on the well logs from wells
(SI, P, NI, EI, WI)

The higher concentrations of radon in wells E2 and E3 should
reflect upgradient emanation conditions where the groundwater has
flowed for a period of days before sampling.   A rough
approximation of the mean distance to the emanation area is given
by [6] as the transport distance corresponding to the mean life of
radon $(1/\lambda)$, which is 5.52 days.   Under natural gradient flow
conditions, the groundwater flow is from south to north at a
velocity of approximately 1 m/d.   The distance upgradient to the

emanation area, based on the mean radon life and the groundwater
velocity is 5.5 meters, which places the emanation region in the
area of well 5. If the aquifer consisting of sands has a greater
emanation rate than the sandy gravel, the sands in the region of
well 5 may be causing the higher radon concentrations observed in
wells E2 and E3.

RESULTS OF EMANATION STUDIES

     Studies were performed to determine whether the sand and
gravel had different emanation characteristics. Emanation measurements were made as a function of particle size, using samples
from a core obtained just west of the S1,S2,S3 series of wells.
Inspection of the core solids indicated that they were representative of the solids in the test zone. The measured emanation
coefficient, E_m (nCi/kg), represents the product of the emanation power and the radium content, the $E_p[C_{Ra}]$ term of Eq. 1.
The emanation coefficients are based on the mass of solids in the
sample.
     Figure 4 presents the emanation coefficient from core solids
as a function of particle size. An increase in emanation
coefficient with a decrease in particle size is shown. There is a
factor of 3 difference between the lowest emanation coefficient
for the gravel fraction and the highest value for the smallest
particle size studied (0.1-0.25 mm), a fine sand. The emanation
coefficient is shown to be fairly constant over a particle size
range of 0.3 to 3 mm, which represents medium and coarse sand.

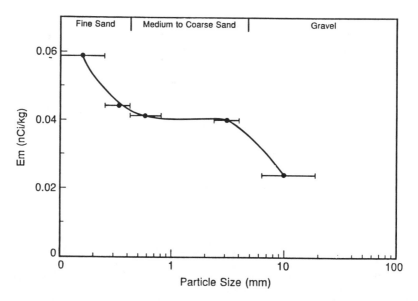

Figure 4.   The mass emanation coefficient as a function of particle
            size.

An increase in emanation coefficient with decrease in particle size may result from several factors. An emanation value would increase with decrease in particle size if the rate is limited by diffusion and the parent radium concentration was constant in the range of particle sizes studied. Thus, smaller size particles would have a higher emanation powers, with a greater fraction of the produced radon escaping due to the shorter diffusion distances. The radium concentration of the solids, however, may not be constant and may increase with a decrease in particle size. Radium concentrations of the solids have not been measured, so the emanation power, which normalizes for radium concentration, was not determined. Studies of soil particles by [20] found that an increase in radium concentration with decrease in particle size was responsible for the higher emanation values of smaller particles size. A more detailed study of emanation and the other characteristics of the aquifer solids of the test zone, including radium concentration, surface area and pore structure as a function of particle size, is required in order to obtain a better understanding of the radon emanation process from these solids.

ESTIMATES OF AQUIFER POROSITY

An estimate of aquifer porosity was made based on the results of the emanation measurements from aquifer solids and the measured radon concentration of the groundwater using Eq. 1. An average value for the emanation coefficient, E_N, was obtained by summing the product of the emanation coefficient for a particle size range by the mass fraction each particle size range represented. The cumulative particle size distributions shown in Figure 1 were used to determine the mass fraction of the particle size range. The values for the sandy gravel were obtained from the curves for samples from wells 4 and 6, while those for the sand were obtained from well 5 distribution. Two different groundwater radon concentrations were used in the estimates: 1) an average concentration of 0.23 nCi/kg represented wells along the south-north transect or emanation from the sandy gravel, 2) a concentration of 0.35 nCi/kg from wells E1 and E2 represented the sand. The average bulk density of the aquifer material, required for the aquifer porosity calculation, was estimated based on the relationship:

$$p_s = (1-\theta)p_s \qquad\qquad (2)$$

where,

p_s = grain density (g/cm^3)

An average grain density value of 2.68 g/cm^3 for gravelly soil was obtained from The Handbook of Physical Properties of Rocks (1984). The radon-based porosity estimate for the sandy gravel aquifer was 31%. The estimated value for the sand aquifer was 28%.

Independent measurements of porosity have not been obtained for aquifer core samples. Both porosity estimates are in the range of values for gravel (25-40) and sand (25-50), as tabulated by [22]. The estimated values are, however, in the low range of reported values, which may result from the aquifer being composed of a broad range of particle sizes. Bouwer (1978), for instance, gives a porosity range for sand and gravel mixtures of 10 to 30%.

The reasonable estimates of aquifer porosity described above, based on the radon measurements, indicate that the emanation studies of aquifer solids performed in the laboratory in conjunction with measurements of groundwater radon concentration are of value in studying interactions of groundwater with the aquifer solid matrix. The emanation data supports the hydrogeology data that suggest higher radon concentrations in the aquifer result from spatial heterogeneities in the solid matrix of the aquifer. Higher radon concentrations result from a zone of higher emanating sands upgradient of observation wells E2 and E3.

The data indicate that the groundwater radon concentration results from emanation from both sands and gravels along the south-north transect.   The data also suggests that contaminants in the groundwater flowing through the test zone will interact with a range of particle sizes. This information has important implications on the fate and transport of contaminants through the test zone.

ACKNOWLEDGEMENTS

This work was conducted at the Stanford experimental site for "In-situ Biodegradation Methodologies for the Restoration of Contaminated Aquifers", sponsored by the Kerr Environmental Research Laboratory of the U.S. Environmental Protection Agency under Cooperative Agreement CR 812220. Several people at Stanford who are involved in this project have contributed to this study: Paul Roberts, Gary Hopkins, Barty Thompson, Helen Dawson, Christoph Buehler, and Lois Epstein.

REFERENCES

1.   Hess, C. T., R. E. Casparius, S. A. Norton, and W. F. Brutsaert. "Investigations of Natural Levels of Radon-222 in Groundwater in Maine for Assessment of Related Health Effects," in Natural Radiation Environment III, T. Gessel and W. Lowder, Eds. Department of Energy, Oak Ridge, TN., (1978), pp. 529-546.

2.   Andrews, J. N., I. S. Giles, R. L. F. Kay, D. J. Lee, J. K. Osmond, J. B. Cowart, P. Fritz, J. F. Barker, and J. Gale. "Radioelements, Radiogenic Helium and Age Relationships for Groundwaters from the Granites at Stripa, Sweden," Geochim. Cosmochim. Acta. 46:1533-1543 (1982).

3. Dyck, W., "The Mobility and Concentration of Uranium and its
    Decay Products in Temperate Surficial Environments." In Short
    Course in Uranium Deposits: Their Mineralogy and Origin, M. M.
    Kimerley, Ed. (Mineralogical Association of Canada, 1978),
    pp. 57-100.

4. Teng, T., "Some Recent Studies on Groundwater Radon Content as
    an Earthquake Precursor," J. Geophysical Res. 85(B6):3089-3099
    (1980).

5. Andrews, J. N., and D. Lee, "Inert Gases in Groundwater from
    the Bunten Sandstone in England as Indicators of Age and
    Paleoclimatic Trends," Jour. Hydrol. 41:233-252 (1979).

6. Tanner, A. B., "Physical and Chemical Controls on Distribution
    of Radium-226 and Radon-222 in Groundwater near Great Salt
    Lake", In The Natural Radiation Environment, J. A. S. Adams
    and W. M. Lowder, Eds. (University of Chicago Press, 1964),
    pp. 253-276.

7. Andrews, J. N., and D. F. Wood, "Mechanism of Radon Release in
    Rock and Matrices and Entry into Groundwaters," Institution of
    Mining and Metallurgy Trans. 81:B198-B209 (1972).

8. Gorgoni, C., G. Martinelli, and G. P. Sighinolfi, "Radon
    Distribution in Groundwater of the Po Sedimentary Basin,"
    Chem. Geo. 35:297-309 (1982).

9. Mitsch, B. F., "Study of Radium-226 and Radon-222
    Concentrations in Ground Water near a Phosphate Mining and
    Manufacturing Facility with Emphasis on the Hydrogeologic
    Characteristics of the Area," Master's Thesis, University of
    North Carolina, Chapel Hill, (1982).

10. Gilkeson, R. H, K. Cartswright, J. Cowart, and R. B. Hotzman,
    "Hydrogeologic and Geochemical Studies of Selected Natural
    Radioisotopes and Barium in Groundwater in Illinois,"
    (Research Report 180, Water Resources Center, University of
    Illinois, Urbana, 1983)

11. Prichard, H. M., and T. F. Gesell, "Radon-222 in Municipal
    Water Supplies in the Central United States," Health Physics
    45(5):991-993 (1983).

12. King, P. T., J. Michel, and S. Moore, "Groundwater
    Geochemistry of ^{224}Ra, ^{226}Ra, and ^{222}Rn," Geochim. Cosmochim.
    Acta., 46: 1173-1182 (1982).

13. Heaton, T. H. E., "Rates and Sources of ^{4}He Accumulation in
    Groundwater," Hydrolog. Sci. Jour. 29(1):29-47 (1984).

14. Hussain, N., and S. Krishnaswami, "U-238 Series Radioactive Disequilibrium in Groundwaters: Implications to Origin of Excess U-234 and the Fate of Pollutants," Geochim. Cosmochim. Acta., 44:1287-1291 (1980).

15. D'Amore, F., J. C. Sabroux, and P. Zettewoog, "Determination of Characteristics of Steam Reservoirs by Radon-222 Measurements in Geothermal Fluids," Paleoph., 177:253-261 (1978).

16. Torgersen, T., "Controls on Pore Filled Concentrations of ^{4}He and ^{222}Rn and the Calculation of ^{4}He/^{222}Rn Ages," J. Geochem. Explor. 13:57-75 (1980).

17. Smith, A. R., H. A. Wollenberg, and D. F. Moosier, "Roles of Radon-222 and Other Natural Radionuclides in Earthquake Prediction," The Natural Radiation Environment III, T. Gessel and W. Lowder, Eds. (Department of Energy, Oak Ridge, TN., 1978), pp. 154-183.

18. Lucas, H. F., " A Fast and Accurate Method for Both Radon-222 and Radium-226," In The Natural Radiation Environment, J. A. Adams and W. M. Lowder, Eds. (University of Chicago Press, 1964), pp. 315- 329.

19. Bouwer, H. Groundwater Hydrology (N.Y. McGraw-Hill Book Company, 1978)

20. Megumi, K. and T. Mamuro, "Concentrations of Uranium Series Nuclides in Soil Particles in Relation to their Size," J. Geophys. Res. 82(2):353-356 (1977).

21. The Handbook of the Physical Properties of Rocks, Volume III, Ed. R. S. Carmichael, (Boca Raton, Florida: CRC Press, Inc., 1984).

22. Freeze, R. A., and J. A. Cherry. Groundwater (Englewood Cliffs, N.J.: Prentice-Hall, Inc., 1979).

An Improved Method for the Simultaneous
Determination of ^{224}Ra, ^{226}Ra and ^{228}Ra
in Water, Soils and Sediments

Henry F. Lucas,
Argonne National Laboratory, Argonne Illinois

INTRODUCTION

The naturally occurring concentrations of radium (^{226}Ra and ^{228}Ra) in public and private water supplies have been of studied for many years. Both general surveys and local studies have established the geographical regions where well waters exceed 3 pCi/L (1-17). In general, the ^{226}Ra level was determined by the emanation method, while the ^{228}Ra level was determined from the beta activity of the ^{228}Ac daughter. In a recent review (18) of the methods used "a number of approved analytic methods can bear improvement, especially the method for ^{228}Ra." The purpose of the work described here was to develop an improved method for the simultaneous determination of ^{226}Ra and ^{228}Ra.

Experimental Method

A selective radium complexer resin was originally developed to remove soluble radium from uranium minewaters (19). This material is unique because it retains only radium in a permanently complexed state (20). We use this resin to concentrate the radium from 20 to 100 L water samples, then place it in a metal can, seal the can, and the resin with a NaI(Tl) gamma spectrometric counting system. The amount of both ^{226}Ra and ^{228}Ra is determined by a least squares method(21,22).

Discussion of Method

The retention of radium on a normal cation exchange resin depends on the volume and hardness of the water, as well as the flow rate. Before the "Radium Selective Complexer" was available, other resins were used to concentrate the radium. The data obtained with water softener grade Dowex 50–X8, 20–50 mesh resin is summarized in Table 1. From water with a hardness of 20 to 34 grains, 95% retention could be routinely obtained at a flow rate of 20 mL/min with a volume of 20 L. Increasing the flow rate to 50 mL/min or the volume to 50 L reduced the retention efficiency to 92% and 61%, respectively.

As shown in Table 2, the retention of radium on the Radium Selective Complexer is considerable better than on the Dowex 50 cation exchange resin. In fact, the retention on as little as 10 mL of resin was equal to that for 200 mL of Dowex 50. While it is expected that conditions will be found in which the radium retention will be reduced, this should not occur for potable waters.

The accuracy and precision of the radium retention were determined by adding varied known amounts of ^{226}Ra and ^{228}Ra to 20 L of tap water. In all cases, the amounts found are within normal statistical limits of the amount added. From this study, we conclude that the accuracy of this method can approach ± 2% and that the limit of sensitivity is about 0.5 pCi/L for ^{226}Ra and ^{228}Ra.

The possible extension of this method to soils, sludges, and other samples was evaluated by adding known amounts of uranium and thorium ore standards to "sea sand". The samples were mixed by stirring and by shaking. The results are also included in Table 3 and indicate that this method can be used to process samples of all types. The reported value for the "Th soil" sample is expected to be reduced by further shaking and mixing.

REFERENCES

1.   John B. Hursh, Radium Content of Public Water Supplies, JAWWA 46, pp.43–54 (1954).

2.   A. F. Stehney, Radium and Thorium–X in Some Potable Waters, ACTA Radiol 43, pp.43–51 (1955).

3.   F. B. Barker and R. C. Scott, Uranium and Radium in Ground Water of the Llano Estacado, Texas and New Mexico, Trans Am Geop Union 39 (3) pp.459–466 (1958).

4.   H. F. Lucas, Jr. and F. H. Ilcewicz, Natural ^{226}Ra Content of Illinois Water Suplies, JAWWA 50 (11) pp.1523-1532 (1958).

5.   R. C. Scott, Radium in Natural Waters in the United States, in Radioecology: Proceedings of the First National Symposium, Colorado State University, Fort Collins, Colorado, Sept. 10-15, 1961, Reinhold Publishing Corporation, New York (1968)

6.   R. C. Scott and F. B. Barker, Ground Water Sources Containing High Concentrations of Radium, U.S. Geological Survey Prof. Paper 424-D, Article 414, pp.357-359 (1961).

7.   F. B. Barker and F. C. Scott, Uranium and Radium in Ground Water from Igneous Terrains of the Pacific Northwest, U.S. Geological Survey Prof. Paper 424-B, Article 128, pp.298-299 (1961).

8.   H. F. Lucas, Jr., Study of Radium-226 Content of Midwest Water Supplies, Radiological Health Data II No. 9, Sept. 1961.

9.   B. M. Smith, W. N. Grune, F. B. Higgins, Jr., and J. G. Terrill, Jr., Natural Radioactivity in Ground Water Supplies in Maine and New Hampshire, JAWWA 53 pp.75-88 (1961).

10.   R. C. Scott and F. B. Barker, Data on Uranium and Radium in Ground Water in the United States 1954-1957, U.S. Geological Survey Prof. Paper 426 (1962).

11.   Anon., Radium-226 and Radon-222 Concentrations in Central Florida Ground Waters, Federal Water Pollution Control Administration, Cincinnati, Ohio, PB-260-211 (1966).

12.   J. L. S. Hickey and S. D. Campbell, High Radium-226 Concentrations in Public Water Supplies, U.S.Public Health Reports 83 (7) pp.551-557 (1968).

13.   M. C. Wukasch and L. M. Cook, High Radioactivity in Drinking Water and Ground Water in South Texas, in Proceedings of the Third International Congress of the International Radiation Protection Association, September 9-14, 1973, Washington D.C., CONF-130907-P1.

14. R. D. Lee and S. W. Fong, An Assessment of Radium in Selected North Carolina Drinking Water Supplies, Health Phys 37 (6) pp.777–779 (1979).

15. J. Michel and W. S. Moore, ^{228}Ra and ^{226}Ra Content of Groundwater in Fall Line Aquifers, Health Phys 38 pp.663–671 (1980).

16. R. E. Rowland, H. F. Lucas, Jr. and A. F. Stehney, High Radium Levels in Water Supplies of Illinois and Iowa, International Symposium on Areas of High Natural Radioactivity, Pocos de Caldos, Brazil, June 16–20, 1975.

17. G. H. Emrich and H. F. Lucas, Jr., Geologic Occurance of Natural Radium-226 in Ground Water in Illinois, Bulletin Internat Assoc Sci Hydrol (Holland) 8 pp.5–19 (1963).

18. R. Blanchard, R. Hahne, B. Kahn, D. McCurdy, R. Mellor, W. Moore, J. Sedlet and E. Whittaker. Radiological Sampling and Analytical Methods for National Primary Drinking Water Regulations, Health Phys 48, pp.587–600 (1985).

19. T. Bouce, and S. Boom, Removal of Soluble Radium from Uranium Minewaters by Selective Complexer, Society of Mining Engineers of AIME, 82–23 (February 1982).

20. R. Rozelle, K. Ma, A New Potable Water Radium/Radon Removal System, Internal Report, The Dow Chemical Company (1983).

21. H. Lucas, Jr., D. Edgington, "Computer Analysis of Gamma-Ray Spectra: Validity of the Results", Modern Trends in Activation Analysis, Proceedings of the 1968 International Conference, Gaithersburg, Maryland, October 7–11, 1968, pp.1207–1214 (1969).

22. H. Lucas, Jr., D. Edgington, The Relative Sensitivity and Accuracy of NaI(Tl) and Ge(Li) Detectors for Minor Components, J. Radioanal. Chem. 15, pp.467–472 (1973).

Table 1.   Radium Retention by Dowex 50 Cation Exchange Resin[a]

Flow Rate (mL/m)	Sample Volume (L)	Retention (%)
20	20	96±2
20	20	95±2
20	20	93±2
50	20	92±2
50	50	63±2

[a]200 mL of Dowex 50-X8, 20-50 mesh.

Table 2.   Radium Retention by Radium Selective Complexer Resin[a]

Flow Rate (mL/m)	Sample Volume (L)	Resin Volume (mL)	Retention (%)
20	20	10	96±2
200	20	10	61±2
20	20	100	100±1
100	100	100	100±1
100	100	100	100±1

[a]Dowex XFS-43230 resin.

Table 3. Accuracy and Precision of the Method

| Experiment | ^{226}Ra (pCi/L) | | ^{228}Ra (pCi/L) | |
	added	found	added	found
WSSP04	2.49	2.6±0.1	0.24	0.2±0.3
WSSP05	2.52	2.7±0.1	0.24	0.3±0.3
WSSP06	0.24	0.4±0.2	2.47	2.7±0.2
WSSP07	0.24	0.6±0.3	2.46	2.4±0.2
WSSP08	0.51	0.4±0.2	0.48	0.6±0.2
WSSP09	0.50	0.5±0.2	0.49	0.5±0.2
WSSP10	0.50	0.6±0.2	0.49	0.5±0.2
WSSP12	105.2	104.8±0.3	0.0	−0.2±0.6
WWSP21	102.5	102.3±0.3	0.0	0.2±0.4
WSSP13	0.0	−0.2±0.3	113.4	102.4±0.4
WSSP22	0.0	−0.6±0.6	92.4	94.0±0.4
U (Soil)	13,158	13,214±77	?	48±36
Th (Soil)	?	1,048±86	15,646	16,297±128

NATIONWIDE DISTRIBUTION OF Ra-228, Ra-226, Rn-222, AND U IN GROUNDWATER

Jacqueline Michel and Mark J. Jordana,
RPI International, Inc., Columbia, South Carolina

INTRODUCTION

As part of the regulatory process for developing National Primary Drinking Water Standards, the U.S. Environmental Protection Agency (USEPA) uses occurrence data to predict human health exposures and the number of public water supplies likely to be affected. Occurrence documents are being prepared for four natural radionuclides, namely, uranium, radium-226, radium-228, and radon-222 (radon). These documents will contain estimates of the number of public water supplies expected to have these radionuclides in the drinking water and the number of people consuming drinking water having the radionuclides present at various levels of concern.

Development of these estimates are difficult for natural radionuclides because of the scarcity and potential for bias of available measurements of public drinking water supplies. One approach funded by the USEPA was the development of a conceptual model for radium-228 which predicted its occurrence in groundwater based on the geochemical and hydrological characteristics of the aquifer [1, 2]. This model was used to prepare the drinking water occurrence document for radium-228 [3, 4]. The same approach was used for the estimation of uranium, radium-226, and radon occurrence in groundwater. It was the objective of this study to prepare nationwide maps, presenting the groundwater occurrence level estimates by county for these natural radionuclides. The occurrence levels are estimated semiquantitatively, using 3-5 categories of occurrence, with estimated concentration ranges associated with each category.

METHODS

Two different approaches were used in determining the distribution of natural radionuclides in groundwater. Detailed
studies were conducted in three hydrogeological provinces--the
Atlantic Coastal Plain, the Piedmont, and the glaciated Central
Platform--to determine whether it was feasible to use aquifer type
and water-quality characteristics to predict radium-228 [1,2].
The aquifer type was determined for all wells or systems for
which radium-228 data were available. Table 1 shows the summary statistics for the specific aquifers analyzed in detail during
these earlier studies. These results were used to develop an
aquifer classification model for predicting radium-228 [5]. However, it was not possible to conduct as detailed an analysis for
the other radionuclides as was done for radium-228. Instead, a
data base was assembled of radon, radium-226, and uranium measurements from public groundwater supplies nationwide. This information was derived from a variety of published sources and
various reports of the U.S. Department of Energy's National
Uranium Resource Evaluation (NURE) Program [1, 2, 6, 7, 8, 9,
10, 11, 12, 13]. Unpublished data were supplied by numerous
state health agencies and midproject results of the USEPA National
Inorganics and Radionuclides Survey (NIRS).

Table 1.   Summary statistics for radium-228 content of groundwater from specific aquifer types.

Aquifer Type	N	$\bar{x}$ (pCi/L)	95% Upper Confidence Limits (pCi/L)
Igneous Rocks (Granites)	42	1.4	16.0
Chemical Precipitates			
° Coastal Plain Aquifers	16	0.1	0.2
° Midwest Aquifers	104	0.6	2.9
Metamorphic Rocks			
° Piedmont Region	75	0.3	2.5
Sand Aquifers			
° Coastal Plain Fall Line Arkoses	89	2.2	17.0
° Coastal Plain – Quartzose	53	0.3	2.4
° Midwest Glacial/Alluvial	135	0.7	4.2
Sandstone Aquifers			
° Midwest Cambrian – Ordivician Aquifer	320	2.1	12.0

The entire data base (exclusive of NURE data) was plotted onto map overlays of individual states at a scale of roughly 1:1,900,000 (1 inch = 30 miles). Distribution of the data was not geographically uniform but rather more closely reflected the usage of groundwater by public water-supply systems and the degree of monitoring performed by various regulatory agencies. Some states had an abundance of data, while other states had a scarcity or were completely devoid of information for one or more radionuclides.

The geology of each state was superimposed onto the map overlays, with special emphasis applied to the distribution and composition or lithology of principal aquifers. The distribution of aquifers within each state was obtained from the United States Geological Survey (USGS) National Water Summary [14], USGS Water Supply Papers, and various state agency publications. Detailed descriptions of aquifer composition or lithology were derived from a wide range of publicly-available literature and occasionally supplemented by communication with state water resources agencies. As most major rock units cross state boundaries, principal aquifers were generally plotted and evaluated on a regional basis.

A classification scheme (Table 2) was devised to categorize the relative abundance of each radionuclide in groundwater supplies. The actual measurements of the data base were then tabulated, by aquifer type, on a state-by-state basis. Upon completion of the tabulation, a radiological categorization of each aquifer type was inferred from the distribution of the data within each of the classification schemes of Table 2. This radiological characterization of aquifer types was then synthesized into a simple geologic model which was used to predict the occurrence of radionuclides in groundwater for areas which lacked actual measurements. The geochemical factors which control the groundwater distribution of each radionuclide are discussed in Hess et al. [15].

Each county in the United States was assigned a relative risk of the potential for radon, radium-226, and uranium in groundwater serving public drinking water supplies in one of two ways. The first method was to check for actual measurements from within each county. If numerous measurements existed, then a statistical interpolation of these values was performed. If a variety of aquifer types existed, then an interpretation was made to determine which aquifer type supplied the majority of the county's public drinking water. The second method, used in areas where no measurements existed, was to rank the county based on the aquifer type according to the results of the geologic model. If the county was served by multiple aquifers, then an interpretation was again made to determine which aquifer was the major source of the county's public water supply.

Following the classification of every county nationwide into its respective ranking, this information was entered into a computer file. This file was merged with a computer mapping program to produce the maps (Figures 1, 2, and 3) which graphically display the results of this study. The distribution map for radium-228 is shown in Figure 4.

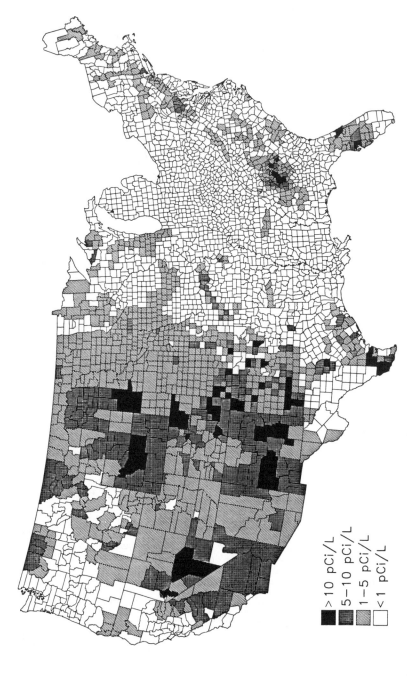

Figure 1.   Generalized map showing the occurrence of uranium by county for selected categories of concentration ranges.

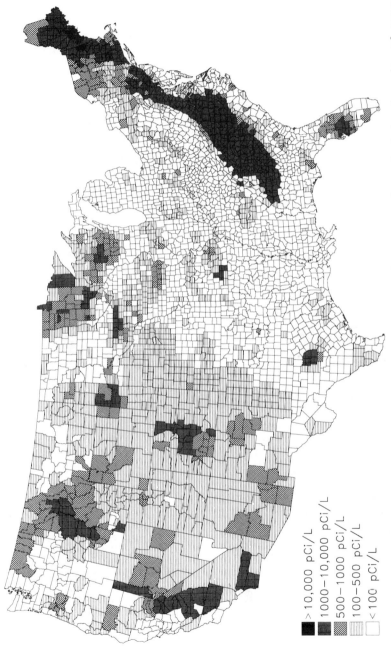

> 10,000 pCi/L
1000–10,000 pCi/L
500–1000 pCi/L
100–500 pCi/L
< 100 pCi/L

Figure 2.   Generalized map showing the occurrence of radon by county for selected categories of concentration ranges.

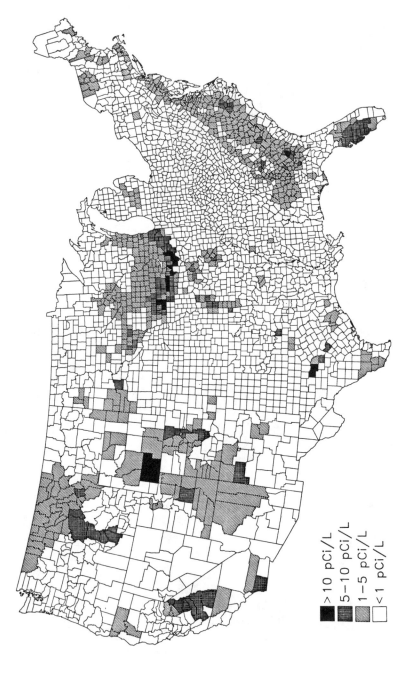

Figure 3.   Generalized map showing the occurrence of radium–226 by county for selected categories of concentration ranges.

> 10 pCi/L
5– 10 pCi/L
1– 5 pCi/L
< 1 pCi/L

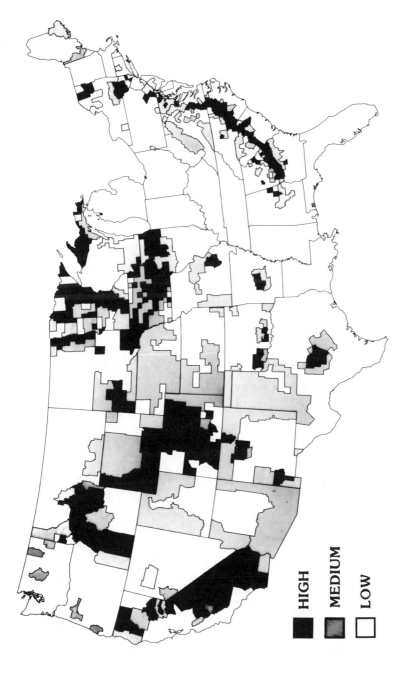

Figure 4.   Generalized map showing the occurrence of radium-228 by county for selected categories of concentration ranges [5].

HIGH

MEDIUM

LOW

Table 2.   Categories of occurrence used to estimate concentration ranges.

	Category	Range (pCi/L)
A) Radon	1	< 100
	2	100 - 500
	3	500 - 1,000
	4	1,000 - 10,000
	5	> 10,000
B) Radium-226	1	< 1
	2	1 - 5
	3	5 - 10
	4	> 10
C) Uranium	1	< 1
	2	1 - 5
	3	5 - 10
	4	> 10

DISCUSSION

Radiological Characterization of Major Rock Types

Igneous Rocks

Igneous rocks seldom supply major municipalities with drinking water but occasionally supply water to smaller public systems. Groundwater derived from silicic igneous rocks, primarily granites, routinely has elevated levels of all the radionuclides investigated in this study.   These rocks are associated with the Appalachian, Rocky, and Sierra Nevada Mountain ranges as well as the Idaho Batholith and the Canadian Shield rocks of the northern Great Lakes region.   Mafic (basaltic) volcanic rocks of the Columbia Plateau region, Hawaii, and other localized occurrences generally have very low concentrations of radionuclides.

Metamorphic Rocks

Widespread regions underlain by metamorphic rocks occur throughout the Piedmont, the Blue Ridge, and the Northeast Highlands Provinces of the Appalachian Mountains, as well as the Canadian Shield of the Great Lakes region.   More localized occurrences exist throughout the Rocky Mountains, Minnesota/South Dakota (Sioux Quartzite), the Uinta Arch of Utah, and numerous other isolated occurrences.   Water derived from these rocks contains a wide range of radionuclide concentrations.   The Sioux Quartzite produces elevated levels of all three radionuclides. Radon values were relatively high throughout the Appalachian region and were generally independent of the degree of metamorphism the rock had undergone [7, 16]. No comprehensive summation of radionuclide concentrations can be made for other metamorphic provinces; therefore, each was evaluated empirically using available data and regional trends.

Carbonate Sedimentary Rocks

Carbonate rocks, primarily limestone and dolomite, serve as principal aquifers in many parts of the country. These areas include the Valley and Ridge Province of the Appalachian Mountains, Paleozoic rocks of the interior plateau in the midwestern United States, the Floridan Aquifer in the southeastern United States, and various localized occurrences in the southwestern and western United States. With the exception of the Floridan Aquifer in central Florida, groundwater derived from these carbonate rocks generally had low concentrations of radionuclides. A multicounty area of central Florida displayed elevated levels of all three radionuclides, which may be associated with phosphate deposits associated with the limestone.

Clastic Sedimentary Rocks

Clastic rocks are utilized as aquifers in nearly every state, occurring as a wide variety of genetic rock units. Unconsolidated sands or sands and gravels occur as coastal plain, glacial, and alluvial valley deposits. The radionuclide concentration of water derived from these unconsolidated deposits generally is related to the proximity and composition of the provenance areas, with deposits adjacent to granites displaying elevated concentrations relative to deposits further removed from granitic source rocks. Consolidated clastic rocks occur in the Interior Lowland Province, the Cretaceous Seaway and Tertiary intermontane basins of the Rocky Mountain region, the Colorado Plateau, the High Plains, and the Triassic rifted basins of the Appalachian Mountain region. Groundwater derived from these rocks display a wide variety of radionuclide concentrations. Paleozoic sedimentary rocks of Iowa, Illinois, and Wisconsin have some of the highest radium-226 and radium-228 levels in the country, while Mesozoic sedimentary rocks of the Colorado Plateau routinely exhibit elevated levels of uranium throughout the four-corner region. The Ogallala Formation of the High Plains region produces groundwater with a wide range of radionuclide concentrations, including some of the highest uranium values in the country.

Regional Hydrogeology

Appalachian Mountains

The Appalachian Mountains are composed of a complex variety of rock units. Igneous and metamorphic rocks form the Piedmont,

Blue Ridge, and Northeast Highlands Provinces and generally display elevated concentrations of radionuclides. Southwestern Maine has the highest radon concentrations in groundwater in the entire United States; it is the only area in the country receiving a class 5 radon ranking (see Figure 2). Localized occurrences of radon concentrations above 10,000 pCi/L occur in the Piedmont of Georgia and the Carolinas but generally are from wells screened in granites which serve a small portion of a county's population. Elsewhere, continental deposits of the Triassic Newark Supergroup display high radionuclide content in New Jersey, Pennsylvania, and Connecticut, and the Devonian Catskill Formation produces elevated levels from continental deposits in Pennsylvania and southeastern New York. The Valley and Ridge Province, primarily folded Paleozoic carbonate rocks and clean sandstones, is generally low in radionuclide content.

Atlantic and Gulf Coastal Plains

The Atlantic and Gulf Coastal Plains are composed of Cretaceous through Recent unconsolidated clastic sediments and carbonate rocks. Atlantic Coastal Plain fall-line sediments display elevated radionuclide content because of their close proximity to granitic source rocks of the Piedmont Province, while groundwater derived from units occurring well seaward of the fall-line and virtually the entire Gulf Coastal Plain (with the exception of Texas) and the lower Mississippi River Valley generally display very low concentrations of radionuclides. Localized occurrences of elevated radionuclide levels occur in the Floridan Aquifer of the southeastern United States.

Interior Lowlands

The Interior Lowlands Province as defined in this report extends from western New York to northern Alabama, and from eastern Oklahoma to southern Minnesota including Michigan. This area is underlain dominantly by Paleozoic clastic and carbonate rocks with a relatively thin cover of glacial deposits over the northern half of the region. Groundwater derived from these units generally displays low concentrations of radionuclides, however, the Wisconsin/Iowa/Illinois region displays elevated radium-226 and radium-228 levels.

Canadian Shield

The Canadian Shield of northern Minnesota, Wisconsin, and Michigan is underlain by Precambrian igneous and metamorphic rocks which are locally overlain by glacial deposits. While these granites are generally not as high in radionuclide content as the granites of the Rocky, Appalachian, and Sierra Nevada ranges, these Canadian Shield granites locally produce elevated concentrations of radionuclides in groundwater. Glacial deposits are used as water sources whenever possible, with crystalline rocks seldom serving major municipalities.

Texas

Texas is underlain by a wide variety of aquifer types. Northwestern Texas derives water from the Ogallala Formation, which occasionally produces very high uranium values in addition to routinely elevated radon concentrations. Granitic rocks and adjacent sedimentary rocks of the Llano Uplift region of central Texas produce elevated concentrations of all three radionuclides in groundwater. Cretaceous carbonate and clastic sedimentary rocks underlie much of central Texas, generally producing relatively low radionuclide concentrations. Cenozoic clastic sediments of the Gulf Coastal Plain Province underlie much of eastern Texas, generally producing low radionuclide values, with the exception of the uranium mineralization belt of southeastern Texas where radionuclide values are elevated.

High Plains

The High Plains physiographic province covers a portion of eight states east of the Rocky Mountains. The primary aquifer is the Ogallala Formation, which represents extensive alluvial deposits of sediments eroded from the Rocky Mountains during the Pliocene. Groundwater derived from this aquifer displays elevated concentrations of radon (the entire region was ranked as Class 2 for radon). While radium-226 levels were generally low throughout the High Plains region, uranium measurements were occasionally quite high with numerous counties ranked as Classes 3 and 4.

Great Plains

Aquifers of the Great Plains region are dominantly clastic sedimentary rocks deposited during the Cretaceous period in a shallow epicontinental sea. With the exception of the region surrounding the Black Hills of southwestern South Dakota and southeastern Montana, radon and radium-226 levels are relatively low. Elevated uranium levels occur in groundwater derived from the Dakota Sandstone and its stratigraphic equivalents in portions of the Great Plains region. In the Black Hills region, elevated concentrations of all three radionuclides occur in groundwater because of the remobilization of uranium derived from Black Hills granites.

Rocky Mountains

The Rocky Mountains are composed of a complex suite of igneous and metamorphic rocks of Precambrian through Cretaceous

age and sedimentary rocks deposited in intermontane basins during the Tertiary period. These rocks include extensive granites
in Colorado, Montana, and Idaho which yield water of relatively
high radionuclide concentrations. This region also consists of
volcanic rocks with felsic compositions which display elevated levels of radionuclides. Since much of this region is sparsely populated with abundant surface water in mountainous regions, limited measurements of radionuclides in groundwater exist. Counties
underlain by granites or sedimentary rocks primarily derived from
granites received higher classifications based on the geologic models.

Colorado Plateau

The Colorado Plateau of the Utah/Colorado/New Mexico/
Arizona region is underlain by Permian through Cretaceous age
sedimentary rocks. These rocks are dominantly clastic rocks
deposited under marine conditions but also include continental deposits and some carbonate rocks. This area constitutes one of
the major uranium mining regions of the United States, with the
occurrence of elevated radionuclide concentrations in groundwater
widespread. Because of scarcity of municipalities, groundwater
resources of the area are generally underutilized.

Basin and Range Province

The Basin and Range Province of Nevada, Utah, Arizona, and
New Mexico consists of intermittent ranges separated by alluvial-
filled basins. The ranges are composed of a wide variety of
rocks including igneous, metamorphic, and sedimentary rocks.
The igneous rocks include units of both intrusive and extrusive
origin and vary in composition from felsic to mafic. The alluvium
which infilled the basins was eroded from the adjacent range;
therefore, the radionuclide concentration of groundwater derived
from the basin is directly dependent upon the composition of the
rocks which form the range.

California

The geology of California is quite complex with numerous
mountain ranges, large and small intermittent basins, deserts,
and extensive volcanic terrain. The Sierra Nevada and portions
of the Coastal Mountain Ranges are composed of granites, while
the Klamath Mountains are composed dominantly of uplifted,
metamorphosed marine sediments. Southeastern California is part
of the Basin and Range Province, while northeastern California is
dominantly volcanic rocks. The San Joaquin Valley in the central
part of the state is a thick accumulation of alluvial sediments
derived from the Sierra Nevada Mountains. The concentration of
radionuclides in groundwater varies markedly from region to region.

Pacific Northwest

Large portions of Oregon, Washington, and western Idaho are underlain by the Columbia River basaltic volcanic rocks. These rocks generally produce groundwater with low concentrations of radionuclides. Northern Washington and Idaho are underlain by igneous and metamorphic rocks, including granites, which contain groundwater with elevated radionuclide levels. The west coasts of Washington and Oregon are lined by uplifted marine sediments, which generally contain groundwater of relatively low radionuclide concentrations.

Alaska and Hawaii

Groundwater is primarily derived from alluvial deposits in Alaska and generally displays low radionuclide concentrations. The exceptions are alluvial deposits adjacent to granites and in southeastern Alaska, where few alluvial deposits exist. Hawaii is underlain by basaltic volcanic rocks, with groundwater derived from these rocks generally quite low in radionuclide content.

REFERENCES

1. Michel, J., and C. Pollman, "A Model for the Occurrence of Ra-228 in Groundwater," Office of Drinking Water, USEPA, Washington, D.C., Environmental Science and Engineering Report No. 81-227-270 (1982).

2. Michel, J., and C. Pollman, "A Model for the Occurrence of Ra-228 in Groundwater, II: Application to the North-Central United States," Prepared by Research Planning Institute, Inc., Columbia, S.C., and Environmental Science and Engineering, Inc., Gainesville, Fla., for the Office of Drinking Water, USEPA, Washington, D.C. (1983).

3. Michel, J., "Aquifer Classification by Relative Risk of Radium-228 Occurrence," Final Report Prepared by Research Planning Institute, Inc., Columbia, S.C., for J.R.B. Associates, McLean, Vir., RPI/R/84/9/26-28 (1984).

4. Johnston, P., and H. Nelsen, "Occurence of Radium-228 in Drinking Water, Food, and Air," Report to the Office of Drinking Water, USEPA, by SAIC/JRB, McLean, Vir. (1985), 39 pp.

5. Michel, J., and C. R. Cothern, "Predicting the Occurrence of ^{228}Ra in Groundwater," *Health Physics*, 51:715-721 (1986).

6. Kaufmann, R. F., and J. D. Bliss, "Effects of Phosphate Mineralization and the Phosphate Industry on Radium-226 in

Groundwater of Central Florida," Environmental Protection Agency Report 520/6-77-010 (1977), 115 pp.

7. Hess, C. T., S. A. Norton, W. F. Brutsaert, R. E. Casparius, E. G. Coombs, and A. L. Hess, "Radon-222 in Potable Water Supplies in Maine: The Geology, Hydrology, Physics and Health Effects," Land and Water Resources Center, University of Maine, Orono (1979), 119 pp.

8. Kriege, C. R., and R. M. A. Hahne, "^{226}Ra and ^{228}Ra in Iowa drinking water," Health Physics, 43:543-559 (1982).

9. Gilkeson, R. H., K. Cartwright, J. B. Cowart, and R. B. Holtzman, "Hydrogeologic and Geochemical Studies of Selected Natural Radioisotopes and Barium in Groundwater in Illinois," University of Illinois Water Resources Center, Urbana-Champaign (1983), 93 pp.

10. Hahn, N. A., Jr., "Radium in Wisconsin Groundwater and Removal Methods for Community Water Systems," Wisconsin Department of Natural Resources (1984), 125 pp.

11. Horton, T. R., "Nationwide Occurrence of Radon and Other Natural Radioactivity in Public Water Supplies," Environmental Protection Agency Report 520/5-85-008 (1985), 208 pp.

12. Lucas, H. F., "Ra-226 and Ra-228 in Water Supplies," Jour. Amer. Water Works Assoc., 77(9):57-67 (1985).

13. Miller, R. L., and H. Sutcliffe, Jr., "Occurrence of Natural Radium-226 Radioactivity in Groundwater of Sarasota County, Florida," USGS Water-Resources Investigations Report 84-4237 (1985), 34 pp.

14. United States Geological Survey, "National Water Summary 1984: Hydrologic Events, Selected Water-Quality Trends, and Groundwater Resources," USGS, Water Supply Paper 2275 (1984), 458 pp.

15. Hess, C. T., J. Michel, T. R. Horton, H. M. Prichard, and W. A. Coniglio, "The Occurrence of Radioactivity in Public Water Supplies in the United States," Health Physics, 48:553-586 (1985).

16. King, P. T., J. Michel, and W. S. Moore, "Groundwater Geochemistry of Ra-228, Ra-226, and Rn-222," Geochem. Cosmochim. Acta., 46:1173-1182 (1982).

RADON MEASUREMENT IN STREAMS TO DETERMINE LOCATION AND MAGNITUDE OF GROUND-WATER SEEPAGE

Roger W. Lee and Este F. Hollyday, U.S. Geological Survey, Water Resources Division, Nashville, Tennessee

ABSTRACT

The location and magnitude of ground-water seepage can be determined by measuring the activity of ^{222}Rn gas in streams. Radon in ground water may be 2 to 4 orders of magnitude greater activity than in surface water. Thus, ground-water seepage to a stream usually increases ^{222}Rn in the streamflow. Downstream of ground-water seepage, ^{222}Rn decreases in the stream as radon escapes to the atmosphere, particularly in turbulent reaches of the stream. The relation between ground-water and surface-water flows can be determined by mass balance assuming no other ^{222}Rn sources and no significant gas loss from the mix of surface water and ground water at the sampling point.

Measurements of ^{222}Rn in water from a 0.75-mile reach of a small bedrock-channel stream in Middle Tennessee ranged from 32 to 196 disintegrations per minute per liter. A sample of ground water from an adjacent spring contained ^{222}Rn activity of 489 disintegrations per minute per liter. Based on ^{222}Rn activities down the sampled reach of the stream, 36 percent of flow leaving the reach was ground-water seepage at a point 0.5 mile downstream from the upstream sampling boundary. Measurements of temperature in the water and bed of the stream verified the point location of ground-water seepage.

INTRODUCTION

Measurements of [222]Rn in water from stream-aquifer hydrologic systems have been successfully used in the characterization of ground-water seepage to rivers and streams [1, 2]. By combining measurements of streamflow with [222]Rn activities in surface water and ground water, quantitative analysis of a ground-water seepage to a stream is possible. Ellins et al. [3,4] have shown the applicability of this approach to stream-carbonate aquifer interactions in Jamaica and Puerto Rico, respectively. The purpose of this investigation was to apply the method in the reconnaissance of small gaining streams in the Ordovician limestone areas of Middle Tennessee. Determination of points of discharge and quantity of ground-water seepage could be useful for locating isolated ground-water reservoirs in these limestone aquifers as well as for quantifying baseflow components of streams. Two surface-water flow measurements and 11 [222]Rn activities were measured in Carters Creek, in Middle Tennessee (Figure 1), during periods of low flow in the summer of 1986.

RADON CYCLING IN GROUND WATER-SURFACE WATER SYSTEMS

Radon-222 occurs naturally in the earth's crust as a daughter of the radioactive decay of radium-226. The concentration of [222]Rn in ground water is primarily related to the amount of [226]Ra present in the aquifer framework, although hydrology and geology may exert some influences. Activities may range from 100 dpm/L (disintegrations per minute per liter), typical of water in clastic sedimentary rocks to greater than 10,000 dpm/L, typical of water in igneous or metamorphic rocks. The amount of [222]Rn in the atmosphere is almost nil; thus, as ground water seeps into surface water bodies, [222]Rn escapes to the atmosphere across the air-water interface, and the concentration of [222]Rn in surface water decreases rapidly. Activities of [222]Rn in water from an aquifer may be 2 to 4 orders of magnitude greater than in an adjacent surface-water body [5]. This makes [222]Rn a suitable geochemical candidate for determining the cycling of water in a ground water-surface water system.

Several factors that control [222]Rn activity in water should be considered for this application:
1. The difference between [222]Rn in ground water and adjacent stream;
2. The magnitude of a ground-water seep in relation to total flow along a reach of stream;

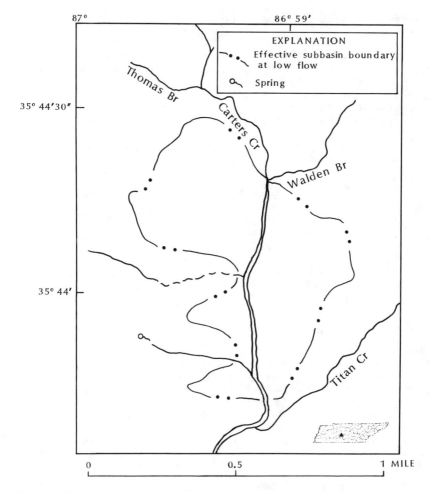

Figure 1.   Location and drainage area of 0.75-mile reach
            of Carters Creek, Maury County, Tennessee.

3.   The rate of gas loss in the stream (gas exchange
     rate) [6];
4.   The length of the stream reach measured and the
     number of points of discharge of ground-water as
     related to scale;
5.   The contribution of tributaries to streamflow and
     ^{222}Rn in the main channel;
6.   The characteristics of ground-water seepage (dis-
     crete or dispersed discharge).
Factors 1 and 2 are most significant in determining
suitability of this method to ground water-surface
water systems.   Factors 3-6 are determined subsequently
by application of the method to the specific system.

CHARACTERIZATION OF BASE FLOW

Ground-water seepage may occur as a dispersed discharge into the stream channel along a given length of
channel; this is typical of channels flowing on thick
granular material. Seepage also may be discrete, seep-
ing into the stream channel at well-defined points along
the stream; this is typical of channels that flow on
soluble bedrock.

The relation between ground-water and surface-water
flows for a reach of stream by use of ^{222}Rn activities
is defined by the mass balance mixing equation

$$Qs \times As + Qgw \times Agw = Qm \times Am \qquad (1)$$

where
Qs	=	rate of streamflow,
As	=	activity of ^{222}Rn at upstream point,
Qgw	=	rate of ground-water seepage,
Agw	=	activity of ^{222}Rn in ground water,
Qm	=	rate of mixed stream/ground-water flow,
Am	=	activity of ^{222}Rn in mixed water down-
		stream,

and

$$Qs + Qgw = Qm. \qquad (2)$$

Equation (1) assumes no other source of ^{222}Rn other than
that conveyed to the stream in ground water and no significant loss of ^{222}Rn from the mix of surface and
ground water at the sampling point. Thus, measurements
of streamflow and ^{222}Rn activity in the stream and in
ground-water sources allow calculation of the magnitude
of ground-water seepage.

In the absence of streamflow measurements, a value
of ground-water seepage may be determined. Substitution
of equation (2) into equation (1) for Qs yields

$$\frac{Qgw}{Qm} = \frac{Am-As}{Agw-As}. \qquad (3)$$

This form of the equation permits use of ^{222}Rn activities to determine the ratio of ground-water seepage to
the total flow at the measured point.

APPLICATION TO CARTERS CREEK, TENNESSEE

Carters Creek is a small stream with base flow of
about 1.0 ft^3/s (cubic feet per second) and average
annual flow of about 17 ft^3/s. The stream is incised
through regolith, and much of the streambed is on thin-
bedded, clay-rich limestone of Ordovician age. At

low-flow conditions, the streamflow is characterized by long, shallow pools and short riffles. The stream channel of the sampled reach is relatively straight, and during higher flows the riffles are submerged and flow is uniform.

A base-flow survey conducted June 26, 1986, in Maury County, Tennessee, indicated that a substantial increase in flow in the uppermost reach of Carters Creek was from ground-water seepage [7]. Subsequently, a reconnaissance sampling under low-flow conditions of ^{222}Rn activities along this reach of Carters Creek was conducted on September 16, 1986 (Figure 2). No significant surface-water inflow from tributaries occurred along the sampled reach of stream.

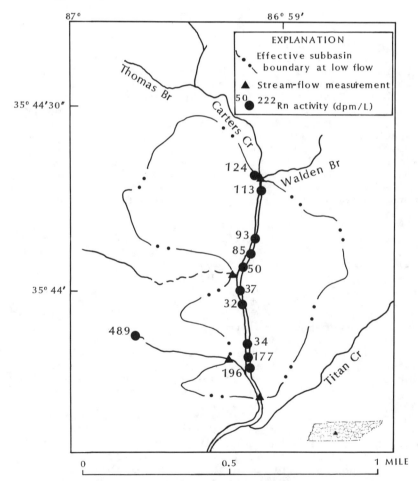

Figure 2.   Locations of streamflow measurement sites and ^{222}Rn activities sampling sites.

METHOD

Water samples of approximately 200 milliliters
volume were collected at 10 surface-water sites and 1
spring. These samples were drawn into evacuated samplers which stripped the water of radon gas and diverted
it into calibrated Lucas cells. The activity of ^{222}Rn
was determined by alpha spectrometry using commercially
available equipment. Sufficient time and repetition of
counting events were performed to achieve an analytical
error of less than 3 percent for all samples.

RESULTS AND DISCUSSION

Measurements of ^{222}Rn activity in water from the
0.75-mile reach of Carters Creek (Figure 2) ranged from
32 to 196 dpm/L. In downstream order, ^{222}Rn initially
was fairly high, at 124 dpm/L, indicating some groundwater seepage upstream of the first sample. The activity of ^{222}Rn decreased consistently downstream to a
low activity of 32 dpm/L. Assuming no dispersed or
discrete ground-water seepage between these points, a
rate of gas loss can be estimated at 50 dpm/L/1,000
feet, an unexpectedly low rate of loss due predominantly
to the low turbulence in the riffles and shallow pools
of the stream. The ^{222}Rn activity increased significantly in a long, shallow pool which began about 0.5
mile below the most upstream sample. Above this long,
shallow pool, ^{222}Rn activity was 34 dpm/L. Just upstream of a riffle near the end of the pool, ^{222}Rn
activity increased to 196 dpm/L. The ^{222}Rn data indicate that nearly all of the ground-water seepage in the
reach studied occurs in this pool (Figure 3).
Simultaneous water temperature measurements were
taken with the ^{222}Rn sampling just above the streambed. Temperatures were around 20 °C, at or near ambient
air temperature throughout most of the stream reach.
In the pool that contains the elevated ^{222}Rn activity,
however, several points measured only 15 °C; an indication of cooler ground-water seepage to the stream. The
temperature gradually increased downstream from this
pool (Figure 3).
Radon-222 activity may vary in an aquifer depending
on the amount of source ^{226}Ra and whether the ^{222}Rn
gas can evade from the saturated zone into the unsaturated zone of an aquifer. Ellins et al. [3] has shown
that springs adjacent to a stream may be most representative of ^{222}Rn activity in ground water that seeps
directly into the stream. A spring was sampled near
the head of a small west bank tributary to Carters

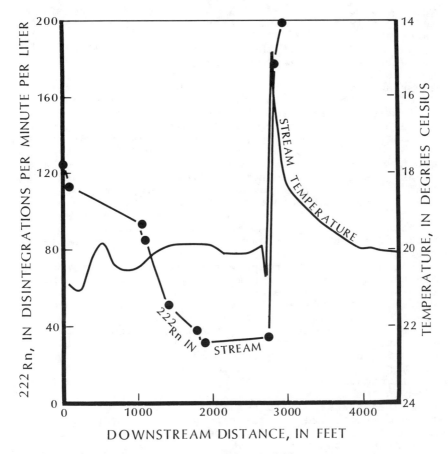

Figure 3.   Stream temperature and ^{222}Rn activity in a 0.75-mile reach of Carters Creek.

Creek (Figure 1). The ^{222}Rn activity was 489 dpm/L, which is presumably representative of ^{222}Rn activity in ground water. Thus,

$$A_s = 34 \text{ dpm/L},$$
$$A_{gw} = 489 \text{ dpm/L},$$
$$A_m = 196 \text{ dpm/L},$$

and substituting these activities into equation (3) yields

$$\frac{Q_{gw}}{Q_m} = 0.36 , \qquad (4)$$

which indicates that about 36 percent of the streamflow leaving this stream reach entered the stream as ground-water seepage in the vicinity of the pool 0.5 mile downstream of the head of the reach for September 16.

Surface-water flow measurements (Figure 2) obtained June 26, 1986, were 1.04 ft^3/s at the upstream end of the reach, and 1.62 ft^3/s at the downstream end of the reach. If the assumption can be made that flow and 222_{Rn} activities in Carters Creek were similar on June 26, and September 16, then an approximate value for ground-water seepage into the lower part of the stream of 0.58 ft^3/s can be calculated from equation (4). The sum of the ground-water seepage and the upstream flow would be 1.62 ft^3/s, equal to the downstream flow. This suggests that all or nearly all of the ground-water seepage to the studied reach occurred in the pool if no major loss of streamflow to the aquifer has occurred. Although inexact, this example implies that 222_{Rn} activity and measurement of streamflow can quantify ground-water seepage.

Because shallow ground-water temperatures are nearly constant at mean annual air temperature, use of temperature to locate seepage needs to be restricted to mid-winter and mid-summer when there is a significant contrast between stream temperature and ground-water temperature. Application of the radon method to the same stream at varying stages of flow could also indicate if ground- water seepages change in terms of number of seepage locations and could indicate variations in the magnitude of ground-water seepage at any given location in the stream channel.

Further research is required at Carters Creek to fully characterize and quantify ground-water seepage beyond the reconnaissance scope of this paper. Further work should include:
1. Combining streamflow measurements with 222_{Rn} measurements synoptically;
2. Collecting more ground-water samples to quantify accurately 222_{Rn} in ground water;
3. Collecting data during higher and lower stages of flow;
4. Determining gas exchange rate during different stages and relating that rate to channel geometry.

SELECTED REFERENCES

1. Rogers, A. S. "Physical Behavior and Geologic Control of Radon in Mountain Streams," U.S. Geological Survey Bulletin 1052-E (1958), 187 pp.

2. Jacoby, G. C., H. J. Simpson, G. Mathieu and T. Torgersen. "Analysis of Ground-Water and Surface-Water Supply Interrelationships in the Upper Colorado River Basin Using Natural Radon-222 As A Tracer," in Report to the John Muir Institute, Inc. (1979), 46 pp.

3. Ellins, K. K., E. Douglas, H. I. Simpson and S. Mathieu. The Relationship Between the Isotope Composition of Precipitation and Hydrologic Systems in Jamaica and the Application of [222]Rn in Measuring Groundwater Discharge to the Martha Brae River, Jamaica. Proceedings, 2d, Caribbean Islands Water Resources Congress (1985), p. 63.

4. Ellins, K. K., A. Roman, and R. W. Lee. Assessment of Ground Water--Surface water relationships using Radon-222 for the Rio Grande de Manati, North Coast of Puerto Rico. U.S. Geological Survey Water-Resources Investigations (in review).

5. King, P. T., J. Michel, and W. S. Moore. Ground Water Geochemistry of [228]Ra, [226]Ra, and [222]Rn. Geochim. Cosmochim. Acta, v. 46, pp. 1173-1182 (1982).

6. Elsinger, R. J., and W. S. Moore. Gas exchange in the Pee Dee River based on [222]Rn evasion. Geophys. Res. Let., v. 10, no. 6, pp. 443-446 (1983).

7. Lowery, J. F., P. H. Counts, H. L. Edmiston and F. D. Edwards. Water resources data, Tennessee, Water year 1986. U.S. Geological Survey Data Report TN-86-1. 411 pp (1987).

POLONIUM IN THE SURFICIAL AQUIFER OF WEST CENTRAL FLORIDA

William C. Burnett, James B. Cowart and Philip A. Chin
Florida State University, Tallahassee, Florida

ABSTRACT

Analysis of water from shallow wells in west central Florida
has revealed the existence of a number of wells with high
concentrations of ^{210}Po, a radioactive daughter of ^{222}Rn. Many
concentrations are considerably higher than the Federal E.P.A.
maximum contaminant levels for total alpha radioactivity (15 pCi
l^{-1}). These high concentrations are surprising in view of the
particle-active nature of polonium. Measured ^{210}Po activities
range from less than 1 to over 500 pCi l^{-1} in filtered samples
and from less than 1 to over 2,500 pCi l^{-1} in unfiltered samples.
Measurements of ^{222}Rn showed a range from less than 100 to over
40,000 pCi l^{-1} and a range of less than 1 to 10 pCi l^{-1} for the
longest lived radon daughter, ^{210}Pb. The chemical
characteristics of the waters high in polonium suggests a
possible association between high ^{210}Po activity and acidic
waters containing sulfide.

INTRODUCTION

A state-wide reconnaissance of radioactivity in water samples
from domestic wells in Florida revealed the existence of several
sites at which the gross alpha measurement exceeded 100 pCi/L yet
both uranium and radium were measured at less than 1 pCi/L.
Further radiochemical analysis at Florida State University
laboratories showed that the radioactivity was due to anomalously
high concentrations of ^{210}Po in the water. This finding led to
the present investigation of the distribution of ^{210}Po in groundwater in west central Florida and of mechanisms which might be
responsible for the mobilization of polonium.

Polonium-210 is a member of the ^{238}U (4n + 2) decay series
and is the final radioactive nuclide in the series. Following
the decay of ^{222}Rn, it is the only remaining alpha emitter in the

series with a half life exceeding a few minutes ($t_{1/2}=138.4$ days); the only nuclide in the series following ^{222}Rn with a longer half life is the grandparent of ^{210}Po, ^{210}Pb, with $t_{1/2}=22.3$ yr. (Figure 1). Probably the measurement of ^{210}Po in groundwater has been only rarely done because 1) it is a very "particle reactive" element and therefore thought likely to be sorbed on the aquifer matrix, 2) its precursor, ^{210}Pb, is not found at very great concentrations in groundwater so that its generation in the "water column" is very small, 3) its half life is relatively short and therefore, even if mobile, it could not travel far in most aquifers, and 4) where it had previously been measured, it had been present at quite low levels.

Disequilibria in the natural ^{238}U decay series within uranium-bearing minerals are caused by internal processes like recoil and diffusion. External processes, such as weathering and leaching, also contribute to the release of radionuclides into the environment in particulate, dissolved and gaseous forms. In the case of ^{210}Po, it is important to determine the distribution of the two precursor nuclides with geologically significant half lives and mobility, ^{222}Rn and ^{210}Pb, to help understand the fractionation mechanism.

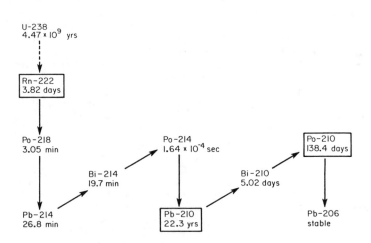

Figure 1.   Abbreviated ^{238}U decay scheme with half-lives of all isotopes from ^{222}Rn through ^{210}Po. There are several intermediate daughters not shown between ^{238}U and ^{222}Rn.

West central Florida, the location of this study, is underlain by the Hawthorn and Bone Valley formations which contain phosphate orebodies containing substantial amounts of uranium. Therefore, there is a very large potential source of ^{238}U series daughter products. The phosphatic horizons in Florida are normally overlain by a sand sheet on the order of 10 m in thickness.

Within this sand sheet is located the surficial water table aquifer which serves as the shallow source of domestic water in the area. It is this surficial aquifer which has provided almost all of the anomalous polonium results. Other important aquifers in the region include the Floridan aquifer which produces from pre-Hawthorn and, at some locations, basal Hawthorn rocks and the intermediate artesian aquifer, which is located within the Hawthorn Formation.

A study by Hinton and Whicker [1] showed that for a reclaimed uranium mill tailing deposit covered by 1.5 m of vegetation or 0.2 m of bare overburden, the ^{222}Rn flux was reduced to background levels. However, an increase in moisture content by precipitation resulted in a 25 to 30% increase in the flux rate. Estimates for how long it would take a ^{222}Rn atom to traverse several meters of bedrock can be approached by comparison to estimates for the equivalent time for radon migration in air or water. On the average, it will take 13 days for a radon atom to diffuse through 5 m of air or 5 cm of water according to the World Health Organization [2]; ignoring convection, it takes approximately 52 days to migrate through 20 m of air or 5200 days to migrate through 20 m of water. This is time enough for complete decay of both ^{222}Rn and ^{210}Po, if unsupported. Since consistently high radionuclide concentrations have been measured in some ground waters from west central Florida, active fractionation and leaching processes of the Hawthorn Formation must be taking place. In general, high radioactivity of short-lived daughters in ground water must necessarily arise from one of the following scenarios: 1) a highly radioactive point source of continuous emission, 2) a persistent flux of these radionuclides emanating from the pathway walls, or 3) emission from particulates carried by the ground water.

The purpose of this paper is to report our observations of anomalously high ^{210}Po in some ground waters from the surficial aquifer in west central Florida. Although it is still unclear why polonium occasionally exists at such high levels, simultaneous measurements of other ^{238}U-series nuclides as well as several chemical parameters have provided some important clues.

EXPERIMENTAL METHODS

(1) Measurement of Polonium-210

The procedure utilized for the analysis of ^{210}Po was modified from that of Oural et al. [3]. Two 900 ml water samples were collected from each station. One sample was immediately filtered through a 0.45 um Millipore filter for analysis of "dissolved" activity and the other sample was collected unfiltered and represents the "total" activity within a sample. Each sample was then immediately fixed by addition of concentrated nitric acid (2% by volume) and spiked with a known activity of yield tracer, [209]Po. Samples were left to equilibrate in the laboratory for at least 24 hours after which time approximately 20 mg of iron carrier solution was added. Solutions were left to equilibrate for a further 48 hours, then the polonium was coprecipated with iron hydroxide by addition of concentrated ammonia. Samples were well mixed, and the precipitate allowed to settle for aproximately 48 hours. The supernate was decanted from the precipitate and discarded. The precipiatate was centifuged and supernate discarded. The precipitate was then dissolved in 6N HCl and transferred to a plating cell.

A modified version of Flynn's [4] procedure was used for the spontaneous plating of [210]Po onto silver disks. The changes incorporated include: (1) the plating solution was adjusted to approximately 0.5N HCl; and (2) the addition of 100mg of ascorbic acid. Yields from simple [209]Po spiked deionized water solutions were typically 80 to 95%, while natural ground water samples were invariably lower, with a range from as low as a few percent to over 60%. The prepared disks were then counted by alpha spectrometry. The tracer, [209]Po, was standardized against the reported ^{210}Pb activity of EPA standard "Climax Sand Tailings".

(2) Measurement of Radon-222

Excess [222]Rn was measured by liquid scintillation counting, using the procedure of Prichard and Gesell [5] on sample volumes of 10 ml. The analyses were all done in triplicate. All results have been corrected for decay to time of sampling. Standards have been prepared from NBS Standard Reference Material 4958, radium solution. Standards prepared ranged from 100-100,000 pCi l^{-1}.

(3) Measurement of Lead 210

The procedure for ^{210}Pb analysis was a modification of those found in Thomson et al. [6] and Koide et al. [7]. A 1.9 liter water sample was collected and immediately fixed with HCl so that the pH was 1.2. On arrival at the laboratory Fe^{+3} and Pb^{+2} carriers were added and, after equilibration, a $Fe(OH)_3$ precipitate containing scavenged lead was formed by addition of ammonia. The precipitate was separated from the supernate by filtering through a 0.45 um Millipore HA filter. The precipitate was dissolved off the filters with 8N HCl. Anion exchange procedures were used to remove Fe and U. The eluate was evaporated to dryness and redissolved in 1.5N HCl. Another anion exchange procedure was used for the separation of Pb. The Pb was eluted with double deionized water. $PbSo_4$ was precipitated from the eluate at pH 2 by addition of saturated sodium sulfate. The

$PbSO_4$ is filtered through a predried and preweighed 0.45 um Millipore filter which, after overnight drying, was weighed for a gravimetric Pb yield. The filter was then combusted and the ash was loaded into vials for counting of the 46.5 KeV photopeak of ^{210}Pb by an intrinsic germanium well-type detector. Standardization of the efficiency and geometry effects were obtained by analyzing standard ^{210}Pb solutions in the same manner.

RESULTS AND DISCUSSION

Regional Results

The distribution of sample collection sites is shown in Figure 2. Tabulated results of filtered and unfiltered ^{210}Po from all these sites is given in Table 1. Although the results are "spotty", there is a rather wide distribution of samples with elevated levels of ^{210}Po, normally a very particle active element. In most of the samples, the concentrations of ^{210}Po commonly range from <0.1 to 10's pCi l^{-1}. Activities of 100's pCi l^{-1} are found in a few cases and more than 2,000 pCi l^{-1} was found in one unfiltered sample.

Of the 80 sites analyzed so far, 25 have ^{210}Po activities which are greater than 15 pCi l^{-1} (Figure 2). Almost all of these wells are shallow, usually on the order of 10 m deep. However, one of the most southerly wells is much deeper (about 85 m), yet contains levels of ^{210}Po of about 90 pCi l^{-1} in the unfiltered water. Anomalously high polonium levels are thus not totally restricted to the surficial aquifer, nor to the immediate vicinity of phosphate mining areas where the phosphate ore is closest to the surface.

Only two other cases are known to us where polonium has been detected in concentrations comparable to what we have measured (up to hundreds of pCi l^{-1}) in ground waters. One well is located in Louisiana in an area of extensive faulting and is believed to overlie a uranium ore deposit [8]. The other well is located in central Florida and serves as a recharge well for a phosphate mine [3].

An analysis of the ^{222}Rn data collected thus far (Table 1), shows that the concentrations tend to be high and extremely variable in the shallow aquifer. The mean value of all stations is 8440 pCi l^{-1}. Almost all samples contain greater than 2,000 pCi l^{-1}, with about 20% of the samples having concentrations above 20,000 pCi l^{-1}. A few samples are much higher. The station with the highest ^{222}Rn concentration, station 45, is close to the station with the highest ^{210}Po concentration, station 18. Although these radon levels are not as spectacular as the >100,000 pCi l^{-1} levels reported in the granitic terrains of New England and Finland, the levels in the surficial aquifer are significantly elevated compared to "normal" ground water.

^{222}Rn will emanate more easily from a medium relative to other members of the uranium series, because it is an inert gas, so reports of high ^{222}Rn concentrations in the water are not

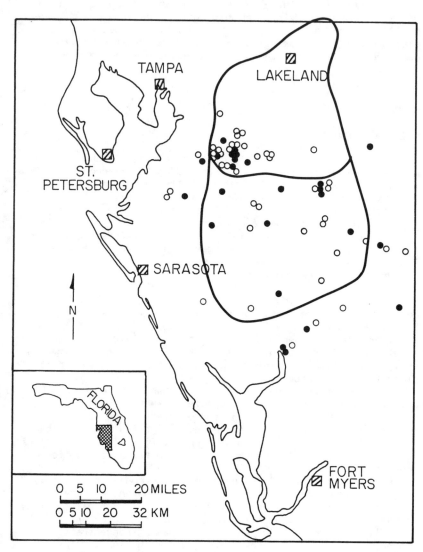

Figure 2.   Index map of southwest Florida showing locations of
sampling sites. Most of these sites are private wells
drawing water from the surficial aquifer.

Table 1.   Chemical and physical characteristics, ^{222}Rn, and filtered and unfiltered ^{210}Po in water samples from wells in west central Florida where all parameters were measured.

Station No.	pH	Conductivity (μmhos)	Sulfide (uM)	^{222}Rn[1] (pCi l^{-1})	^{210}Po Filtered (pCi l^{-1})	^{210}Po Unfiltered (pCi l^{-1})
**** JUNE 16-17, 1985 ****						
1					0.6	1.6
2					0.1	0.2
3					1.4	2.4
4					0.3	1.3
5					0.9	1.3
6					0.12	3.0
7					0.3	0.3
8					17.3	17.0
9					109.1	145.2
**** AUGUST 29-30, 1985 ****						
10					29.0	112.7
11					145.6	189.2
12					74.1	92.2
12*					NA[3]	3.9
13					0.8	1.4
14					62.9	114.8
15					1.2	3.6
16					1.2	4.2
17					0.8	9.2
**** NOV 14-15, 1985 ****						
18					338.3	NA
19					23.5	NA
20					5.0	NA
21					2.5	NA
21*					3.2	NA
22					2.6	NA
23					31.0	NA
24					9.4	0.6
25					0.6	NA
26					0.5	NA
27					.0	NA
28					20.4	31.1
29					6.5	4.3
30					47.3	NA
**** JAN 30-31, 1986 ****						
31					NA	86.9
32					NA	8.0
33					NA	5.1
**** MAY 1-2, 1986 ****						
10	4.26	169	NA	3470 ± 114	36.5	27.4
11	4.64	239	NA	10690 ± 320	60.7	99.0
18	4.80	43	NA	16100 ± 1008	181.8	282.5

Table 1. (continued)

Station No.	pH	Conductivity (µmhos)	Sulfide (uM)	^{222}Rn[1] (pCi l^{-1})	^{210}Po Filtered (pCi l^{-1})	^{210}Po Unfiltered (pCi l^{-1})
MAY 1-2, 1986 (continued)						
34	5.34	62	NA	7064 ± 88	0.7	0.3
35	6.54	223	NA	2840 ± 112	1.1	0.4
36	5.13	90	NA	2263 ± 35	1.0	0.5
37	6.45	291	NA	26480 ± 254	0.4	0.3
38	4.56	43	NA	13500 ± 254	2.8	2.7
39	4.93	294	NA	10110 ± 581	21.9	0.3
40	4.76	550	NA	19190 ± 262	0.7	0.6
41	4.85	44	NA	21550 ± 459	12.9	20.7
**** MAY 23-24, 1986 ****						
11	5.15	173	NA	10570 ± 224	87.4	340.9
12	4.60	52	NA	24810 ± 344	378.7	141.6
18	3.80	40	NA	16480 ± 663	456.0	2569.4
42	4.43	93	NA	6435 ± 95	NA	27.5
42*	NA	190	NA	6008 ± 41	3.6	5.1
43	4.69	88	NA	1273 ± 24	1.5	2.6
44	4.52	45	NA	22910 ± 308	37.2	50.6
45	4.95	55	NA	41440 ± 925	0.7	1.5
46	5.03	38	NA	17340 ± 327	3.7	2.7
47	4.70	90	NA	8500 ± 514	1.1	0.8
48	5.86	148	NA	NA NA	2.4	0.2
49	5.86	102	NA	2552 ± 39	1.1	2.1
50	7.05	315	NA	3457 ± 44	1.2	55.4
51	4.26	48	NA	1195 ± 26	20.0	38.3
52	7.15	312	NA	3064 ± 61	2.2	BD[3]
53	4.76	65	NA	3810 ± 138	3.1	10.5
54	4.23	134	NA	13730 ± 299	58.2	57.9
**** JUNE 20-21, 1986 ****						
18-	5.13	40	38	17070 ± 799	446.0	468.3
18-	5.19	40	35	18170 ± 285	461.5	481.9
18-	4.99	40	38	18160 ± 500	480.7	424.4
18-	5.08	40	40	18870 ± 173	433.3	416.0
18-	5.00	40	41	18450 ± 318	384.1	369.6
18-	5.06	40	34	19020 ± 372	380.0	323.4
18-	5.06	40	34	NA NA	NA	651.1
18-	NA	NA	NA	18520 ± 371	NA	NA
55	5.20	44	6	120 ± 26	6.3	1.3
56	5.24	33	BD	7030 ± 29	9.0	6.2
57	7.47	132	BD	2560 ± 377	0.7	0.9
58	5.77	150	BD	4580 ± 105	2.6	20.7
59	5.14	72	15	636 ± 20	4.0	3.1
60	5.69	142	BD	28320 ± 550	9.3	6.4
61	5.01	30	22	5352 ± 3	16.5	18.3
62	5.04	62	BD	6730 ± 563	19.1	1.0
63	4.92	58	25	1095 ± 37	15.7	24.1

Table 1. (continued)

Station No.	pH	Conductivity (µmhos)	Sulfide (uM)	^{222}Rn (pCi 1^{-1})		^{210}Po Filtered (pCi 1^{-1})	^{210}Po Unfilt (pCi 1^{-1})
			**** JULY	31- AUG 1,	1986 ****		
14	5.28	67	40	7157	± 37	244.3	257.4
18	4.61	48	46	16160	± 371	289.4	358.6
64	7.05	473	NA	59	± 27	19.1	21.8
65	7.19	790	214	2555	± 24	8.5	92.9
66	7.35	1000	229	1088	± 56	8.5	9.2
67	7.10	720	90	9180	± 275	7.4	67.3
68	4.59	710	4	4750	± 216	1.4	3.2
69	4.68	780	BD	4420	± 156	10.1	1.6
70	5.22	447	BD	35000	± 1713	2.7	17.8
71	7.12	360	4	2476	± 80	0.9	21.7
			**** AUGUST	28-29,	1986 ****		
10	4.58	140	43	2590	± 109	141.4	177.2
18	4.80	46	35	15140	± 472	566.8	538.6
19	5.75	312	BD	7120	± 367	4.0	2.3
20	6.12	310	BD	3900	± 221	2.1	24.7
21	4.95	219	BD	3410	± 258	1.3	4.2
22	6.43	400	BD	8380	± 333	10.1	2.7
25	6.05	365	BD	4127	± 58	2.4	0.6
31	3.85	40	BD	13960	± 388	35.5	33.2
32	6.12	230	BD	3180	± 100	6.2	18.4
72	6.00	193	BD	2780	± 244	9.4	0.4
73	6.25	300	BD	4690	± 473	1.6	2.6
74	6.45	183	BD	2346	± 65	0.6	0.5
75	6.40	312	BD	10780	± 286	12.0	0.7
76	6.13	301	BD	5000	± 225	1.3	18.2
			**** OCT	23-24, 1986	****		
14	5.11	97	22	8260	± 65	63.7	83.3
18	5.00	52	20	17290	± 354	425.9	544.7
65	7.30	830	230	3119	± 23	34.5	23.9
77	7.21	940	220	3171	± 40	4.0	2.0
78	7.20	600	NA	NA	NA	3.5	2.2
79	7.22	1200	149	1441	± 60	0.4	3.0
80	6.82	1100	163	2210	± 118	1.8	8.2

[1] Standard deviation of ^{222}Rn concentrations based on triplicate measurements
[2] BD = Below detection
[3] NA = Not analyzed
* Station numbers with "*" suffix refer to stations where water purification apparatus was installed and samples retrieved both before and after purifer

Note: For samples collected from June 16, 1985 to January 31, 1986, only polonium data is available.

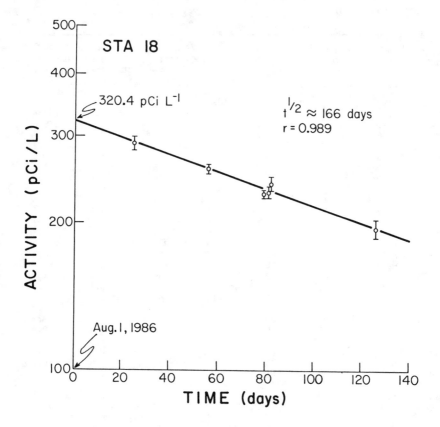

Figure 3. ^{210}Po activity measured at various times from one
large, filtered and acidified water sample. The
slope of this curve indicates an apparent half-life
of about 166 days, reasonably close to the expected
half-life for ^{210}Po (138.4 days).

uncommon [9,10] and the mechanisms responsible are recoil and
diffusion [11].

In a few cases, we have relatively complete analyses of well
water for several members of the ^{238}U decay series (Table 2).
Apparently, subsurface leaching and/or recoil and transport by
ground waters has led to marked disequilibria between members of
the ^{238}U-series. The concentrations of ^{238}U in all wells
measured were in the ppb range, while the ^{234}U/^{238}U ratio is
greater than unity in almost all cases. There is an inverse
correlation between the uranium activity ratio and concentration,

a situation which has been previously observed by Osmond and Cowart [12].

A comparison of the polonium and radon results (Table 1) shows that wells possessing high polonium activities also have high excess radon. However, low polonium wells occur in cases where there is low or high radon activities. When there is high ^{222}Rn and low ^{210}Po in the water, all the ^{210}Pb and ^{210}Po are apparently strongly adsorbed onto particle surfaces. Analyses for the intermediate daughter, ^{210}Pb, have ranged from below detection to 10 pCi l^{-1}. The anomalously high ^{210}Po, therefore, appears to be largely unsupported.

Much work has been concentrated at the site of the highest ^{210}Po values, station 18, in an effort to identify the unique characteristics of this environment in terms of why it has such high levels of radioactivity in its water compared to ground-waters from other, apparently similar, environments.

Station #18 Results

Replicate samples of water collected on October 24, 1986 from a large water sample from station #18 were analyzed for ^{210}Po (Table 3). The filtered samples were filtered sequentially from the bulk sample in the order of A to E without replacing the Millipore filter. Both filtered and unfiltered ^{210}Po activities lie within two standard deviation counting errors of the respective average activities and most are within one standard deviation counting error. These results imply that the average filtered and unfiltered activities are significantly different. However, filtering in the field for other samples has occasionally resulted in higher filtered (versus unfiltered) concentrations.

Results of different pretreatment (oxidizing) conditions on unfiltered samples for ^{210}Po from station #18 are given in Table 4. The results show a wide variation in results, from 323-805 pCi l^{-1}. The average activities appear to be significantly different for filtered and unfiltered samples. Both the filtered and unfiltered ^{210}Po activities lie within two standard deviations of the respective activities and most are within one standard deviation counting error. Assuming that the higher values are more correct, it may be preferable to fix the sample with either a mixture of acids or ammonium hydroxide. If acid fixation is used, a digestion procedure is required, which apparently is not the case for the ammonia treated sample. Acid fixation followed by digestion or ammonium fixation gave similar results, which are higher for ^{210}Po than just following a HNO_3 fixation followed by iron coprecipitation.

Sequential analysis of a large, filtered water sample from station #18 collected on August 1, 1986, is given in Table 5. These results show that the ^{210}Po activity has decreased over time since the collection date of the bulk sample and the half-life of the decay curve of this sample (Figure 3) is somewhat greater but still close to the half life of ^{210}Po. This implies that polonium is largely unsupported, in concurrence with our

Table 2. ^{238}U-series radionuclides in unfiltered groundwater samples
from west central Florida.

Station No.	U (pCi l^{-1})	^{234}U/^{238}U	^{226}Ra (pCi l^{-1})	^{222}Rn (pCi l^{-1})	^{210}Pb[1] (pCi l^{-1})	^{210}Po (pCi l^{-1})
8	0.04	9.96	1.0	NA[2]	NA	17
9	0.007	2.01	NA	NA	NA	145
10	0.005	4.67	2.8	2590-3470	NA	27.4-178
12	0.005	3.59	1.1	24800	2.75	92-142
14	0.16	1.07	NA	7160-8260	NA	83-260
15	2.77	0.97	NA	NA	NA	3.62
16	0.007	2.71	NA	NA	NA	4.21
17	0.25	2.03	NA	NA	NA	9.20
18	0.003	2.91	0.73-1.1	16100-19000	2.0-4.4	260-2570
23	0.31	0.67	NA	244	NA	0.59
26	0.008	2.55	NA	NA	NA	.503
37	NA	NA	8.5-9.5	26500	NA	0.26
56	NA	NA	NA	7030	2.43	6.19
62	NA	NA	NA	6730	7.70	0.95

[1] ^{210}Pb in filtered water samples.
[2] NA = not analyzed.
[3] ^{210}Po in filtered water sample.

Table 3.   Replicate analyses of samples from a homogeneous bulk sample
collected from station #18 on October 24, 1986. Errors
quoted are ±1σ based on counting statistics.

Sample	^{210}Po Filtered[1]	Unfiltered
	--- pCi l^{-1} ---	
A	430 ± 20	513 ± 22
B	340 ± 23	578 ± 23
C	436 ± 20	512 ± 20
D	437 ± 34	564 ± 36
E	487 ± 21	557 ± 33
Mean[2]	447 ± 27	545 ± 31

[1] Samples were filtered in sequential order from A to E.
[2] This is the mean and standard deviation of 4 replicate analyses. The
filtered sample B was considered an outlier and excluded from the
calculation.

Table 4.   Results of various pretreatments of unfiltered water from
station #18. All samples collected on June 21, 1986.

Notes	Experimental Treatment	^{210}Po (pCi l^{-1})
(1)	1 ml $HClO_4$ + 1 ml HNO_3	805 ± 25
	1 ml $HClO_4$ + 1 ml HNO_3	700 ± 16
	1 ml HCl + 1 ml HNO_3	665 ± 20
	1 ml $HClO_4$ + 1 ml HNO_3 + 1 ml H_2O_2	697 ± 18
	1 ml H_2SO_4 + 1 ml HNO_3	679 ± 25
(2)	1 ml $HClO_4$ + 5 ml HNO_3 + 3 ml H_2SO_4 + 3g Na_2SO_4	687 ± 23
(3)	18 ml HNO_3	323 ± 13
(4)	18 ml NH_4OH	651 ± 27

----

(1) 100 ml of sample was fixed in the field and ^{209}Po tracer added.
On return to the laboratory the acids were added and the sample
fumed to approximately 5 mls in teflon beakers. Acids were added
a second time and fumed to 1 ml. The sample was then prepared for
spontaneous plating.

(2) 100 ml of sample was fixed in the field and ^{209}Po added. On
return to the laboratory the acids were added and the sample fumed
to a Na_2SO_4 cake in a teflon beaker. Acids were added again and
fumed to a Na_2SO_4 cake a second time. Sample was diluted to a HCl
molarity of 0.1 M and prepared for spontaneous plating.

(3) 900 ml of sample was collected in the field and ^{209}Po added. Upon
return to the laboratory, Fe^{3+} was added and polonium was co-
precipitated by addition of Na_4OH. The sample was then prepared
for spontaneous plating.

(4) Procedure same as that in note 3.

direct measurements for ^{210}Pb.

    Activities of ^{238}U-series radionuclides in station #18 water
samples show extreme disequilibrium between all nuclides measured
(Figure 4). The disruption of equilibrium in the radon-lead-
polonium chain may occur by: (1) adsorption of dissolved lead
onto particles or surfaces such as the bedrock of the reservoir;
or (2) the authigenic precipitation of lead from the dissolved
phase as a lead sulfide. Either of these removal mechanisms must
then be followed by a partial release of polonium into the water
or requires removal of lead while leaving significant amounts of
polonium in solution. The latter mechanism is possible under
these conditions. A study by Jean and Bancroft [13] showed a pH
of at least 6 is required before adsorption takes place upon a
sulfide mineral surface; the pH we found at station #18 has
ranged from 3.8 to 5.19 during the course of this study. In
addition, sulfide concentrations are always significant in all
groundwaters containing elevated polonium. We have also noted
that, upon addition of Pb^{2+} as $PbNO_3$, precipitation of lead
sulfide occurs immediately, implying that these waters are at or
near saturation with respect to PbS. Ingrowth of ^{210}Po occurs
into the ^{210}Pb and a significant faction may be subsequently
released into the aqueous media by recoil, diffusion, and perhaps
other processes.

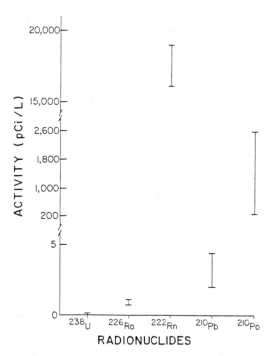

Figure 4.   Concentration ranges of ^{238}U, ^{226}Ra, ^{222}Rn, ^{210}Pb, and
            ^{210}Po in Station 18 water samples. Note two changes
            in scale.

Table 5. Results of long-term experiment to determine if ^{210}Po is supported by successive measurements from one large-volume sample, collected August 1, 1986. Activities in pCi l^{-1}.

Date Measured	HNO_3- Fixed	NH_4OH- Fixed
Aug. 1, 1986[1]	320.4	–
Aug. 26, 1986	289.4 ± 9.4	295 + 11
Sept. 24, 1986	–	303.6 + 6.5
Sept. 26, 1986	257.9 ± 6.8	–
Oct. 19, 1986	229.6 ± 5.1	–
Oct. 20, 1986	–	271.0 + 7.2
Oct. 21, 1986[2]	231.5 ± 6.7	–
Oct. 22, 1986[3]	240.6 ± 7.3	–
Oct. 30, 1986[4]	198.8 ± 8.0	230.3 + 3.9
Dec. 5, 1986	194.6 ± 8.9	235.9 + 8.9

[1] Values are calculated assuming ^{210}Po is unsupported based on extrapolated value from Fig. 5.4.
[2] This sample represents a separate aliquot taken on August 1, 1986, precipitated and processed just before counting.
[3] Separate aliquot also taken on August 1, 1986, precipitated and processed just before counting.
[4] Recount of ^{210}Po planchet from Aug. 26, 1986.

Table 6. ^{238}U-series radionuclides in unfiltered, radon-rich ground water from New Hampshire. The ^{222}Rn concentrations were measured at the University of New Hampshire. Note that the units for ^{222}Rn are in nanocuries per liter, while all other parameters are in picocuries per liter.

Sample	^{238}U	^{234}U	^{226}Ra	^{222}Rn	^{210}Po	^{231}Pa	^{232}Th	^{230}Th
	---- pCi l^{-1} ----			(nCi l^{-1})	-------- pCi l^{-1} -------			
D1	9.44	9.70	6.74	122	7.33	BD*	0.01	0.04
D19	NA	NA	BD	5	0.22	NA	NA	NA
D21	NA	NA	0.17	118	4.77	NA	NA	NA

*BD = Below Detection. Approximate detection limits as follows: ^{231}Pa = 0.003 pCi l^{-1}; and ^{226}Ra = 0.01 pCi l^{-1}.

Analysis for activities in a scale sample taken from the pressure tank at station #18 (Figure 5) showed a marked disequilibrium pattern similar to that found in the water (Figure 4). Although the absolute activities of the radionuclides in the scale were considerably higher than that in the water, the ratios are similar. The scale thus serves as a useful monitor of the conditions within the ground water. Although lead and polonium are particle reactive under the conditions of low pH, high sulfide and low Eh found in this environment, significant amounts of ^{210}Po are still available in solution.

Radioactivity in New Hampshire Groundwater

Three ground water samples from New Hampshire with ^{222}Rn activities ranging from 5,000 to 122,000 pCi l^{-1}, were provided to us by colleagues from the University of New Hampshire for analysis of other decay-series isotopes. The ^{210}Po activities in these samples ranged from 0.2 to 8.0 pCi l^{-1} (Table 6). Thus, even though these waters contain extraordinary amounts of ^{222}Rn, their ^{210}Po levels were reasonably low. No chemical parameters were measured in these waters but we noted the absence of any sulfur smell. We propose that these samples from New Hampshire represent a case of high excess ^{222}Rn and low ^{210}Po in a system with fast moving ground water associated with relatively low activities of ^{210}Pb and ^{210}Po on particulates.

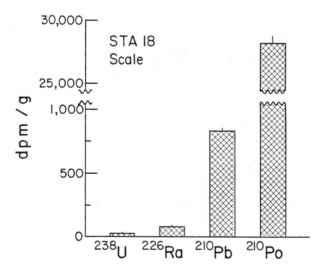

Figure 5.   Radiochemical results of an analysis of solid scale removed from the inside of a water storage tank. Activities given in dpm/g. Note change in scale.

CONCLUSIONS

(1) Some ground water in the surficial aquifer of west central Florida contains significantly elevated levels of ^{210}Po, a daughter of ^{222}Rn.

(2) High values of ^{210}Po exist in both filtered and unfiltered samples although filtered samples are significantly lower in some cases.

(3) The waters with anomalously high values of ^{210}Po are characterized by moderately high ^{222}Rn, low pH, and the presence of sulfide.

(4) Waters from a granitic terrain with exceedingly high ^{222}Rn have low ^{210}Po, supporting the interpretation that the presence of high ^{222}Rn alone will not necessarily lead to high polonium in ground water.

ACKNOWLEDGEMENTS

This work was funded by the Florida Institute of Phosphate Research (Contract No. 83-05-016S) and by the State of Florida Department of Environmental Regulation (Contract No. WM 147). Special thanks go to Sheila Heseltine who typed and assisted with the preparation of this manuscript.

REFERENCES

1.   Hinton, T.G. and F.W. Whicker. "A Field Experiment of Radon
     Flux from Reclaimed Uranium Mill Tailings," Health Physics
     48(4):421-427 (1985).

2.   Environmental Health Criteria No. 25: Selected Radionuclides
     – Radon (Geneva: United Nations Environment Programme and
     The World Health Organization, 1977).

3.   Oural, C.R., H.R. Brooker and S.B. Upchurch. "Source of
     Gross Alpha Radioactivity Anomalies in Recharge Wells,
     Central Florida Phosphate District," Florida Institute for
     Phosphate Research Publication No. 05-014-034, 83 pp.
     (1986).

4.   Flynn, W.W. "The Determination of Low Levels of ^{210}Po in
     Environmental Materials," Anal. Chim. Acta 43:221-227
     (1968).

5.   Prichard, H.M. and T.F. Gesell. "Rapid Measurements of
     ^{222}Rn Concentrations in Water with a Commercial Liquid
     Scintillation Counter," Health Physics 33:577-581 (1977).

6.   Thomson, J. and K.K. Turekian. "^{210}Po and ^{210}Pb
     Distributions in Ocean Water Profiles from the Eastern
     South Pacific," Earth Planet. Sci. Lett. 32:297-303 (1976).

7.   Koide, M., A. Soutar and E.D. Goldberg. "Marine
     Geochronology with ^{210}Pb," Earth Planet. Sci. Lett. 14:442-
     446 (1972).

8.   Mullin, A. "Abnormally High Alpha Activity in a Louisiana
     Drinking Water Supply," paper presented at 27th Annual
     Meeting of the Health Physics Society, Las Vegas, NV, 27
     June-1 July, 1982.

9.   Brutsaert, W.F., S.A. Norton, C.T. Hess and J.S. Williams.
     "Geology and Hydrologic Factors Controlling Rn-222 in
     Ground Water in Maine," Ground Water 19(4):407-417 (1981).

10. Asikainen, M. "State of Disequilibrium Between ^{238}U, ^{234}U,
     ^{226}Ra and ^{222}Rn in Groundwater from Bedrock," Geochim.
     Cosmochim. Acta 43:1681-1686 (1981).

11. Rama and W.S. Moore. "Mechanism of Transport of U-Th-Series
     Radioisotopes from Solids into Ground Water," Geochim.
     Cosmochim. Acta 48:395-399 (1984).

12. Osmond, J.K. and J.B. Cowart. "The Theory and Uses of
     Natural Uranium Isotopic Variations in Hydrology," Atomic
     Energy Rev. 14(4):621-679 (1976).

13. Jean, G.E. and G.M. Bancroft. "Heavy Metal Adsorption by Sulphide Mineral Surfaces," Geochim. Cosmochim. Acta 50:1455-1463 (1986).

A TECHNIQUE FOR THE RAPID EXTRACTION OF RADON-222
FROM WATER SAMPLES AND A CASE STUDY

William M. Berelson and Douglas E. Hammond,
University of Southern California, Los Angeles, California
Andrew D. Eaton, J.M. Montgomery Engineers, Pasadena, California

INTRODUCTION

The radon concentration in our environment has become a topic
of recent attention and medical concern. Federal and state
governments are in the process of assessing safe limits of radon
exposure, and the public and private sector are working on
developing fast and accurate methods of detecting and measuring
this element. The first part of this paper is a description of
our adaptation of previously used techniques for extraction and
analysis of dissolved gases to make quick and accurate
measurements of radon in water samples. In the second part of
this paper we present some data from a case study of radon
measurements made using this technique and other radioactivity
measurements (total alpha activity, uranium concentration) made in
ground waters near Los Angeles, California

RAPID RADON EXTRACTION SYSTEM (RRES)

RRES Design

The rapid radon extraction system (RRES), shown schematically
in Figure 1, was developed to extract radon from a water sample
and transfer it to a counting cell in a single step. It utilizes
two devices previously developed for gas extractions and radon
measurements, a Swinnerton-type stripper [1] and a Lucas-type
counting cell [2]. Radon and other dissolved gases are extracted
from a water sample injected into an evacuated system, and a flow
of helium carrier gas is used to complete the extraction and
transfer the radon to a scintillation counting cell where its
concentration can be determined from the measured count rate of
radon and its daughters, polonium-218 and polonium-214.

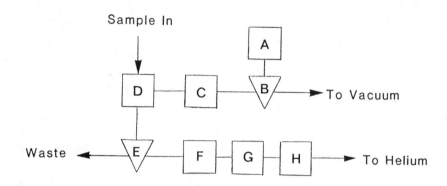

Figure 1.   Schematic of RRES. A is a counting cell, B is a 3-way
valve, C is a CaSO$_4$ column, D is a stripping column, E
is a 3-way valve, F is a pressure gauge, G is a bleeder
valve, H is a helium toggle.

Samples are collected in a manner similar to that used for
dissolved oxygen. Water is introduced to a 300ml glass BOD bottle
through a polyethylene tube, and care is taken to avoid bubbling
the sample while filling. After two or more bottle volumes are
overflowed, the bottle is sealed with a ground glass stopper.
Although any glass bottle with a gas-tight top could be used as a
sample vessel, it is important to consider the diffusive loss of
radon out of the sample bottle, especially if samples are to be
stored for longer than a few hours. Samples should be collected
free of bubbles, although often small (<1cc) bubbles form due to
changes in temperature, and their volume can be estimated.
     Before extracting a sample, the system is evacuated through
the three-way valve (B in Fig. 1) and "rinsed" several times with
helium to insure a low system blank. A short leak test is
performed by evacuating the line with a cell attached (A),
isolating the line, and observing the change in pressure (F) over
a 5 minute period. During the leak test, a sample can be prepared
for analysis.
     Before drawing a sample syringe, the sample bottle is shaken
to equilibrate the bubble gas with the dissolved gas. The
measured radon concentration can be corrected for the amount
present in the bubble based on published radon solubility data
[3]. Immediately after opening the water sample bottle, a 20 ml
aliquot is withdrawn into a glass syringe through a piece of
flexible tubing inserted to the bottom of the sample bottle. The
precision of measuring volume from graduations on the syringe

barrel is about 1%. The syringe barrels can be weighed before and after filling to reduce this uncertainty. After the syringe is filled, a 1.5 inch, 22 gauge needle is attached to the tip and the sample is ready for injection. The syringe filling procedure takes about 3 minutes.

With the extraction board evacuated, helium toggle (H) open, bleeder valve (E) closed and counting cell (A) in place, the sample is ready for extraction. First, the bleeder valve is cracked open to allow a very small flow of helium into the cell. Immediately, the syringe needle is inserted into a septum at the top of the stripping chamber (D) and the pressure gradient draws the sample out of the syringe. As the sample enters the stripping chamber (chamber volume is 35 ml) it de-gases with the net flow of gases moving into the counting cell (cell volume is approximately 100 ml). After the sample is completely injected (30 seconds), the syringe is removed and the bleeder valve opened to allow a greater flow of helium through the sample and into the cell. The bleeder valve is opened so that the cell fills at a constant rate and reaches ambient pressure (1 atmosphere) in 1-2 minutes. At this time, the counting cell is removed from the extraction line by way of a quick-connect fitting and is ready for counting. Water is purged from the stripping chamber by letting the helium pressure increase over ambient in the line, then turning the three-way stopcock (E) to connect the stripper with a waste line. The sample may be collected and saved for further analyses if desired. The RRES is re-evacuated, leak tested, and ready for use with another sample. The entire system fits on a board 40x30x10 cm.

RRES Quality Control

The accuracy of this technique was verified by running samples both on the RRES and on our standard radon extraction boards [4,5]. The average standard deviation of four samples run with both techniques was ± 5%. The standard technique has been calibrated with radium-226 sub-standards prepared from NBS standards by G. Mathieu of Lamont-Doherty Geological Observatory. By analyzing radon standards with the same counting cells for many years, we know the efficiency of the standard extraction system and the cells to 1-5%. From running standards on different boards with the same counting cell it is possible to deduce that the efficiency measured (between 70-90%) is due to peculiarities of the counting cell geometry and not to the extraction technique. Thus, we used the cell efficiencies determined on the standard system in calculating the radon activity in samples extracted with the RRES. However, the counting efficiency of a radon cell will be a function of the type of gas present in the cell. We have done experiments showing that the effect of CO_2 as a matrix gas is to reduce the counting efficiency of a cell by about 20%. Assuming that a water sample were saturated with CO_2 (at STP), there would be <<1% CO_2 in the cell after filling with helium. Alkaline samples with a high partial pressure of CO_2 could cause higher CO_2 fractions in the gas phase, although such samples are

likely to be rare. Water vapor can also affect the efficiency of
a counting cell, but water vapor in the RRES is trapped on a $CaSO_4$
column (C in Figure 1) before it reaches the cell. Other gases
dissolved in the sample (N_2, O_2) will not be present in great
enough abundance to affect the counting efficiency of radon.

One question concerning the effectiveness of this RRES was
whether the radon in the sample is quantitatively removed to the
counting cell. We checked this question by making multiple
extractions on the same sample. In these experiments, a cell was
filled by the above described procedure, removed from the line and
an evacuated cell was connected to the line and filled. We found
that the second extraction had less than 2% of the activity of the
first.

The precision of the RRES was tested by running replicates of
water samples from the same bottle. In this case a syringe was
filled, the sample extracted, a second syringe filled and sample
extracted. The time between the first and second run was less
than 10 minutes which is too short for a significant fraction of
the dissolved radon to enter the head space in the sample bottle
created by the withdrawal of the first sample. The results of
these experiments are plotted in Figure 2 and the fit of this data
indicates our precision is about ± 6%.

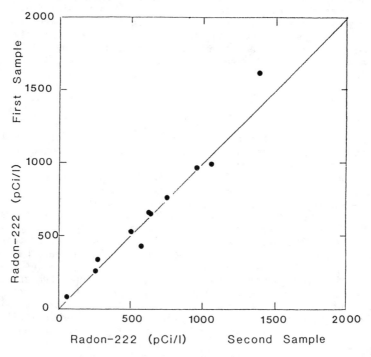

Figure 2. Plot of replicate samples run with the RRES. The
line represents the 1:1 relationship.

The sensitivity of this technique depends on the background activity of the system and the cells used, the desired counting statistics and the desired counting interval. We find that the radon blank, determined by filling a cell with helium with no sample injected, is <.05 dpm, a limit whose detectibility is less than we can measure with our current counting cells. The radon counting cells that were used in this study are 5-10 years old and thus have backgrounds that are relatively high, 0.3-1.5 cpm. With the uncertainty in the background values at about ± 30%, it is important that the sample activity be at least 1.5 times the cell background. Thus, the lowest activity sample we could count using the RRES is about 5 pCi/l (11 dpm/l). However, this sample would have to count for 33 hours in order to get 5% counting statistics. The ground water samples analyzed for this study have activities >100 pCi/l, thus counting these samples takes <2 hours. High activity samples can be stored to allow the radon activity to decrease, or smaller volumes can be injected into the RRES. The rate limiting step in the procedure, providing that counting cells are available, will be counting time. One sample can be processed in less than 10 minutes.

CASE STUDY

Background

A groundwater storage program has been proposed as a long term solution to periodic water shortages in southern California. In order to evaluate this proposal, the Metropolitan Water District of Southern California contracted with James M. Montgomery Consulting Engineers (JMM) and Camp, Dresser, and McKee (CDM) to investigate current water quality in the recharge and storage areas of a potential storage site (the Chino Basin in Los Angeles and San Bernardino Counties, California). This involved collection of approximately 150 samples. A number of parameters were analyzed in these samples including such known problems as TDS and Nitrate, a wide variety of organic compounds including various classes of pesticides and volatiles, and radiological parameters including gross alpha and beta activity, uranium, and radon. EPA has indicated that much greater emphasis will be placed on these parameters in the future and it was, therefore, deemed important to obtain good baseline data. EPA has not yet determined a Maximum Contaminant Level (MCL) for radon, but possible values range from 500 to 5000 pCi/l.

Sampling Protocols

In order to carry out the sampling and analysis it was necessary to develop careful protocols for all phases of the program. Most of the analytical work was done by JMM's laboratory, but the radon analysis was performed by the University of Southern California (USC) Department of Geological Sciences.

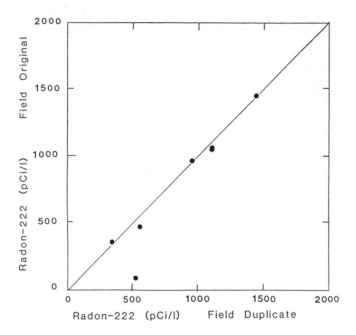

Figure 3.   Plot of field replicates of radon samples. Line
represents 1:1 relationship.

Samples were delivered to the JMM laboratory each day and
delivered to USC within 3 days of collection after examination for
leaky bottles or other problems which would affect sample
integrity for radon analysis.

Precision of Radon Analysis

Approximately 5% of the samples were collected in duplicate in
the field to test the precision of field sampling techniques.
Figure 3 summarizes the precision of field duplicates. With one
exception, the precision of field duplicates are comparable to
that of lab duplicates (Fig. 2) indicating that the sampling
protocols and sample transport procedures were reliable. The one
exception appears to be due to an inadvertent aeration in sampling
with consequent loss of radon, since the duplicate result is
significantly lower than any other results measured during the
survey.

Other Radiological Parameters

In addition to radon, we also obtained samples for gross alpha
and beta activity and uranium analysis. Similar protocols of

field and laboratory duplicates were used for both of these parameters. Gross alpha and beta measurements were made with a Tennelec Low Background Gas Flow proportional counter using EPA method 900.0 (evaporation on planchets and counting the deposit for periods of 1-3 hours). Uranium measurements were made using a VG Instruments Inductively Coupled Plasma/Mass Spectrometer (ICP-MS) which gives detection limits of <0.1 ug/l, well below probable MCL's of 10-20 ug/l. For gross alpha activity the precision of field duplicates was within the counting error of individual samples. For uranium activity, precision of both field and lab duplicates was 6% for samples with concentrations above 1 ug/l.

Results of Field Survey

Figure 4 shows the localized nature of radon distributions in the Chino Basin. Areas with radon greater than 1000 pCi/l are found in a random pattern, in contrast to other measured parameters such as TDS, which increases down gradient. The high radon activities in wells near the Jurupa and San Gabriel Mountains may indicate an association of radon with local geology. Activities vary from <100 pCi/l to >2000 pCi/l, with a median

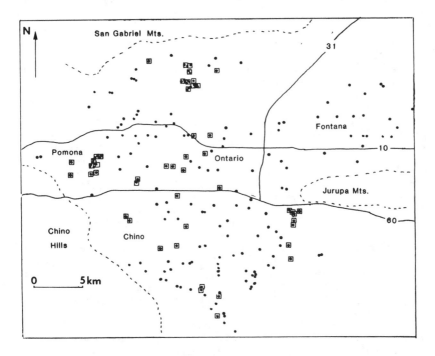

Figure 4. The distribution of sample wells (circles) in the study area. Sites with radon-222 activities >1000 pCi/l are indicated by a square. Numbers indicate highways.

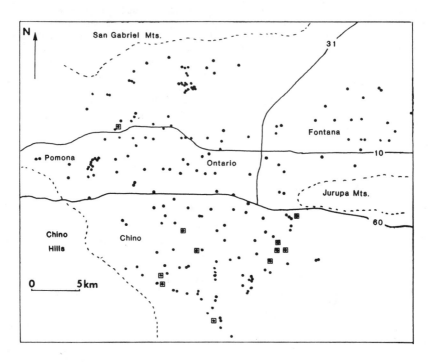

Figure 5.   The distribution of sample wells (circles) in the study
            area. Sites with gross alpha activities >5 pCi/l are
            indicated by a square.

value of 980 pCi/l. In only a few locations in the basin are
values either below 100 pCi/l or above 2000 pCi/l. Figure 5
shows the distribution of gross alpha activity in the basin.
Elevated gross alpha activity is much more localized. Values
above 5 pCi/l are found in only a few sites in the basin.
Gross alpha activity varied between <1 pCi/l and 10 pCi/l.
Uranium values ranged from <0.1 ug/l (0.07 pCi/l assuming
U-234/U-238 equilibrium) to 30 ug/l (20 pCi/l assuming
equilibrium).

Although the highest radon activity is located in areas of
granitic terrain, which also contain high levels of gross alpha
activity and uranium, there is no simple correlation between these
parameters. When the data are examined in detail using scatter
plots of radon versus gross alpha (Figure 6) and radon versus
uranium (Figure 7), it is apparent that radon concentrations are
generally independent of other radioactivity measurements in this
basin. Thus it is not possible to use gross alpha screening
measurements to predict those areas where radon concentrations
might be high enough to be above action levels for drinking
water.

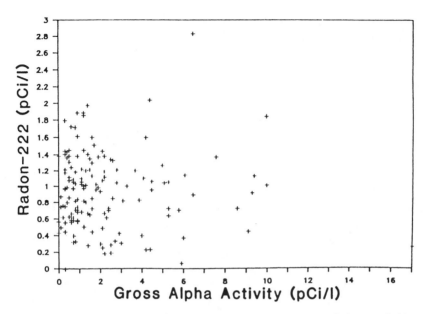

Figure 6.   Scatter plot of radon-222 versus gross alpha activity.
Radon activity in thousands.

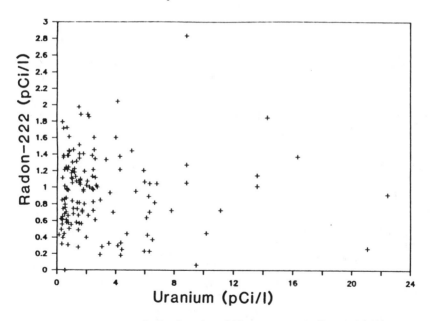

Figure 7.   Scatter plot of radon-222 versus uranium activity.
Radon activity in thousands.

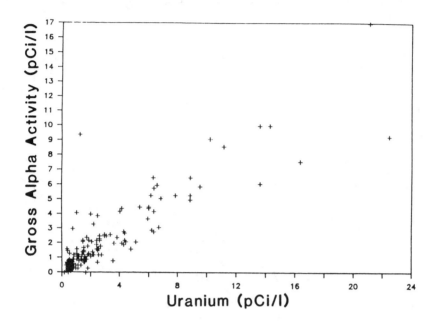

Figure 8.   Scatter plot of gross alpha activity versus uranium.

In contrast to the lack of correlation of gross alpha and radon, most of the high gross alpha activity is due to uranium concentrations, as can be seen from the good correlation shown in Figure 8. When uranium activities (in pCi/l) are subtracted from gross alpha activity for the same sites, there is essentially no remaining alpha activity for most of the samples.

CONCLUSIONS

The rapid measurement technique for radon analysis has been demonstrated to be useful in the analysis of well water samples for this study of the chemistry of a groundwater basin. This procedure has been shown to be accurate, reliable and takes a minimum of sample preparation or analytical work. Because the system is portable, it may be used in field areas when logistics prohibit the transport of samples to a laboratory.

Results of the screening program for the Chino groundwater basin indicate that radon concentrations are independent of other radiological measurements in this basin. If EPA sets the limit for radon in drinking water below 1000 pCi/l a significant portion of the basin may be out of compliance, without treatment.

ACKNOWLEDGEMENTS

We would like to acknowledge the financial assistance of the Metropolitan Water District of Southern California, which supported the lab work reported herein, Dr. Joe LeClaire of CDM, Inc. who provided geographic plots of sample locations and assisted with data manipulation, and Mr. Michael Hamilton who assisted with radon analyses.

REFERENCES

1. Swinnerton, J.W., V.J. Linnenbom and C.H. Cheek "Revised sampling procedure for determination of dissolved gases in solution by gas chromatography," Anal. Chem. 34:1509 (1962).

2. Lucas, H.F. "Improved low level alpha scintillation counter for radon," Rev. Sci. Instrum. 28:680-683 (1957).

3. Jenkins, A.C. and G.A. Cook "Gas phase properties," in Argon, Helium and the Rare Gases, G.A. Cook ed. (New York: Interscience Publications, 1961) p. 175.

4. Mathieu, G. "Radon-222/Radium-226 technique of analysis," in Annual Report to ERDA, Transport and Transfer Rates in Waters of the Continental Shelf, Contract EY 76-S-02-2185, (1976) 30 pp.

5. Hammond, D.E. and C. Fuller "The use of radon-222 to estimate benthic exchange and atmospheric exchange rates in San Francisco Bay," in San Francisco Bay, the Urbanized Estuary, T.J. Conomos, ed. (A.A.A.S., 1979) pp. 213-230.

RELATION BETWEEN NATURAL RADIONUCLIDE ACTIVITIES AND
CHEMICAL CONSTITUENTS IN GROUND WATER IN THE
NEWARK BASIN, NEW JERSEY

Zoltan Szabo and Otto S. Zapecza,

U.S. Geological Survey, West Trenton, New Jersey

ABSTRACT

    The U.S. Geological Survey, in cooperation with the
New Jersey Department of Environmental Protection,
Division of Water Resources, is conducting a study to
determine the occurrence and distribution of naturally
occurring radionuclides in ground water in the Newark
Basin, New Jersey, and to identify other aqueous
chemical constituents that may be associated with these
radionuclides. Water samples were collected by the
U.S. Geological Survey in 1985 and 1986 from 260
ground-water sites areally distributed throughout the
Newark Basin. Specific conductance, pH, alkalinity,
dissolved oxygen, Eh (oxidation-reduction potential),
and water temperature were measured on site.
Subsequently, these samples were analyzed in the
laboratory for gross alpha- and gross beta-particle
activity. If gross alpha radiation exceeded 5 pCi/L
(picocuries per liter), uranium, radium-226, major ion,
and trace metal concentrations were also determined.
In addition, sampling density was increased in these
areas, and selected wells were sampled for dissolved
radon-222.
    Gross alpha-particle radiation activities ranged
from less than 0.1 to 124 pCi/L; uranium activities
ranged from 0.1 to 40 pCi/L; radium-226 activities

ranged from less than 0.01 to 22.5 pCi/L; and radon-222 activities ranged from 71 to 15,900 pCi/L. Where gross-alpha radiation in reducing waters was high, radium-226 activities were elevated; however, where gross-alpha radiation in oxidizing waters was high, uranium activities were elevated. These results suggest that oxidation-reduction potential is an important control on radium-226 and uranium concentrations in the ground water. Elevated activities of radium-226 or uranium were not found exclusively in any one chemical class of ground water; however, elevated levels of radium-226 are absent from sulfate-dominated waters. In several instances, radium-226 activities in excess of 1 pCi/L were associated with anomalously high levels of iron (11,000 μg/L [micrograms per liter]), manganese (1,600 μg/L), and barium (1,300 μg/L). Uranium activities in excess of 4 pCi/L were locally associated with strontium concentrations as high as 5,100 μg/L. No definitive relation could be determined between concentrations of radon-222 and radium-226 in the ground water.

INTRODUCTION

Ingestion of water containing radium and/or uranium can result in the accumulation of these radioactive elements in bone tissue with resultant bone and sinus cancers [1]. Because ingested radium poses a health hazard, the U.S. Environmental Protection Agency (USEPA) has established the following maximum contaminant levels (MCL): (1) Radium-226 and radium-228 (total radium, combined), 5 pCi/L (picocuries per liter); (2) Gross alpha-particle activity (including radium but excluding uranium and radon), 15 pCi/L; (3) Gross beta-particle activity, 4 millirems per year (mrem/yr) [2]. In addition, a 10 pCi/L MCL for uranium in water [3] and a 10,000 pCi/L MCL for radon-222 in water [4] have been suggested by other investigators. The USEPA established a sampling requirement that composite water samples from the distribution system of all public water suppliers be analyzed once every 4 years for activities of gross alpha- and gross beta-particle, and radium-226.

Data from various sources indicates that elevated levels of gross alpha-particles, radium-226, and uranium occur in ground water of the Newark Basin. Based on the occurrence of uranium minerals [5] and favorable geologic settings [6] , the Newark Basin in New Jersey was selected for inclusion in the National Uranium Resource Evaluation (NURE) program of the U.S.

Department of Energy. Analysis of water from more than 300 wells in the New Jersey part of the Newark Basin revealed several areas where the uranium concentration in the ground water was above 10 μg/L micrograms per liter [7]. The USEPA-mandated analysis for radioactivity in public water-supply systems showed that several samples had radionuclide activities near the MCL levels [8].

The U.S. Geological Survey, in cooperation with the New Jersey Department of Environmental Protection (NJDEP), Division of Water Resources, is conducting a study of naturally occurring radionuclides in ground water of the Newark Basin in New Jersey (fig. 1). The principal objectives of the study are to (1) determine the occurrence and distribution of naturally occurring radionuclides; (2) identify nonradioactive chemical constituents in the ground water that may indicate the presence of these radionuclides; and (3) identify geological and geochemical factors controlling radionuclide concentrations, distribution, and migration in ground water. For this study, the U.S. Geological Survey collected 260 ground-water samples for chemical and radiochemical analysis.

The purpose of this paper is to present relations observed between radionuclides and nonradioactive chemical constituents in ground water. Indicators of the possible presence of radionuclides in ground water in the study area are suggested herein on the basis of correlations between nonradioactive and radioactive chemical constituents. These correlations are based on analyses completed to date; consequently, they may be modified as additional information becomes available. Possible geochemical reactions that control the association of radium-226 and uranium with the chemical constituents identified in this report are discussed. A companion paper [9] discusses the areal distribution of radioactivity in ground water of the Newark Basin and its relation to the geologic framework.

THE STUDY AREA

The Newark Basin in New Jersey (fig. 1) comprises a 10-county area of approximately 2,000 square miles. It is exposed at land surface in a band 16 to 32 miles wide trending southwestward from the New York-New Jersey border to the Delaware River. It accounts for one-fifth of the State's land area as well as approximately two-thirds of its population. The study area is bordered on the northwest by Precambrian and Paleozoic crystalline and sedimentary rocks of the New Jersey Highlands, also known as the Reading Prong, and

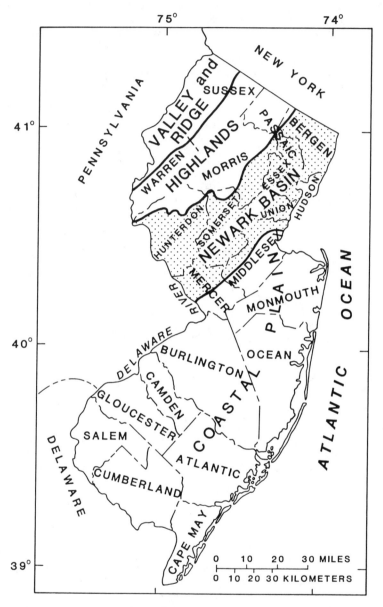

Figure 1. Location of study area, Newark Basin,
New Jersey.

on the southwest by unconsolidated deposits of the
Coastal Plain. The Newark Basin is filled with a thick
sequence of late Triassic and early Jurassic
continental sediments interbedded with basaltic lava
flows and cut by diabase intrusives. For purposes of
this report, these sediments have been classified from
oldest to youngest, into the following three major
formations: Stockton Formation, Lockatong Formation,
and Passaic Formation. Sedimentary deposits overlying
the Passaic Formation in this report are termed
"Undifferentiated".

The Stockton Formation consists primarily of gray,
buff, and red, arkosic conglomerate and sandstones.
The Lockatong Formation is composed of gray, green,
black, and red mudstones, siltstones and minor
carbonates. The Passaic Formation is composed
primarily of red siltstone and sandstone. The lower
part of the Passaic Formation consists largely of
alternating black and red mudstones, a lithology
similar to that of the Lockatong Formation. The
undifferentiated deposits overlying the Passaic
Formation are predominantly red, gray, and black
mudstones and siltstones with local beds of sandstone
and conglomerate. A more detailed discussion of the
geologic framework of the study area is presented in
the companion paper [9], and in references contained
therein.

SAMPLING AND ANALYTICAL PROCEDURES

Wells sampled for this study were selected to
provide complete geographic coverage of the study area
and representative samples of each geologic formation.
Potential sampling sites were selected by reviewing
well records in the files of the U.S. Geological Survey
and the NJDEP. This approach insured that information
about well construction, particularly that on depth and
contributing aquifer(s), was available.

Wells finished in the rocks of the Newark Basin
usually are cased from the land surface, through the
weathered regolith zone, into the unweathered rock
below. The rest of the well, often open to several
hundred feet of fresh rock, is left uncased so that
water from the fractures intersected by the borehole
drain freely into the well. Individual fracture zones
within the same well may yield water of somewhat
different chemical compositions [10]. These open-hole
wells can be visualized as penetrating a thick aquifer
in which water quality is vertically inhomogeneous.
Water from these wells has a gross chemistry that is a
composite of the different water chemistries

contributed to the well by the individual fracture
zones.

Sampling

Sampling was divided into two phases:
reconnaissance phase at 211 wells, and an intensive
phase at an additional 49 wells located in areas where
elevated radionuclide concentrations were found in
ground water in the reconnaissance phase.
Characteristics of the reconnaissance-sampling network
were: (1) maintenance of widespread geographic
distribution, although additional sites were sampled in
areas where MCL violations were demonstrated or
suspected; (2) inclusion of wells finished in all
geologic formations within the basin; and (3) emphasis
on public supply wells, because water from these wells
is ingested by large numbers of consumers.
Characteristics of the intensive-sampling network were:
(1) approximately uniform areal distribution of
sampling sites around wells with anomalous radionuclide
concentrations; (2) where possible, wells drilled to
similar depths along strike in the same formation were
selected because ground water flows primarily in
fractures parallel to strike [11]; and (3) no
preference as to type of well (public supply, domestic,
and industrial wells were all given the same
preference).
All samples consisted of untreated water collected
near the well head; a few passed through a pressure
tank or small holding tank before collection. Where
possible, these tanks were drained before sampling. A
minimum of one well-casing volume of water was pumped
from the well, and three well-casing volumes of water
were removed whenever possible. Stabilization of
specific conductance, pH, and water temperature was
necessary before a water sample was collected.
Specific conductance, pH, alkalinity, dissolved oxygen,
Eh (oxidation-reduction potential), and water
temperature were measured on site according to the
guidelines of Wood [12]. Dissolved-oxygen
concentration and Eh were measured in water passing
through a flow-through chamber. Samples were collected
and preserved following U.S. Geological Survey
procedures [13,14]. Dissolved radon-222 gas samples
were collected according to USEPA methods [15].

Analysis

All water samples were analyzed for gross alpha-particle and gross beta-particle activity. If gross alpha radiation exceeded 5 pCi/L, radium-226, uranium, major ion, and trace-metal concentrations were determined. In addition, uranium and radium-226 analyses were performed on selected samples to insure representative areal coverage. Gross alpha-particle, gross beta-particle, and radium-226 activities were determined by the NJDEP Radiation Protection Laboratory with planchet-counting methods following USEPA procedures [16]; radon-222 activity was determined by liquid scintillation. Uranium concentration was determined by the U.S. Geological Survey National Water Quality Laboratory using the fluorometric method [13], and was reported in micrograms per liter. Assuming that uranium isotope ratios in the Newark Basin equal natural abundances, the measured concentrations were multiplied by a factor of 0.68 pCi/μg (picocuries per microgram) to convert them to pCi/L values [13,3]. Concentrations of major cations and anions, and trace-metals were determined by the U.S. Geological Survey National Water Quality Laboratory using standard methods [14]. Blanks, standards, and duplicate samples were submitted to each laboratory for each parameter being analyzed, so that analytical quality could be monitored.

RADIONUCLIDE ACTIVITIES AND DISTRIBUTION IN GROUND WATER

The range and median value of gross alpha- and gross beta-particle, radium-226, uranium, and radon-222 activities are listed in table 1. Approximately 5 percent of the analyzed samples exceed the MCL established by the USEPA. Most of the ground water samples have levels of natural radioactivity well below USEPA required and recommended limits. The activities of these radioactive constituents, which are approximately lognormally distributed, are skewed strongly toward low values, suggesting low activities of these constituents. In general, the levels of radioactivity in ground water measured by this study are consistent with, although locally somewhat higher than, measurements made previously in ground water in the study area [7,8,17].

Table 1. Range and median values, in picocuries per liter (pCi/L),
and number of samples exceeding maximum contaminant
levels (MCL) for radionuclide analyses of ground water in
the Newark Basin, New Jersey, completed as of March 6,
1987.

Radionuclide	No. samples analyzed to 3-6-87	Highest value (pCi/L)	Lowest value (pCi/L)	Median value (pCi/L)	No. samples exceeding USEPA MCL***
Gross alpha	259	124	<0.1	3.2	15*
Gross beta	259	49	< .1	3.1	--**
Radium-226	177	22.5	.1	.2	7***
Uranium	55	40	.1	2.1	4****
Radon-222	23	15,900	71	1420	3****

(*)   Uranium and radium-226 activities have not been subtracted
        from gross alpha-particle activities
(**)   Gross beta-particle activity was not measured in mrem/yr
        units
(***)   The U.S. Environmental Protection Agency maximum
        contaminant level (MCL) calls for measurement of the
        activities of both radium-228 and radium-226. Addition of
        radium-228 activities to the radium-226 activities already
        measured may increase the number of samples over the MCL.
(****) MCL are only recommended by USEPA

Radium-226 and Uranium

Elevated activities of radium-226 (greater than 1
pCi/L) and uranium (greater than 4 pCi/L) are present
predominantly in ground water from wells near or along
the geologic contact of the Lockatong Formation and
Passaic Formation (fig. 2 and 3). Elevated activities
of these radionuclides also are present locally in
other parts of the Lockatong Formation and Stockton
Formation. Large areas of the Passaic Formation, the
basalt and diabase units and the area shown as
undifferentiated, have ground water with low radium-226
and uranium activities (figs. 2 and 3).

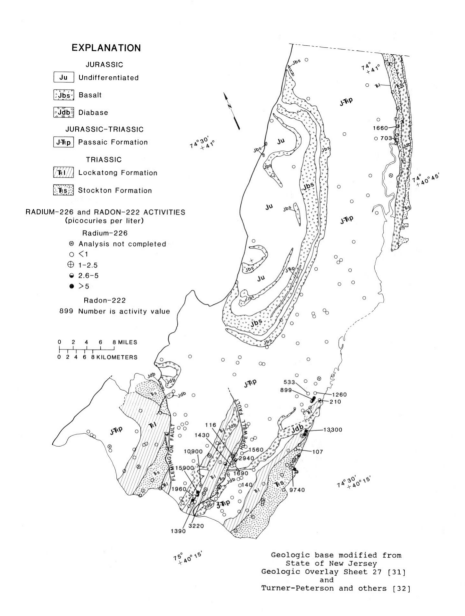

Figure 2. Radium-226 and radon-222 activities
in ground water, Newark Basin, New Jersey.

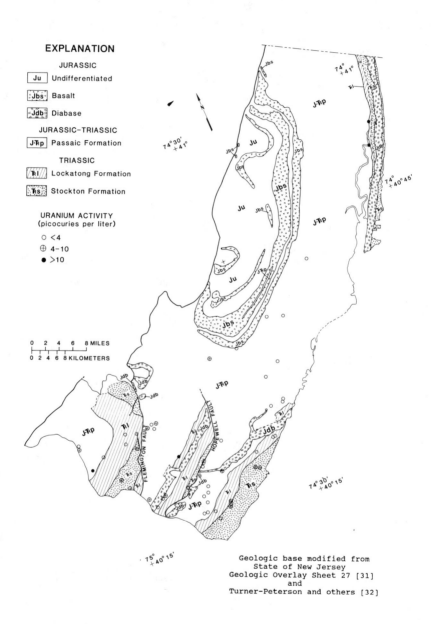

Figure 3. Uranium activities in ground water,
Newark Basin, New Jersey.

The areas of elevated activities of radium-226 and uranium in ground water correspond to areas of elevated gross alpha activity in ground water and uranium enrichment in the underlying rock [9]. Although elevated activities of radium-226 and uranium in ground water of the study area are spatially related, elevated activities of both radionuclides are rarely found concurrently in the same ground-water samples. The lack of concurrence within a given water sample is due to different geochemical behavior of the two radionuclides.

Radon-222
_____

Radon-222 activity in the ground water correlates with radium-226 activity. However, some of the water with the highest radium-226 activity also contained the lowest radon-222 activities (fig. 2). Radon-222 activity in ground water is controlled by the concentration of radium-226 [18] in the aquifer matrix. The amount of radon-222 that can enter the ground water depends on the emanating coefficient of the different radium-226-containing materials in the aquifer matrix [19]. Therefore, elevated radium-226 activities in ground water may be useful as a possible indicator of elevated radon-222 activities.

RELATIONS BETWEEN RADIONUCLIDES AND INORGANIC CHEMICAL CONSTITUENTS

The generalized chemical characteristic of ground water in the study area are summarized below. Ground water in the study area is primarily a calcium-magnesium-bicarbonate type although some calcium-sulfate type water is also present. These results are consistent with previous findings in the Newark Basin [20,21,22]. All major cations and anions increase as total dissolved solids increase. Total dissolved solids range from 103 to 1,300 mg/L (milligram per liter) with a median value of 259.5 mg/L.

The ground water ranges from slightly acidic to slightly alkaline, with pH values ranging from 5.5 to 9.8. Ninety percent of the pH values range from 6.4 to 8.1; however, 50 percent of the pH values range from 7.2 to 7.8. The median pH is 7.6.

The ground water is mostly oxidizing, dissolved oxygen values ranged from less than 0.1 mg/L to 9.8 mg/L, with a median concentration of 2.8 mg/L. Approximately 20 percent of the sites sampled had dissolved oxygen concentrations of less than 1 mg/L.

Eh values ranged from +24 to +827 mV (millivolts), with a median value of +328 mV. Field measurements of Eh were used as a tool to classify ground water broadly from the study area as being relatively oxidizing or relatively reducing. Eh values below +200 mV are considered reducing in this report, whereas ground water with measured Eh values above +200 are considered oxidizing. Because shallow ground water systems, in general, are not in chemical equilibrium [23], Eh values measured in the field are not used to calculate exact concentrations of specific dissolved constituents in the ground water.

For this report a nonparametric statistical technique, the Spearman rank correlation, was used to correlate the concentration of inorganic chemical constituents with activities of radium-226 and uranium [24]. The minimum detection limit (MDL) of radium-226 and uranium was 0.10 pCi/L and 0.1 µg/L, respectively. When radium-226 or uranium was reported as less than the MDL, the value of the MDL was assigned to these constituents for the purpose of statistical calculation and graphical representation. The Spearman rank correlations significant at the 0.05 level are listed in table 2. In addition, Spearman correlation coefficients are listed for barium, bicarbonate, strontium, total dissolved solids, and pH.

In table 2, where the Spearman correlation coefficient is positive, the concentration of the chemical constituents indicated will tend to increase with increasing radionuclide activity. Conversely, where the Spearman correlation coefficient is negative, the concentration of the chemical constituents indicated will tend to decrease with increasing radionuclide activity. Therefore, these constituents can be used as an indicator to the possible presence or absence of elevated levels of the radionuclide. Radium-226 has a positive correlation with gross alpha- and gross beta-particle activity, iron, manganese, potassium, and silica, and possibly with barium, and bicarbonate concentrations. Radium-226 has a negative correlation with dissolved oxygen concentration, Eh, pH, and possibly with uranium concentration. Uranium has a positive correlation with gross alpha- and gross beta-particle activity, Eh, dissolved oxygen, and possibly strontium concentrations. Uranium has a negative correlation with iron and manganese. Abundant scatter in the data may partially be due to mixing and averaging of water from several different fracture zones within the well bore. This mixing may have reduced the values of some of the correlation coefficients.

Significant correlation alone does not indicate cause and effect relations between radionuclide activities and concentrations of other dissolved chemical constituents. Relations between radium-226 and uranium and individual chemical constituents are discussed below in terms of possible geochemical reactions that may control the presence of elevated activities of these radionuclides with high concentrations of the previously mentioned inorganic chemical constituents.

Table 2. Spearman Rank correlations for gross alpha, gross beta, total dissolved solids, pH, Eh, and selected inorganic chemical species with radium-226 and uranium at the 0.05-significance level. Where the significance level is greater than 0.05, it is listed next to the correlation coefficient in parentheses.

Radium-226 (92 samples)		Uranium (49 samples)	
Chemical Species	Correlation Coefficient	Chemical Species	Correlation Coefficient
Gross Alpha	.35	Gross Alpha	0.51
Gross Beta	.44	Gross Beta	.40
Uranium	-.22 (0.12)	Radium 226	-.22 (0.12)
pH	-.17	pH	-.02 (0.88)
Eh	-.38	Eh	.31
Dissolved Oxygen	-.35	Dissolved Oxygen	.37
Total Dissolved Solids	.05	Total Dissolved Solids	-.07
Iron	.45	Iron	-.49
Manganese	.53	Manganese	-.32
Potassium	.37	Strontium	.26
Silica	.24		
Barium	.18 (0.12)		
Bicarbonate	.12 (0.12)		

Dissolved Oxygen and Eh

Dissolved oxygen concentrations and Eh (oxidation-reduction potential) are important controls on the distribution of radium-226 and uranium in ground water. In areas with elevated activities of gross alpha in ground water, such as near the geologic contact of the Lockatong Formation and Passaic Formation [9], elevated radium-226 and uranium activities are present but usually are not present in elevated concentrations in the same ground water. Dissolved-oxygen concentrations in ground water are shown in figure 4. Ground water with dissolved oxygen concentrations of less than 1 mg/L are considered relatively reducing and correspond with areas of elevated radium-226 activities in the study area (fig. 2). Uranium activities (fig. 3) increase significantly only in areas where the dissolved-oxygen content of ground water is above 1 mg/L (fig. 4).

The contrasting geochemical behavior of radium-226 and uranium in the presence or absence of dissolved oxygen is shown in figure 5. Generally, radium-226 activities only exceed 1 pCi/L in ground water that contains less than 1 mg/L dissolved oxygen. Conversely, uranium concentrations are less than 1 µg/L most often in ground water that contains less than 1 mg/L of dissolved oxygen. In general, ground water samples determined to be oxidizing or reducing by measurement of dissolved oxygen content corresponded to those determined to be oxidizing or reducing using Eh measurements. The contrasting geochemical behavior of radium-226 and uranium are also evident with respect to Eh (fig. 6a and 6b). Radium-226 activities in ground water increase as Eh decreases, whereas uranium concentrations increase in the ground water as Eh increases.

Iron and Manganese

Radium-226 activities increase with iron (fig. 6c) and manganese (table 2) concentrations in ground water in the Newark Basin. Radium-226 activities above 1 pCi/L are accompanied by iron concentrations as high as 11,000 µg/L and manganese concentrations as high as 1,600 µg/L. In addition, elevated concentrations of iron and/or manganese are present in almost all ground water samples that contain elevated radium-226 activities. Iron and manganese hydroxides, which are abundant in the rocks of the Newark Basin [6], may exert significant control on radium-226 activity in the ground water.

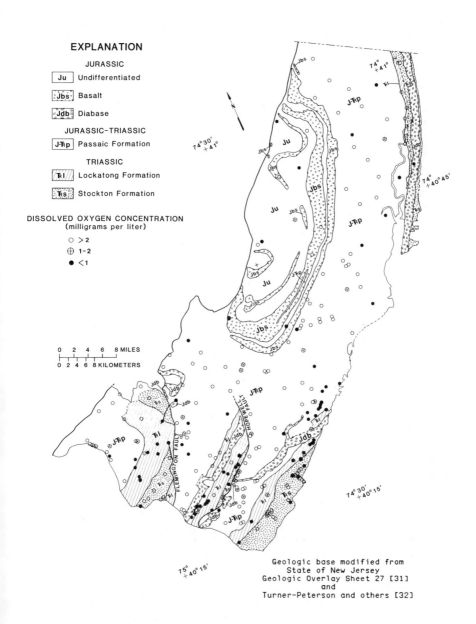

Figure 4. Dissolved oxygen concentrations in ground water, Newark Basin, New Jersey.

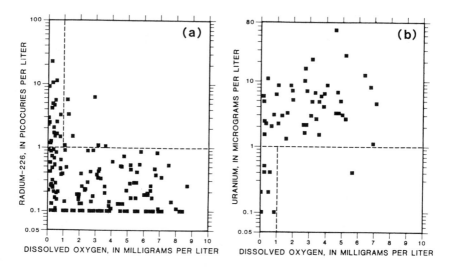

Figure 5. Graphs of (a) radium-226 activities, and
(b) uranium concentrations versus dissolved
oxygen concentrations in ground water,
Newark Basin, New Jersey. Vertical dashed
lines indicate dissolved oxygen concentration
of 1 milligram per liter. Horizontal dashed
lines indicate: (a) radium-226 activity of
1 picocurie per liter, and (b) uranium
concentration of 1 microgram per liter.

It is well known that iron and manganese hydroxides
adsorb radium very strongly [18,19,25]. Radium-226 is
produced by decay from its parent radionuclide,
thorium-230.   Thorium-230 decays to radium-226 by
ejecting an alpha-particle. The momentum of the
ejected alpha-particle causes the newly created radium-
226 radionuclide to recoil in the direction opposite to
the one in which the alpha-particle departed. This
mechanism is known as alpha-recoil. Although the
distance a radium-226 radionuclide travels due to
alpha-recoil is very small [26], a small fraction of
radium-226 at the edge of a host mineral grain may be
ejected from the host into an adjoining pore space or
into a microfracture within the mineral grain [26]. In
this way, radium-226 is continuously released to ground
water contained in the pore space. However, because
radium is so strongly adsorbed by iron and manganese
hydroxides, a radium-226 radionuclide ejected into a
pore space (or microfracture) would quickly be adsorbed
onto the iron or manganese hydroxides.

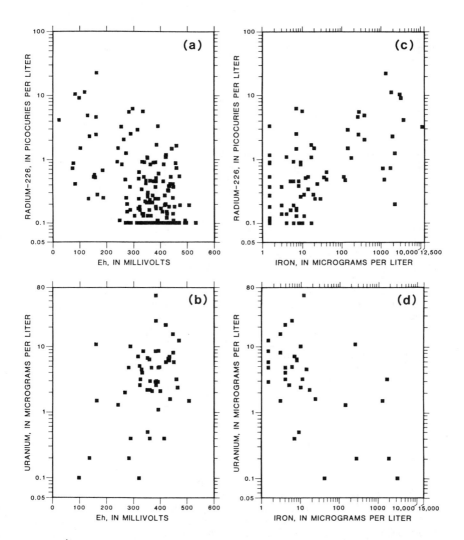

Figure 6. Graphs of (a) radium-226 activities,
and (b) uranium concentrations versus
Eh and (c) radium-226 activities and
(d) uranium concentrations versus
iron in ground water, Newark Basin,
New Jersey.

Iron and manganese hydroxides are soluble where ground water pH values decrease below 7.0 and Eh values decrease below approximately +200 mV [25].   In such waters, iron and manganese hydroxide coatings are less abundant on mineral grains or along microfractures, because they are unstable in such environments and would tend to dissolve.   Therefore, radium-226 released by alpha-recoil will have a smaller likelihood of being adsorbed onto iron and manganese hydroxides in such waters, and will have a greater likelihood of migrating into larger pore spaces and major fractures. Concentrations of iron, manganese, and radium-226 increase together in the ground water with decreasing dissolved oxygen concentrations, decreasing Eh, and decreasing pH.   These relations are observed throughout the study area.

The negative correlation between uranium and iron (fig. 6d) and manganese (table 2), reflects the fact that in oxidizing water, uranium is present in the very soluble +6 valence state [27,28]; whereas iron and manganese are present in the essentially insoluble +3 and +4 valence states, respectively [25].   Conversely, in reducing water, uranium is present in the very insoluble +4 valence state, whereas iron and manganese each are present in the soluble +2 valence state. Uranium is present predominantly as amorphous uranium oxides [6], the solubility of which probably controls the abundance of uranium in ground water.

Sulfate and Barium

Based on completed chemical analyses, most of the ground water in the study area is of a calcium-magnesium-bicarbonate type, but some calcium sulfate-type water is present (figs. 7 and 8).   Concentrations of sulfate greater than 100 mg/L were measured in this study, and have been reported by others [20,22].   Water samples containing radium-226 activities above 1 pCi/L are almost exclusively bicarbonate-type waters (fig. 7), whereas samples containing uranium activities greater than 4 pCi/L encompass both these classes of water (fig. 8).   The absence of significant radium-226 activities from calcium sulfate-type water may be accounted for by the precipitation of sulfate minerals. Although radium sulfate is essentially insoluble [29], it is not abundant enough in nature to precipitate. However, radium co-precipitates with barium sulfate (barite) [29,16,13].   Sulfate minerals are reported in the rocks of the Newark Basin [21].

In the Newark Basin, as radium-226 activities increase above 1 pCi/L in ground water, corresponding barium concentrations are as high as 1,300 μg/L (fig. 9a). However, several ground water samples with radium-226 activities above 1 pCi/L do not contain any barium. This fact, along with the fact that scatter is produced in the data by mixing of waters in the well from different fracture zones, may explain why barium concentration does not correlate with radium-226 activity at the 0.05 significance level.

Thermodynamic calculations indicate that sulfate is reduced to sulfide in ground water at a pH of 7.0 and an Eh value of approximately -100 mV [25]. Radium-226 ejected into the ground water from the aquifer minerals or from amorphous uranium oxides by alpha-recoil would not be removed by barite precipitation. This would allow radium-226 activities in the ground water to increase; with possible dissolution of barite from the aquifer material, barium concentrations also would increase. Sulfide odor was noted at several well sites in this study where low Eh values (+20 to +90 mV) were measured and high concentrations of barium and radium-226 were reported from the ground water. Gilkeson and others [30] report elevated barium and sulfide concentrations with elevated radium-226 activities in ground water from Illinois.

Potassium, Silica, and Strontium

Correlation of radium-226 activities with potassium and silica concentrations may represent desorption reactions by these three constituents from clay minerals in the aquifer material. Clay minerals strongly adsorb radium-226, especially in alkaline waters [25] and therefore, may also control radium-226 activities in ground water. Determination of the types of clay minerals in the aquifer matrix is needed to improve understanding the role of clay minerals in controlling radium-226 activity, and potassium and silica concentrations in ground water.

Uranium concentrations show a positive correlation with strontium concentrations (table 2). Strontium concentrations up to 5,100 μg/L were observed accompanying uranium activities in excess of 4 pCi/L (fig. 9b). The significance of this relation is difficult to interpret, because strontium is a common constituent of most minerals that are present in Newark Basin rocks, including clays, carbonates, sulfates, and feldspars.

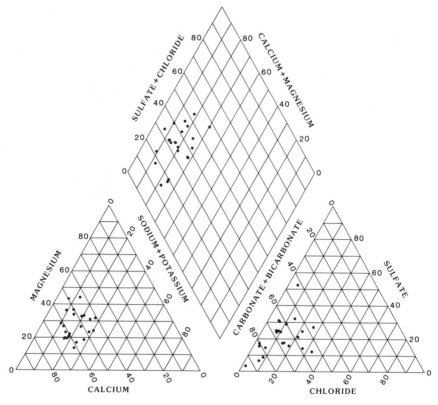

CATIONS   Percent milliequivalents per liter   ANIONS

Figure 7. Piper diagram of ground water having
          radium-226 activities in excess
          of 1 picocurie per liter, Newark Basin
          New Jersey.

CONCLUSIONS

    Approximately 5 percent of 260 ground-water samples
collected for this study in the Newark Basin, New
Jersey exceeded USEPA required or recommended MCL for
either gross alpha-particle, radium-226, and uranium
activity.   Elevated activities of radium-226 and
uranium were found predominantly in ground water near
the geologic contact of the Lockatong Formation and
Passaic Formation.   Elevated activities of radium-226
and uranium also were found locally in the Stockton
Formation.   Most of the ground water sampled in this
study contain levels of natural radioactivity well
below USEPA limits.

Several chemical constituents may serve as indicators of the presence of elevated uranium or radium-226 in ground water. Where gross alpha-particle activity in reducing waters was high, radium-226 activity was elevated; however, where gross alpha-particle activity in oxidizing waters was high, uranium activity was elevated. With a few exceptions, elevated uranium and radium-226 activities were not present concurrently in ground water samples. Radon-222 activity shows some correlation with radium-226

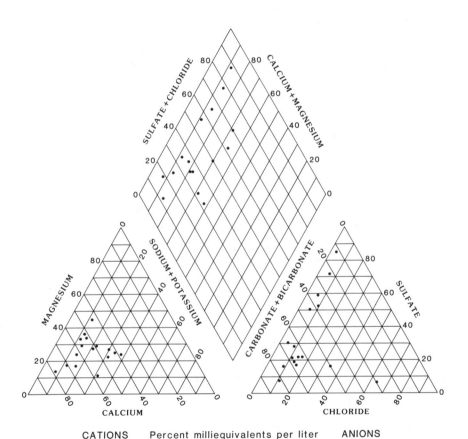

CATIONS   Percent milliequivalents per liter   ANIONS

Figure 8. Piper diagram of ground water having
          uranium activities in excess of
          4 picocuries per liter, Newark Basin,
          New Jersey.

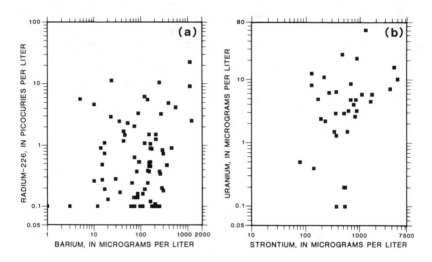

Figure 9. Graphs of (a) radium-226 activities
vs barium concentrations and (b) uranium
vs strontium concentrations in ground water,
Newark Basin, New Jersey.

activity. However, water from some of the wells with
the highest radium-226 activities contained radon-222
activities that fell below the median value. Elevated
radium-226 activity is associated with elevated
concentrations of iron, manganese, and barium. Radium-
226 also correlates positively with potassium and
silica. Radium-226 activities above 1 pCi/L were not
found in waters where the major anion is sulfate.

Uranium is negatively correlated to iron and
manganese. Positive correlation with strontium is
noted, but its significance is difficult to interpret.

Radium-226 activities in the ground water of the
Newark Basin are controlled by the simultaneous
occurrence of at least three important geochemical
processes: (1) adsorption onto iron and manganese
hydroxides, (2) adsorption onto clay minerals, and (3)
precipitation as sulfate minerals. These reactions, in
turn, may be controlled by the pH and Eh of the ground
water. Uranium concentrations may be controlled by the
solubility of the +6 uranium ion.

Additional research is needed to define the
distribution and concentration of uranium and radium in
varying mineral phases in the rock matrix in the
different formations in the Newark Basin. The
importance of individual geochemical processes
controlling the levels of these radionuclides in ground
water needs to be evaluated by using thermodynamic
calculations; elevated levels of radionuclides in
ground water can then be estimated more accurately.

REFERENCES

1. Mays, C. W., R. E. Rowland and A. P. Stehney.
    "Cancer risk from the lifetime intake of Ra and U
    isotopes," Health Physics 48(5):635-647 (1985).

2. "National interim primary drinking water
    regulations," Washington DC, Office of Water
    Supply, U.S. Environmental Protection Agency Report
    EPA-570/9-76-003 (1976).

3. Cothern, R. C., W. L. Lappenbush and J. A. Cotruvo.
    "Health effects guidance for uranium in drinking
    water," Health Physics 44(1):377-384 (1983).

4. Cross, F. T., N. H. Harley and W. Hoffman. "Health
    effects and risks from Rn-222 in drinking water,"
    Health Physics 48(5):649-670 (1985).

5. Bell, C. "Radioactive mineral occurrences in New
    Jersey," New Jersey Geological Survey Open-File
    Report 83-5 (1983), p.21.

6. Turner-Peterson, C. E. "Sedimentology and uranium
    mineralization in the Triassic-Jurassic Newark
    Basin, Pennsylvania and New Jersey," in Uranium in
    sedimentary rocks: Application of the facies
    concept to exploration, Soc. Economic
    Paleontologists and Mineralogists Short Course
    Notes, Denver, CO (1980), pp. 149-175.

7. Heffner, J. D. "Newark 1°x2° Area New Jersey,
    New York, and Pennsylvania, Data Report, National
    Uranium Resource Evaluation Program,
    Hydrogeochemical and stream sediment
    reconnaissance," Grand Junction, CO, U.S. Dept.
    Energy Report DPST-70-146-9 (1980).

8. Saroya, S. New Jersey Dept. Environmental
    Protection. Personal communication (1984).

9. Zapecza, O. S. and Zoltan Szabo. "Source and
    distribution of natural radioactivity in ground
    water in the Newark Basin, New Jersey," in
    Proceedings of Radon, Radium and Other
    Radioactivity in Ground Water Conference:
    Hydrogeologic Impact and Application to Indoor
    Airborne Contamination.

10. Claassen, H. C. "Guidelines and techniques for
    obtaining water samples that accurately represent
    the water chemistry of an aquifer," U.S. Geological
    Survey Open-File Report 82-1024 (1982), p. 49.

11. Vecchiolli, J., L. D. Carswell and H. F. Kasabach. "Occurrence and movement of ground water in the Brunswick Shale at a site near Trenton, New Jersey," U.S. Geological Survey Professional Paper 650-B (1969), pp. 154-157.

12. Wood, W. W. "Guidelines for collection and field analysis of ground-water samples for selected unstable constituents," U.S. Geological Survey Techniques of Water-Resources Investigations, Book 1, Chapter D2 (1976), p. 24.

13. Thatcher, L. L., V. J. Janzer and K. W. Edwards. "Methods for determination of radioactive substances in water and fluvial sediments," U.S. Geological Survey Techniques of Water-Resources Investigations, Book 5, Chapter A5 (1977), p. 95.

14. Fishman, M. J., and L. C. Friedman. "Methods for determination of inorganic substances in water and fluvial sediments," U.S. Geological Survey Open-File Report 85-495 (1985), p. 709.

15. "Radon in water sampling program," USEPA EPA/EERF-Manual-78-1 (1978), p. 11.

16. Krieger, H. L., and E. L. Whittaker. "Prescribed procedures for measurement of radioactivity in drinking water," USEPA Manual EPA-600/4-80-032 (1980), p. 111.

17. Anderson, S. B. "Levels of Ra-226 and Rn-222 in well water of Mercer County, New Jersey," Senior Thesis, Princeton University, Princeton, N.J. (1983), p. 59.

18. Tanner, A. B. "Physical and chemical controls on distribution of radium-226 and radon-222 in ground water near Great Salt Lake, Utah," in The Natural Radiation Environment, J. A. S. Adams and W. M. Lowder, Eds. (Chicago, IL: Univ. Chicago Press, 1964), pp. 253-278.

19. Korner, L. A., and A. W. Rose. "Rn in streams and ground waters of Pennsylvania as a guide to uranium deposits," U.S. Energy Research and Development Assoc., Grand Junction, CO, Open-File Report GJBX-60(77) (1977), p. 151.

20. Nemickas, B. "Geology and ground-water resources of Union County, New Jersey" U.S. Geological Survey Water-Resources Investigations 76-73 (1976), p. 103.

21. Herpers, H., and H. C. Barksdale. "Preliminary report on the geology and ground-water supply of the Newark, New Jersey area," New Jersey Dept. Conservation and Economic Development Div. Water Policy and Supply Special Report 10 (1951), p. 52.

22. Longwill, S. M., and C. R. Wood. "Ground-water resources of the Brunswick Formation in Montgomery and Berks Counties, Pennsylvania," Pennsylvania Geologic Survey Ground Water Report 22 (1965), p. 59.

23. Lindburg, R.D., and D. Runnels. "Ground water redox reactions: An analysis of equilibrium state applied to Eh measurements and geochemical modeling, "Science 225:925-927 (1984).

24. Conover, W.J. Practical nonparametric statistics, 2nd ed. (New York, NY: John Wiley & Sons, Inc., 1980), p. 493.

25. Hem, J.D. "Study and interpretation of the chemical characteristics of natural water," U.S. Geological Survey Water-Supply Paper 2254 (1985), p. 264.

26. Rama, and W. S. Moore. "Mechanism of transport of U-Th series radioisotopes from solids into groundwater," Geochimica Cosmochimica Acta 48:395-399 (1984).

27. Hostetler, P. B., and R. M. Garrels. "Transportation and precipitation of uranium and vanadium at low temperatures, with special reference to sandstone-type uranium deposits," Economic Geology 57:137-167 (1962).

28. Langmuir, Donald. "Uranium solution-mineral equilibria at low temperatures with applications to sedimentary ore deposits," Geochimica Cosmochimica Acta 42:547-569 (1978).

29. Langmuir, D. and A. C. Riese. "The thermodynamic properties of radium," Geochimica Cosmochimica Acta 49:1593-1601 (1985).

30. Gilkeson, R. H., K. Cartwright, J. B. Cowart and R. B. Holtzman. "Hydrogeologic and geochemical studies of selected natural radioisotopes and barium in ground water in Illinois," Bureau Reclamation, U.S. Dept Interior, Final Technical Completion Report Proj. No. B-108-ILL (1983), p. 93.

31. New Jersey State Geologic Overlay Sheet 27.

32. Turner-Peterson, C. E., P. E. Olsen and V. F. Nuccio. "Modes of uranium occurrence in the Newark Basin, New Jersey and Pennsylvania," in "Proceedings of the Second U.S. Geological Workshop on the Early Mesezoic Basins of the Eastern United States", G. R. Robinson, Jr. and A. J. Froelich, Editors, U.S. Geologic Survey Circular 946 (1985), pp. 120-124.

Session V: Radon and Radium in Water Supply Wells

Moderators: Michael Weber, U.S. Nuclear Regulatory Commission

James B. Cowart, Florida State University

Kevin L. Dixon and Ramon G. Lee,
American Water Works Service Co., Marlton, N.J.

ABSTRACT

An analytical survey designed to determine the distribution and concentration of Radon in more than 370 wells utilized as water supply sources for 49 companies within the American Water Works Company system was conducted in 1986 and 1987. The survey encompassed well supplies located in 15 states across the continental U.S. Selected wells were monitored to assess the variation in Radon levels with pumpage. Additionally, samples were collected to assess the removal efficiency of selected water treatment processes, i.e. GAC adsorption, aeration (packed tower and tray aeration). A limited number of samples were also collected to monitor the fate of Radon during storage and transit within distribution systems.

INTRODUCTION

The American Water Works System is comprised of over 100 separate water supply systems serving 5 million consumers in 21 states.

Individual operating companies within the American Water Works System are grouped into six geographical regions. These regions and the 15 states in which operating companies that participated in the Radon Survey are located, are listed in Table 1. The results of the survey will be presented according to this regional grouping.

Table 1.   Regional Organization of the American Water
           Works Company and Corresponding List of State
           Locations for Operating Companies
           Participating in the Radon Survey

Geographical Region	Operating Company State Location
New England	Massachusetts
	New Hampshire
	Rhode Island
Eastern	Connecticut
	New Jersey
Pennsylvania	Pennsylvania
Southern	Virginia
	West Virginia
Mid-America	Ohio
	Indiana
	Illinois
	Iowa
Western	New Mexico
	Arizona
	California

The operating companies participating in the survey
utilize varying percentages of groundwater to meet
their total system demand.   Some companies utilize
groundwater to meet as little as 1 percent of their
distribution demands while other companies are entirely
dependent upon groundwater to supply their customers.
Systemwide, groundwater is drawn from approximately 500
wells and constitutes approximately 20 percent of the
total amount of water provided to customers of the
American Water Works Company. This equates to over 50
billion gallons of groundwater being pumped each year.
     It is immediately evident that the forthcoming
drinking water regulations governing Radon could
significantly impact the American System. This fact,
coupled with the need to provide a timely response to
the concern felt by our customers as a result of
statements issued to the national news media by the
U.S. EPA Office of Drinking Water in August 1986
relative to the health risks associated with Radon,
prompted the initiation of a systemwide analytical
survey of American System well supplies.
     In September, 1986, the System Water Quality
Department of the American Water Works Service Company
began conducting a Radon Survey which was designed to
yield data and information on: (1) the occurrence,

distribution and magnitude of Radon levels in American well supplies, and (2) the impact of operational factors that could be developed to deal with Radon in the groundwater supplies.

SURVEY OBJECTIVES

Following were the major objectives of the American Water Works Service Company Radon Survey:

1. analytical determination of the levels of Radon in American System groundwater supplies.

2. provision of information to American System operating companies and community relations people for development of factual responses to inquiries from the concerned public.

3. assessment of the Radon removal efficiency of existing treatment processes within the American System such as GAC and various forms of aeration.

4. assessment of the effects of pumping time, blending, and storage on Radon concentrations.

SURVEY DESIGN

In order to fulfill the listed objectives, the survey was designed to incorporate a two phase approach. Phase one, henceforth designated as the Occurrence Phase, was designed to result in the collection of replicate samples of groundwater from each of the wells included in the survey. Sample kits, supplied by the University of Maine, containing the equipment necessary to collect replicate samples from each well, were mailed to designated personnel experienced in the technique of collecting samples for VOC analysis. It was felt that a familiarity with a somewhat elaborate sample collection technique (such as for VOC analysis) would lead to properly-collected Radon samples, and thus, the most accurate analytical results possible.

Sample collection personnel were instructed to ensure that: the water samples collected were representative of fresh formation water (not stagnant casing water); the sample information section of the instruction sheet which was included in the sample kit, was filled out immediately after sample collection; and, the samples were mailed to the University of Maine no later than the Wednesday of the sample collection week in order to minimize the time of decay prior to analysis. This was found to be particularly important for supplies with low levels of Radon.

Phase two of the survey, henceforth designated as the Operational Phase, was designed to result in the collection of selected samples which would yield data and information relative to each of the following operational/water quality areas of concern.

Pumping Time

The concentration of Radon in a given well supply may be significantly impacted by the length of time that water is drawn from the well. Variation in concentration as a function of pumping time may be attributable to aquifer yield capacity, the hydrogeologic nature of the aquifer strata, aquifer material geo-chemistry, and the magnitude and configuration of the cone of depression resulting from the pumpage. To assess the effect of pumping time on Radon concentrations, a series of Radon samples were collected at selected time intervals during a 5 day pumping test at a well supply known to contain a significant level of Radon.

Blending

A significant number of American System operating companies rely on both surface water and groundwater as their source of supply. Blending of the surface water, which is presumed to have very low levels of Radon, with Radon-laden groundwater, offers an operating company an operational mechanism to utilize well water with elevated levels of Radon and maintain compliance with future Radon standards, simply by virtue of dilution. To verify the dilution effect of blending, a series of samples were collected at a selected site prior to and following the blending of treated surface water with groundwater known to contain a substantial level of Radon.

Distribution System Storage

Radon has a short half-life of 3.8 days. Thus, water stored in a vessel as part of a water supply distribution system, would be expected to contain less Radon after a given period of storage than at the time the water was introduced into the storage vessel. Additionally, there will be a release of Radon to the air space in the tank.

Depending upon the length of the storage period and the amount of ventilation associated with the storage vessel, a substantial decrease in the Radon concentration could result. Modification of storage patterns to the extent possible, may offer an operational tool to deal with water supplies containing elevated levels of Radon.

In order to document the degree of effect of storage and ventilation on Radon levels, a series of samples were collected from a storage vessel within the distribution system of an operating company.

Treatment

American System operating companies employ diverse methods of treatment to provide potable water to their customers. A series of samples (generally influent and effluent samples) were collected at seven sites employing the following methods of treatment of groundwater, to assess the resulting effect on Radon levels:

    granular activated carbon
    packed tower aeration

In assessing the Radon removal efficiency of various treatment methods, an evaluation of the treatment options available to a given operating company (based on the levels of Radon in their well supplies) becomes possible. This is important from the standpoint of achieving compliance with future radionuclide regulations while concurrently keeping capital/operational costs at a minimum.

As was the case with the Occurrence Phase of the survey: samples were collected by personnel familiar with intricate sampling techniques; personnel were instructed to collect samples representative of fresh formation water when sampling at the well head; sample instruction/information sheets, specific to the type of samples being collected, were provided to collection personnel; and, samplers were instructed to mail the samples to the University of Maine no later than 1 day after collection.

EXPERIMENTAL

All samples were collected in the following manner. A 10 ml sample of water, representative of a selected water source, was drawn into a plastic syringe, without aeration. The sample was injected into a 22 mL borosilicate glass scintillation vial containing 5 mL of a mineral oil/scintillation fluor. Injection of the sample was performed beneath the surface of the scintillation fluor. The vial was sealed tightly with a poly-cone insert cap. A second sample was then collected in the same manner, to provide a replicate-oriented data base. Important information about the water sample (time and date of collection, name and location of well, etc.) was then entered onto a sample collection instruction/information sheet.

The vials were mailed to the laboratory at the University of Maine, where the replicate samples were

counted on a Hewlett-Packard Model 3255 scintillation
counter, utilizing a modification of the liquid scintillation procedure described by Pritchard and Gesell
(1).

Occurrence Phase samples were counted for a period
of 20 minutes while Operational Phase samples were
counted for 60 minutes (to minimize and better define
the counting error associated with the method). The
Radon concentration, as well associated error, (a
two-sigma confidence interval) was reported in
picocuries per liter (pCi/L).

Based upon extensive discussion with Jerry D.
Lowry, associate professor of civil engineering at the
University of Maine, it was deemed that all samples
containing 0 to 99 pCi/L of Radon be reported as <100
pCi/L, due to a number of factors relative to the
method of sample collection, sample shipment time and
the analytical technique utilized (2). For purposes of
discussion, 100 pCi/L is considered to be the quantification limit (QL) for the Radon Survey results.

The results presented in this report represent an
arithmetic mean value of the concentration of Radon
detected in the set of replicate samples collected at
each well.

OCCURRENCE OF RADON IN AMERICAN SYSTEM WELL SUPPLIES

The occurrence phase of the Radon Survey resulted
in the collection of samples from 377 well supplies in
15 States. Table 2 presents a summary of the survey
results on a regional basis.

The range of detected Radon concentrations throughout the American System was <100 pCi/L to 4622 pCi/L.
The systemwide arithmetic mean Radon concentration was
686 pCi/L +/- 104 pCi/L (associated error). The Median
Radon concentration was 320 pCi/L. Table 3 presents a
summary of the statistical analysis of the data, on a
regional basis.

Table 4 presents a summary of the statistical
analysis of the data on a state-by-state basis.

Geographically, the survey revealed the highest
levels of Radon in individual wells in the states of
NH, MA, RI, NJ, PA, CT, and CA.

Following is a detailed presentation of the Occurrence Phase data, on a regional/state basis. Frequent
reference will be made to Figure 1 and Table 5, in an
effort to relate the detected Radon levels to the
predominant geological patterns of each state (3).
Frequent reference is also made to a U.S. EPA map (not
included in this report) outlining areas of the United
States susceptible to potentially high Radon levels,
based on the distribution of known/potential uranium
deposits(4).

Table 2.   Regional Summary of Results of the Radon Survey.

Region	Number of Wells Sampled	Range of Detected Radon Concentrations (pCi/L)						
		<100	100-499	500-999	1000-1999	2000-2999	3000-3999	4000-5000
New England	43	1	3	14	20	1	3	1
Eastern	116	55	35	11	7	7	1	0
Pennsylvania	64	10	12	8	9	14	7	4
Southern	9	4	5	0	0	0	0	0
Mid-America	60	10	41	9	0	0	0	0
Western	85	8	47	24	5	1	0	0
Totals	377	88	143	66	41	23	11	5
Percent of Total Wells Sampled	23.3%	23.3%	37.9%	17.6%	10.9%	6.1%	2.9%	1.3%

Table 3.   Regional Summary of Detected Radon Concentrations and the Corresponding
Mean Radon Levels from the Radon Survey.

Region	Number of Wells Sampled	Range of Detected Radon Concentrations (pCi/L)	Mean Rn Concentration (pCi/L)	Associated Error (+/-pCi/L)
New England	43	61–4609	1250	112
Eastern	116	<100–3805	405	72
Pennsylvania	64	<100–4622	1570	89
Southern	9	<100–468	176	35
Mid-America	60	<100–714	297	99
Western	85	<100–2003	446	166

Table 4. Summary of the Range of Detected Radon Concentrations and the Corresponding Mean Radon Levels for States Participating in the Radon Survey.

State	Number of Wells Sampled	Range of Detected Rn Levels (pCi/L)	Mean Rn Conc. (pCi/L)	Associated Error (+/-pCi/L)
AZ	5	434 – 681	582	105
CA	44	<100 – 2003	589	91
CT	3	757 – 984	841	98
IA	6	911 – 100	12	75
IL	16	182 – 714	449	115
IN	28	<100 – 624	324	106
MA	28	<100 – 3288	1145	101
NH	12	880 – 4609	1716	134
NJ	113	<100 – 3805	394	79
NM	36	<100 – 678	253	266
OH	10	<100 – 343	148	116
PA	64	<100 – 4622	1570	89
RI	3	640 – 787	702	61
VA	2	465 – 468	467	53
WV	7	<100 – 281	93	42

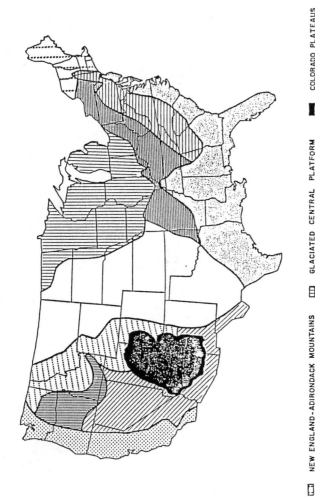

NEW ENGLAND-ADIRONDACK MOUNTAINS

APPALACHIAN HIGHLANDS - PIEDMONT

APPALACHIAN AND INTERIOR PLATEAUS

COASTAL PLAIN

GLACIATED CENTRAL PLATFORM

WESTERN CENTRAL PLATFORM

ROCKY MOUNTAIN SYSTEM

COLORADO PLATEAUS

BASIN AND RANGE

COLUMBIA PLATEAUS

PACIFIC MOUNTAIN SYSTEM

Figure 1. Map of geological provinces of the United States according to Beddinger

Table 5.   Summary of Potential Host Rocks and Geologic
           Framework in the Provinces of the Contiguous
           United States as Applied to States Partici-
           pating in the Radon Survey.

Province	Geological Framework
New England Adirondack Mountains	New England--complexly faulted metamorphic and metasedimentary rocks intruded by large masses of granite. Adirondacks--mountains composed of marble and schist intruded by granites, anorthosite, and gabbro.
Appalachian Highlands- Piedmont	Appalachian Highlands--mountain belt of granites and metamorphics thrust westward over Paleozic rocks. Piedmont-non-mountainous belt of highly complex metamorphic rocks with abundant granites.
Appalachian & Interior Plateaus	Gently dipping, gently folded, sandstones, shales, carbonates, and evaporites.
Coastal Plain	Seward dipping thickening wedge of sand, sandstones and shales with some evaporites and limestones; underlain by a basement of metamorphic rocks.
Glaciated Central Platform	Igneous and metamorphic rocks on the northwest overlain by sandstones, carbonates, shales, and evaporites.
Western Central Platform	Horizontal to gently dipping sandstones, deep sedimentary basins. Capped with sands and gravels.
Colorado Plateaus	Flat-lying to gently warped layers of sandstones, shales, limestones and evaporites.
Pacific Mountain System	Large uplifted and tilted blocks of granite with inliers of metasediments; folded and faulted folded and faulted sedimentary rocks; deep elongate troughs filled with fluvial sediments.

New England Region

The range of Radon levels in the 43 wells tested in the New England Region was <100 pCi/L to 4609 pCi/L. The average Radon concentration for the region, which for survey purposes consisted of the states of MA, NH and RI, was 1250 pCi/L +/- 1122 pCi/L (associated error). Table 6 presents a summary of the data from the region for states included in the survey.

Figure 1 and Table 5 indicate that the well supplies for the New England Region companies are located in the New England-Adirondack Mountain Province. This area exhibits a geological framework which is intruded by large masses of granite(3). Granitic formations are often associated with the presence of Uranium deposits and could thus be expected to yield higher than normal Radon levels in the groundwater. The U.S. EPA map of potentially high Radon areas tends to bear this out, indicating that Southwestern New Hampshire and Eastern Massachusetts, where the affected wells are located, display granitic rock formations with 4 to 9 ppm Uranium. Eastern RI exhibits a near-surface distribution of potential Uranium resources(4).

The geological information presented here seems to correlate well with the levels of Radon detected in the well supplies of these areas.

Radon concentrations in the 28 wells samples in Eastern MA ranged from <100 pCi/L to 3288 pCi/L. The mean Radon concentrations was 1145 pCi/L +/- 101 pCi/L (associated error).

Radon levels ranged from 880 pCi/L to 4609 pCi/L in the samples drawn from 12 wells in Southeastern NH. The mean Radon concentration detected in these wells was 176 pCi/L +/- 134 pCi/L.

In the 3 wells sampled from Eastern RI, Radon concentrations ranged from 640 pCi/L to 787 pCi/L, with a mean concentration of 702 pCi/L +/- 61 pCi/L.

Eastern Region

The levels of Radon detected in the 116 wells tested in the Eastern Region ranged from <100 pCi/L to 3805 pCi/L. The mean Radon concentration for the region, which for purposes of the survey consisted of the states of CT and NJ, was 405 pCi/L +/- 72 pCi/L. Of the 116 wells tested, 55 (48%) had Radon concentrations below the QL of 100 pCi/L.

Because of the variability in the predominant geological patterns within the region (particularly within N.J.) analysis of the data will be presented for the following geographical areas: southwestern CT, eastern-central to southeastern N.J., southwestern N.J., northeastern N.J., and northwestern N.J.

Table 6. Summary of Results of Radon Survey for the New England Region.

State	Number of Wells Sampled	Range of Detected Radon Concentrations (pCi/L)						
		<100	100-499	500-999	1000-1999	2000-2999	3000-3999	4000-5000
MA	28	1	3	10	10	1	3	0
NH	12	0	0	1	10	0	0	1
RI	3	0	0	3	0	0	0	0
Totals	43	1	3	14	20	1	3	1
Percent of Total Wells Sampled		2.3%	7%	32.6%	46.5%	2.3%	7%	2.3%

Table 7 presents a summary of the survey data from well supplies in the Eastern Region.

The occurrence patterns noted in CT and NJ seem to correlate well with the geological information presented in Figure 1 and Table 5 (3).

The well supplies located in southwestern CT as well as northeastern and northwestern N.J. are situated on or near the Appalachian Highlands Province, which is composed principally of granitic and metamorphic rocks (3). This area has a high potential for distribution of Uranium resources near the surface (4). In fact, the wells in northern N.J. are believed to fall within the boundaries of the highly publicized Reading Prong, thus accounting for the higher Radon levels which were detected.

Specifically, the Radon concentrations in the 3 wells tested in southwestern CT ranged from 757 pCi/L to 984 pCi/L with a mean concentration of 841 pCi/L +/- 98 pCi/L.

Radon levels in Northern N.J. ranged from 505 pCi/L to 3805 pCi/L for the 19 wells analyzed in the eastern part of the state and from 401 pCi/L to 2235 pCi/L in the 4 wells analyzed in the western part of N.J. Mean Radon concentrations for the northeastern and northwestern geographic areas were 1608 pCi/L +/- 95 pCi/L and 880 pCi/L +/- 130 pCi/L, respectively.

The well supplies included in the survey from eastern-central N.J. to southeastern N.J. are located within the boundaries of the Coastal Plain Province. This province is known for sand, sandstones, shales & limestones and is thought to have a low potential for the presence of Uranium deposits (3,4). This would tend to explain the low Radon concentrations detected in the 46 wells tested in this area. Radon levels ranged from <100 pCi/L to 264 pCi/L with a mean Radon concentration of 86 pCi/L +/- 80 pCi/L. Thirty (65%) of the wells had Radon concentrations below the QL of 100 pCi/L.

The southwestern N.J. well supplies included in the survey are distributed on the Coastal Plain Province and extend up to the border of the Appalachian Highlands Province (3). For the most part, these wells exhibit low Radon concentrations with the exception of two well supplies in the northern-most section of this area, which appear to fall within the influence of the Reading Prong. These wells had Radon concentrations of 1074 pCi/L +/- 93 pCi/L and 1020 pCi/L +/- 90 pCi/L. The remainder of the 42 wells in this area had Radon concentrations which ranged from <100 pCi/L to 351 pCi/L, with a mean Radon concentration of 104 pCi/L +/- 62 pCi/L. Twenty five (57%) of the wells tested in this area had Radon concentrations below the QL of 100 pCi/L.

Table 7. Summary of Results of Radon Survey for the Eastern Region.

State/Geographical Area	Number of Wells Sampled	Range of Detected Radon Concentrations (pCi/L)						
		<100	100–499	500–999	1000–1999	2000–2999	3000–3999	4000–5000
CT	3	0	0	3	0	0	0	0
NJ								
Eastern-Central to Southeastern	46	30	16	0	0	0	0	0
Southwestern	44	25	17	0	2	0	0	0
Northeastern	19	0	0	7	5	6	1	0
Northwestern	4	0	2	1	0	1	0	0
Totals	116	55	35	11	7	7	1	0
Percent of Total Wells Sampled		47.4%	30.2%	9.5%	6.0%	6.0%	0.9%	0

The mean Radon concentration for all well supplies in N.J. was 394 pCi/L +/- 79 pCi/L.

Pennsylvania Region

The concentrations of Radon detected in the 64 wells tested in the Pennsylvania Region ranged from <100 pCi/L to 4622 pCi/L. The mean Radon concentration for the region was 1570 pCi/L +/- 89 pCi/L.

Based upon predominant geological characteristics, the state is divided into two geographical areas, east and west.

Table 8 presents a summary of the survey data from well supplies in the Pennsylvania Region.

The occurrence patterns noted in PA appear to correlate well with the geological information presented in Figure 1 and Table 5 (3).

The well supplies located in eastern PA are situated within the Appalachian Highlands Province. This province is characterized by a mountainous belt of granitic and metamorphic rocks (3). The EPA map of potentially high Radon areas indicates that the wells in eastern PA are in an area with a near-surface distribution of potential Uranium resources (4). This area is known as the Reading Prong. The Radon levels detected in groundwater from this formation reflect the presence of the Uranium deposits.

In the 42 wells sampled in eastern PA, Radon levels ranged from <100 pCi/L to 4622 pCi/L with a mean Radon concentration of 2273 pCi/L +/- 56 pCi/L.

Well supplies in western PA are located in the Appalachian Plateau Province (Figure 1, Table 5), an area consisting of folded sandstones, shales, & carbonates(3). The potential distribution of Uranium deposits in this area is low and the levels of Radon detected in the groundwater bear this out (4). Radon concentrations in the 22 wells tested in western PA ranged from <100 pCi/L to 724 pCi/L, with a mean Radon concentration of 281 pCi/L +/- 76 pCi/L. Eight (38%) of the wells tested exhibited Radon concentrations below the QL of 100 pCi/L.

Southern Region

The range of Radon levels detected in the 9 wells tested in the Southern Region was <100 pCi/L to 468 pCi/L. The mean Radon concentration for the region, which for survey purposes consisted of the states of VA and WV, was 176 pCi/L +/- 35 pCi/L.

For purposes of discussion the state of WV is divided into three geographical areas: southern-central, southeastern and southwestern WV.

Table 9 presents a summary of the survey data for the Southern Region.

Table 8. Summary of Results of Radon Survey for the Pennsylvania Region.

State/Geographical Area	Number of Wells Sampled	Range of Detected Radon Concentrations (pCi/L)						
		<100	100-499	500-999	1000-1999	2000-2999	3000-3999	4000-5000
PA								
Western	22	8	8	6	0	0	0	0
Eastern	42	2	4	2	9	14	7	4
Totals	64	10	12	8	9	14	7	4
Percent of Total Wells Sampled		15.6%	18.8%	12.5%	14.1%	21.9%	10.9%	6.2%

Table 9. Summary of Results of the Radon Survey for the Southern Region.

State/Geographical Area	Number of Wells Sampled	Range of Detected Radon Concentrations (pCi/L)						
		<100	100-499	500-999	1000-1999	2000-2999	3000-3999	4000-5000
VA	2	0	2	0	0	0	0	0
WV								
Southern-Central	2	2	0	0	0	0	0	0
Southeastern	3	2	1	0	0	0	0	0
Southwestern	2	0	2	0	0	0	0	0
Totals	9	4	5	0	0	0	0	0
Percent of Total Wells Sampled		44.4%	55.6%	0%	0%	0%	0%	0%

The occurrence patterns noted in VA and WV appear to correlate reasonably well with the geological information presented in Figure 1 and Table 5.

The well supplies located in southeastern WV and northeastern VA are situated near the Appalachian Highlands-Piedmont Province, an area of abundant granitic rock formations (3). However both well systems border on adjacent provinces (southeastern WV-Appalachian Plateau; northeastern VA-Coastal Plain) not characterized by the presence of granitic rock. The EPA map of potentially high Radon areas indicates that these sites exhibit a low potential for distribution of Uranium deposits and should thus be expected to exhibit low concentrations of Radon in the groundwater (4). The analytical results of the survey bear this out. In the 2 well supplies of northeastern VA, Radon concentrations ranged from 465 to 468 pCi/L. In southeastern WV, Radon concentrations in the 3 wells supplies tested there, varied from <100 pCi/L to 105 pCi/L with a mean Radon concentration of 50 pCi/L +/-43 pCi/L, a level well below the QL of 100 pCi/L.

The well supplies of southern-central WV and southwestern WV are located within the province of the Appalachian Plateaus (Figure 1; Table 5), an area consisting of dipping and folded sandstones, shales & carbonates (3). This area exhibits no evidence of the distribution of Uranium in these locations (4). The Radon data reflect this as all levels detected in southern-central WV were <100 pCi/L and the levels in southwestern WV ranged from 222 pCi/L to 280 pCi/L with a mean concentration of 252 pCi/L +/- 39 pCi/L in the 2 wells tested.

Overall, the Southern Region exhibited the lowest mean Radon concentrations of the 6 regions within the American System.

Mid-America Region

The range of Radon levels detected in the 60 wells tested within the Mid-America Region was from <100 pCi/L to 714 pCi/L. The mean Radon concentration for the region, which for purposes of the survey consisted of well supplies in the states of Ohio, Indiana, Illinois and Iowa, was 297 pCi/L +/- 99 pCi/L. Approximately 17 percent (10 wells) exhibited Radon levels below the QL of 100 pCi/L.

Table 10 presents a summary of the survey data from the Mid-America Region.

The occurrence patterns noted throughout the Mid-America Region correlate reasonably well with the geological information presented in Figure 1 and Table 5. All of the well supplies included in the survey are located within the boundaries of the Glaciated Central

Table 10. Summary of Results of Radon Survey for the Mid-America Region.

State/Geographical Area	Number of Wells Sampled	Range of Detected Radon Concentrations (pCi/L)						
		<100	100-499	500-999	1000-1999	2000-2999	3000-3999	4000-5000
IA	6	6	0	0	0	0	0	0
IL	16	0	10	6	0	0	0	0
IN								
Northern-Central	13	0	12	1	0	0	0	0
Eastern-Central	8	0	6	2	0	0	0	0
Western-Central	7	1	6	0	0	0	0	0
OH	10	3	7	0	0	0	0	0
Totals	60	10	41	9	0	0	0	0
Percent of Total Wells Sampled		16.7%	68.3%	15.0%	0%	0%	0%	0%

Platform Province. This province is characterized by sandstones, carbonates, shales and evaporites (3).

EPA's map of potentially high Radon areas suggests that the wells in northern-central IN may be affected by high Radon levels due to the presence of significant uraniferous black shales in that area, however the analytical data do not support this. In the 13 wells tested in that area, Radon concentrations ranged from 136 pCi/L to 584 pCi/L with a mean concentration of 297 pCi/L +/- 117 pCi/L.

Well supplies in the eastern-central part of IN had Radon concentrations in the range of 295 pCi/L to 624 pCi/L while the western-central section of IN exhibited levels of <100 pCi/L to 343 pCi/L. The mean Radon concentration for each of these areas was 445 pCi/L +/- 97 pCi/L and 235 pCi/L +/- 102 pCi/L, respectively.

In the 10 well supplies of central OH, Radon levels ranged from <100 pCi/L to 343 pCi/L with a mean concentration of 148 pCi/L +/- 116 pCi/L.

The 16 wells tested in northern-central IL demonstrated Radon levels which ranged from 182 pCi/L to 714 pCi/L with a mean Rn concentration of 449 pCi/L +/- 115 pCi/L.

The 6 well supplies located in eastern-central IA all had Radon concentrations below the QL of 100 pCi/L.

Western Region

The range of Radon levels detected in the 85 wells sampled in the Western Region was from <100 pCi/L to 2003 pCi/L. The mean concentration for the region, which for survey purposes included the states of AZ, NM and CA, was 446 pCi/L +/- 166 pCi/L.

For purposes of discussion, the State of CA has been divided into the geographical areas of western-central CA and southern CA.

Table 11 presents a summary of the analytical data from the states participating in the survey.

The Radon occurrence patterns noted throughout the Western Region tend to deviate from the patterns predicted by the geological information presented in Figure 1 and Table 5.

The two geographical areas displaying deviation were for well supplies located in southern-central AZ and supplies located in the western-central CA.

The wells in southern-central AZ are located within the Colorado Plateau Province (Figure 1; Table 5). This area is characterized by warped layers of sandstones, shales and limestones (3). It is an area that has a low potential distribution of Uranium, however Radon concentrations in the 5 wells tested ranged from 434 pCi/L to 681 pCi/L, with a mean concentration of 582 pCi/L +/- 105 pCi/L.

Table 11. Summary of Results of Radon Survey for the Western Region.

State/Geographical Area	Number of Wells Sampled	Range of Detected Radon Concentrations (pCi/L)						
		0-99	100-499	500-999	1000-1999	2000-2999	3000-3999	4000-4999
NM	36	7	26	3	0	0	0	0
AZ	5	0	1	4	0	0	0	0
CA								
Western-Central	13	1	5	4	2	1	0	0
Southern	31	0	15	13	3	0	0	0
Totals	85	8	47	24	5	1	0	0
Percent of Total Wells Sampled	9.4%	9.4%	55.3%	28.2%	5.9%	1.2%	0%	0%

The 13 wells located in western-central CA exhibited Radon levels of <100 pCi/L to 2003 pCi/L with a mean Radon concentration of 656 pCi/L +/- 92 pCi/L. The EPA map of potentially high Radon areas indicates this site has a low potential for distribution of Uranium deposits but clearly the wells with Radon levels greater than 1000 pCi/L are indicative of a significant source of Uranium within the cone of depression of this system.

The well supplies of eastern-central NM and southern CA appear to correlate well with the geological information of Figure 1 and Table 5.

The eastern-central NM wells are situated within the Western Central Platform Province, an area composed of sandstones capped with sands and gravels (3). This area has a low potential for Uranium distribution and the analytical results seem to reflect this. In the 36 wells sampled, the Rn concentrations ranged from <100 to 678 pCi/L with a mean concentration of 253 pCi/L +/- 266 pCi/L. Note that the associated error was higher than the mean detected Radon concentration. This is indicative of lengthy holding periods from the time of sample collection to the time of analysis. This phenomenon is particularly aggravated when dealing with water samples that contain low levels of Radon.

The 31 wells tested in southern CA are located in the Pacific Mountain System Province (Figure 1; Table 1). This province is characterized by several complex rock formations including granites, sedimentary rocks, and fluvial sediments (3). This is an area with a near-surface distribution of potential Uranium deposits. The survey results tend to confirm this as Radon levels ranged from 246 pCi/L to 1212 pCi/L, with a mean Radon concentration of 553 pCi/L +/- 93 pCi/L.

OPERATIONAL ASPECTS CONCERNING RADON

The Operational Phase of the Radon survey resulted in the collection of samples, each designated by one or more of the following categories, intended to yield data and information on a very specific topic:

Category	Purpose of Sample
1	assessment of effect of storage
2	assessment of effect of blending
3	assessment of effect of pumping time
4	assessment of effect of treatment

Discussion of the Operational Phase results will be presented according to each of the above-listed categories.

Category #1 - Effect of Storage

Category #1 samples were collected to determine the effect of storage within a water supply distribution system on Radon levels. In order to make this assessment samples were collected from a storage vessel in the distribution system of an American System operating company. The sampling points chosen to assess the effect of storage were: at the well; at the point of introduction of water into the storage vessel, and; the storage vessel effluent after selected periods of storage time.

The storage vessel involved in this assessment was a 0.32 MG capacity steel standpipe. The standpipe had dimensions of a height of 45 feet and a diameter of 35 feet. Samples were collected at this site on each of two days.

During the sample collection process, the well which supplied water to the standpipe was operated only to the extent to fill the standpipe. Once full, the well pump was shut off, thereby allowing for maximum holding time of a given quantity of water within the standpipe. Prior to initiation of the sampling process, the standpipe was emptied to the lowest practical level, so as to eliminate, to the greatest extent possible, dilution of freshly-pumped well water with water that had been previously stored in the standpipe.

On each sampling day, samples were collected at the well at a time designated as Time 0. Samples were then collected as the standpipe was filling (8 to 11 minutes later). The tank was filled to near-capacity and allowed to empty for a period of 1 hour, at which time a sample of the tank effluent was collected. The tank was then allowed to empty to the lowest level possible and another effluent sample was collected.

Table 12 provides a summary of the types of samples collected, important increments of time associated with the storage-effects assessment and the corresponding concentrations of Radon.

From the data in Table 12, approximately 18% of the original Radon concentration had dissipated by the completion of the storage test, on Day 1 and Day 2. Closer analysis of the data, utilizing the decay curve for Radon reveals that a small fraction of the 18% decrease noted each day is actually attributable to decay. The factor apparently responsible for the largest increment of decrease in the Radon concentration is the loss of Radon to the atmosphere as a result of agitation of the water and ventilation in the

Table 12.   Summary of Data from Assessment of Effect of Storage on Rn Levels

Type of Sample	Time of Sampling	Total Elapsed Time (minutes/hours)	Mean Rn Conc. (pCi/L)	Associated Error (+/-pCi/L)	Decrease in Rn Conc. (pCi/L)
		DAY 1			
well supply	0835	0	4636	75	--
tank influent	0843	8 min./0.13 hr.	4104	103	532
tank effluent	0945	62 min./1.03 hr.	3927	69	177
tank effluent with tank @ lowest water level	1920	645 min./10.75 hr.	3794	95	133
		DAY 2			
well supply	0840	0	4610	95	--
tank influent	0851	11 min./0.18 hr.	4467	71	143
tank effluent	0953	62 min./1.03 hr.	3919	60	548
tank effluent with tank @ lowest water level	1600	440 min./7.33 hr.	3799	62	120

standpipe. Following is a more detailed analysis of
each data set.

Day 1

The largest decrease noted in Day 1, a decrease in
the Radon concentration of 532 pCi/L (63%), occurred as
the water was pumped from the well into the standpipe.
This is likely attributable to the agitation of the
water and the subsequent liberation of the volatile
Radon gas to the atmosphere.   Inside the standpipe the
decrease in the Radon concentration, during the time
interval between 1 hour and 10.75 hours when the
samples were collected, is attributable largely to
ventilation.   Decay, as calculated from the Radon decay
curve, can be responsible for a decrease of only 46
pCi/L.   This amounts to only 35% of the total decrease
observed.   Liberation of the Radon gas to the atmosphere within the tank must be responsible for the
remaining fraction of the noted decrease.   Similarly,
only 34% of the total observed decrease in Radon is
attributable to decay.   Of the total decrease of 842
pCi/L, 555 pCi/L of that quantity must be attributed to
volatilization of the Radon gas from the water.

Day 2

The results from Day 2 are similar to those of Day
in that only 26% (207 pCi/L) of the total decrease in
Radon concentration from Time 0 to 7.3 hours is attributable to decay.   A much larger fraction of the Radon
(604 pCi/L) was lost through volatilization of the gas
from the water.

Also similar was the fact that only a minor amount
of the observed decrease in Radon, while in storage, is
attributable to decay.   In the 6.3 hours of storage,
decay accounts for only 23% (30 pCi/L) of the noted
decrease.   90 pCi/L (77%) lost during storage, must be
attributed to ventilation within the standpipe.

The results from Day 2 were different from those of
Day 1 in that a decrease of only 143 pCi/L was noted
between the time the water was pumped from the well to
the time it was introduced into the standpipe.   Further, loss during the first hours of storage was
significantly higher on Day 2 than on Day 1 (Day 1
decrease = 177 pCi/L; Day 2 decrease = 548 pCi/L). The
reason for these differences is not known at this time.
The important trends to note, however, are that the
observed decreases in the levels of Radon, while in
storage, were similarly proportional to the time of
storage on both days and that volatilization of the
Radon gas was a far more important factor in the noted
decreases than was the decay of Radon into its daughter
elements.

Category #2 - Effect of Blending

Category #2 samples were collected to determine the effect of blending essentially Radon-free surface water with Radon laden groundwater. It was assumed prior to conducting this assessment, that the net effect of the blending would be a decrease in the Radon concentration of the blended water as a simple function of dilution on a volume basis. The data presented in Table 13 bears this out.

Table 13.   Effect on Radon Levels of Blending Surface
            Water with Groundwater.

Type of Sample	Mean Rn Conc.(pCi/L)	Associated Error(+/-pCi/L)	Volume of Water (MG)
well water	1079	80	6.34
surface water	0	NA	12.0
blended water	226	52	18.34

Dilution calculations on a strictly volumetric basis (a ratio of 1.9 to 1; surface water to groundwater) would predict a blended Radon concentration of 243 pCi/L. From Table 13, the detected Radon concentration of the blended water was 226 pCi/L +/- 52 pCi/L. Given the associated error, it is valid to conclude that the resultant blended water Radon concentration is simply a function of volumetric dilution.

Category #3 - Effect of Pumping Time

Category #3 samples were collected to document the effect which pumping time at a selected well site had on the levels of Radon in the groundwater drawn from the well.

To make this assessment, a series of 13 sets of samples were collected at predetermined time intervals over the course of a planned 5 day pump test. A summary of the results of that pump test are presented in Table 14. An important point to note about the data is that on Day 4 a break occurred in the water main leading from the well, necessitating a 5 hour shutdown of the well and an interruption of the pump test. As such, to account for the interruption and the subsequent pumping time, data is shown for a 6th day.

The well selected for this assessment is a gravel-pack well screened in coarse sand and gravel. The well has a safe yield of 600 gpm and was pumped at 510 gpm during the pump test. The aquifer from which the well draws its water is a water table aquifer with no known significant confining layers. It is composed

Table 14.   Summary of Data from Radon Survey Pump Test.

Pumping Time	Measured Water Level	Mean Rn Conc.(pCi/L)	Associated Error(+/-pCi/L)	Difference in Rn Conc. from Time 0 (pCi/L)	Difference in Rn Conc. from Previous Sample(pCi/L)
0	43' 9"	1957	73	--	--
15 min	42' 10"	2186	112	+229	+229
30 min	41' 3"	2071	79	+114	-115
45 min	41' 1.5"	2136	74	+179	+65
1 hr.	41' 1.5"	2510	83	+553	+374
2 hr.	41' 3"	2517	85	+560	+7
3 hr.	41' 3.5"	2485	80	+528	-32
6 hr.	41' 5"	2440	75	+483	-45
day 1	41' 7.5"	2412	102	+455	-28
day 2	41' 6"	1949	57	-8	-463
day 3	41' 4"	1920	52	-37	-29
day 4	no data due to main break				
day 5	43' 0"	2098	43	+141	+178
day 6	41' 3"	2055	40	+98	-43

of stratified drift (sand and gravel) bounded by glacial till and bedrock.

A number of trends are obvious from the data. Through the first 2 hours of pumpage, Radon levels increased from an initial concentration of 1957 pCi/L +/- 73 pCi/L (@ Time 0) to a concentration of 2517 pCi/L +/- 85 pCi/L (@ hour 2). From that point through Day 3, the Radon levels dropped off steadily from a concentration of 2517 pCi/L +/- 85 pCi/L (@ hour 2) to 1920 pCi/L +/- 52 pCi/L (Day 3). During the course of pumping on Day 4, a nearby mainbreak resulted in an interruption of the pump test. Upon repair of the break, the well was turned back on and sampled after approximately 19 hours of pumping time (Day 5) and again on Day 6.

The pump test data for the period from Time 0 to Day 3 tends to suggest that the Radon source is within the boundaries of the cone of depression created by 2 hours of pumpage. As the cone expands with continued pumpage, water bearing lower concentrations of Radon is drawn into the well, thereby diluting the highly Rn-laden water and resulting in a concurrent decrease in the Radon concentration of the water drawn from the well. This occurrence appears to have repeated itself on Days 5 and 6, following the shut down of the well after the main break on Day 4.

The pattern of Radon concentration varying with time in this well is attributed to the site-specific geology of this location. It is likely that most wells would be expected to show some variation in Rn concentration, as a result of pumping time, depending upon the geology of the water bearing formation.

Category #4 - Effect of Treatment

Category #4 samples were collected to assess the efficiency of various water treatment methods in the removal of Radon from groundwater. The two general categories of treatment which will be discussed are filtration through Granular Activated Carbon (GAC) and aeration.

Granular Activated Carbon

GAC can effectively reduce Radon concentrations in drinking water to very low levels. However, the required amount of detention time within a GAC bed, or the Empty Bed Contact Time (EBCT), is quite lengthy from an operational point of view. For example, a 99 percent reduction in Radon concentration would require an EBCT of 130 minutes (5). In order to achieve this EBCT, medium-large water systems would be required to operate multiple contactor units in series. This would not be feasible from a cost standpoint.

Radon removal efficiency by GAC was evaluated on a limited basis at three American System operating companies.

The first site consists of two down-flow, pressure GAC contactors, each containing 10,000 pounds of GAC to provide a bed depth of 10 feet. The EBCT of each contactor is 10.5 minutes at a design loading rate of 7.1 gpm/ft². The contactors were designed for the removal of tetrachloroethylene and trichloroethylene.

Each GAC contactor is equipped with 3 sampling taps at various bed depths. This provided an opportunity to measure Radon removal efficiency per specific fractions of the total bed depth. Following is a summary of the Radon data received from sampling at this site.

Type of Sample	Mean Rn Conc.(pCi/L)	Associated Error(+/-pCi/1)	Percent Red. of Influent Conc.
Run #1			
raw water	401	110	----
contactor effluent	226	99	43.6%
Run #2			
raw water	312	59	----
tap 1	237	54	24%
tap 2	251	56	19.6%
tap 3	209	53	33%
contactor effluent	200	52	35.9%

The reductions achieved in each run generally agree with test data reported in the literature. For an EBCT of 10.5 minutes, a reduction in the Radon concentration of approximately 35 percent can be expected (5).

It is interesting to note that the detected Radon concentration from tap 2 (mid-bed depth) was higher than that of tap 1 (near surface of the bed). While possibly attributable to the range of associated analytical error, this finding may also be indicative of channeling within the carbon bed.

At the second site, one pressure GAC contractor contains 8640 pounds of GAC with a bed depth of 27 inches over 12 inches of graded gravel. The EBCT of the contactor is approximately 2 minutes at an estimated design loading rate of 10.5 gpm/ft². The contactor was designed for removal of hydrogen sulfide from the groundwater. Following is a summary of the Radon data received from sampling at this site.

Type of Sample	Mean Rn Conc.(pCi/L)	Associated Error(+/-pci/L)	Percent Reduction
raw water	840	155	--
contactor effluent	751	153	10.6%

The reduction achieved by adsorption onto the carbon bed agrees fairly well with published test data which indicates that a reduction in the Radon concentration of approximately 12 percent can be expected with an EBCT of 2 minutes (5).

At site #3, the pressure GAC contains about 8800 pounds of GAC with a bed depth of 31.5 inches over 12 inches of graded gravel. The EBCT of the contactor is calculated to be approximately 2 minutes at an estimated design loading rate of 11 gpm/ft^2. This contactor was also designed for removal of hydrogen sulfide from the groundwater. Following is a summary of the analytical data for samples collected at this site.

Type of Sample	Mean Rn Conc.(pCi/L)	Associated Error(+/-pCi/L)	Percent Reduction
raw water	675	146	___
contactor effluent	823	153	none

It had been expected that a reduction in the Radon concentration in the range of 10-15% would be noted at this site, based on the EBCT of 2 minutes. The increase in the Radon concentration of the contactor effluent over the raw water concentration may be attributable to one of a number of factors, the most likely being: during the sample collection process, the raw water sample was improperly handled resulting in volatilization of a fraction of the Radon gas; given the rather high associated analytical error (due to the extended period of time between sample collection and sample analysis), the results may be masking an actual reduction of the Radon concentration, or; Radon may have desorbed from the carbon and returned to the water phase, possibly as a result of a recent backwashing of the carbon bed. Without sufficient data, the exact cause of this occurrence cannot be identified at this time.

The data reported here tends to confirm the information reported in the literature. In order to achieve substantial reduction (>90%) in the levels of Radon from groundwater, GAC systems of impractical EBCTs for medium or large water supply systems would have to be designed. From a cost and operational standpoint, more efficient removal technology than GAC is available to water purveyors.

Aeration

Because Radon is a highly volatile gas, it was expected that water treatment methods which employ some form of aeration at several American System operating companies, might be capable of achieving significant reductions in the levels of Radon in the well supplies of those companies. The Radon removal efficiency of various forms of aeration was evaluated on a limited basis at four American System operating companies.

The first site utilizes a packed tower aeration column with a diameter of 4.5 feet and height of 16 feet, designed to remove a minimum of 70 percent of the volatile organic contaminants detected in the well supply owned by the company. The air to water ratio of the column is 50:1.

Few packed tower aeration studies have been conducted for Radon removal. Because Radon behaves similarly to some highly volatile organic compounds and packed columns have been shown to be very efficient at VOC removal, packed tower aeration should be an effective technology for Radon removal. The data obtained from this site bear this out.

Type of Sample	Mean Rn Conc.(pCi/L)	Associated Error(+/-pCi/L)	Percent Reduction
Run #1			
raw water	783	95	--
aerator effluent	20	55	97%
Run #2			
raw water	649	32	--
aerator	21	14	97%
Run #3			
raw water	646	32	--
aerator	24	15	96%
Run #4			
raw water	646	21	--
effluent	34	11	95%

The packed tower aerator consistently removed 95% or more of the Radon from the well water.

Samples were collected from three sites (#2, #3, and #4) which utilize tray aerators for the purpose of oxidizing iron and manganese. Following is the data from each site.

Type of Sample	Mean Rn Conc.(pCi/l)	Associated Error(+/-pCi/L)	Percent Reduction
Site #2			
west aerator influent	327	63	--
west aerator effluent	50	23	85%
east aerator influent	342	47	--
east aerator effluent	37	32	89%
Site #3 - Run #1			
raw water	465	76	--
aerator effluent	108	64	77%
Run #2			
raw water	521	60	--
aerator effluent	46	43	91%
Site #4 - Run #1			
aerator influent	269	45	--
aerator effluent	54	36	80%
Run #2			
aerator influent	260	63	--
aerator effluent	50	36	81%

The data indicate that tray aerators are able to remove greater than 75% of the Radon concentrations in the well water supplies. This is an important finding because tray aeration devices may offer a water utility a rather simple, non-capital intensive means of removing Radon from well waters.

CONCLUSIONS

Following are the major conclusions drawn from the Radon Survey:

1. American System operating companies in the northeastern part of the country experienced the highest levels of Radon in their groundwater supplies. The highest detected concentrations of Radon were in individual wells in the states of NH, MA, CT, RI, NJ, PA and CA.

2. The analytical results from the Occurrence Phase of the Radon survey seemed to correlate well with the known geology of the aquifer materials from which samples of groundwater were drawn. The highest observed Radon levels were associated with formation of Uranium-bearing granitic rocks.

3. The observed decrease (approximately 18%) in the levels of Radon in water stored within a water supply distribution system was attributable largely to volatilization of the gas due to pumping and agitation of the water and ventilation within the storage vessel. Decay of the Radon accounted for only approximately 30% of the noted decrease in the Radon concentration, while in storage.

4. Blending of essentially Radon-free surface water with Radon-laden groundwater results in a blended water Radon concentration that is a function of volumetric dilution.

5. During a 5 day pump test, Radon levels fluctuated by more than 20% and were at their highest after 2 hours of pumpage. The Radon concentrations dropped steadily, to a level below that detected at the initiation of the pump test, during the remainder of the test.

6. GAC can effectively reduce Radon concentrations in drinking water supplies to very low levels, however the amount of contact time within the carbon bed, required to do so would be prohibitive to many water utilities from an operational and economic standpoint

7. Aeration is very effective in the removal of Radon from drinking water. Packed tower aerators achieved greater than 95% reduction in Radon concentrations and conventional tray aerators achieved greater than 75% reduction in Radon concentrations.

8. The precision and accuracy of the quantification of Radon was affected by sample collection procedures, sample holding times, the length of counting time with the scintillation counter and the counting equipment.

REFERENCES

1. Pritchard, H.M., and T.F. Gesell. "Rapid Measurements of Radon-222 Concentrations in Water with a Commercial Liquid Scintillation Counter," Health Physics, 33:577-581 (1977).

2. Lowry, J. Personal Communication (1986).

3. Hess, C.T., J. Michel, T.R. Horton, H.M. Prichard and W.A. Coniglio. "The Occurrence of Radioactivity in Public Water Supplies in the United States", U.S. Environmental Protection Agency National Workshop on Radioactivity in Drinking Water, Easton, MD (1983) p.1.

4. U.S. Environmental Protection Agency. Unpublished Results (1986).

5. Dyksen, J.E., D.J. Hiltebrand and R. Guena. "Treatment Facilities and Costs for the Removal of Radon from Groundwater Supplies", in Proceedings of the 1986 Environmental Engineering Specialty Conference (Ohio: Published by the American Society of Civil Engineers, 1986), pp. 510-521.

A CONNECTICUT RADON STUDY- USING LIMITED WATER SAMPLING
AND A STATEWIDE GROUND-BASED GAMMA SURVEY TO HELP GUIDE
AN INDOOR AIR TESTING PROGRAM. A PROGRESS REPORT:

Margaret A. Thomas, Natural Resources Center, Department of
Environmental Protection, Hartford, Connecticut

ABSTRACT

The Connecticut Geological Survey within the State Department
of Environmental Protection (DEP) is working with the State Department of Health Services (DOHS) investigating the occurrence of
radon in Connecticut. In 1985 and 1986, approximately 300 private
and public water supply wells from 20 geological areas were tested
for radon by the Toxic Hazards and Public Water Supply Sections of
DOHS. Highest ground water radon was 130,240 pCi/1 from the
Nonewaug Granite, a two-mica granite (range 10,720-130,240 pCi/1).
Elevated radon was found in wells within several granitic gneisses:
the Glastonbury Gneiss (3070-80,900 pCi/1), the Canterbury Gneiss
(10,010-64,510 pCi/1), and the Hope Valley Alaskite Gneiss (4060-
59,180 pCi/1). These Paleozoic and PreCambrian age rocks underlie
~5% of the State. Intermediate radon levels were found in water
from PreCambrian and Paleozoic age stratified metamorphic rocks
where radon levels > 10,000 pCi/1 were widely scattered in these
surveys. Relatively low ground water radon values were found in
central Connecticut Mesozoic age sedimentary rock wells (390-8490
pCi/1) and in Paleozoic age carbonate rock wells (200-4130 pCi/1).
Preliminary results from the ground (carborne) gamma radiation
survey generally show a positive correlation with radon water
analyses, enabling characterization of geological areas in
Connecticut as radiation sources. DOHS is currently conducting
2,200 air tests in homes located in geologic areas selected from
water analyses and the ground survey. All data is compiled on
1:24,000 maps to be included in multiple natural resource spatial
analyses using an automated Geographic Information System. The
analyses will explore relations between the distribution of radon
levels and earth materials to better define the geologic areas and
possible origins of radon in ground water and in indoor air.

INTRODUCTION

In the early part of 1986 the Connecticut Department of Environmental Protection (DEP) and the Department of Health Services (DOHS) began working together to address the radon issue. Available geological information for Connecticut and other states of the northeast indicates that elevated concentrations of radon are possible in many areas of Connecticut [13, 15, 31, 32, 18, 19, 14, 23, 2]. Connecticut is geologically diverse, with many geological regions having suspected potential for elevated radon. These include areas of: Mesozoic sedimentary rocks of central Connecticut, granitic rocks (particularly pegmatites) that occur in parts of western, coastal, and eastern Connecticut, high grade metamorphic rocks of eastern and western Connecticut, carbonates of western Connecticut, and 'Reading Prong-type' Proterozoic Massifs of western Connecticut (Fig 1).

In order to identify the occurrence and magnitude of radon concentrations statewide the DEP and DOHS began an exploratory ground water sampling program, followed by indoor air testing. DEP has been using this information and conducting a car-borne gamma radiation survey in order to identify the geologic and hydrogeologic relationships of radon occurrence. These activities will enable better targeting of future home testing and help determine geographic areas at risk.

The DOHS and DEP collective endeavor is ongoing, slated to be continued for the next several years. This paper gives an overview and progress report of the Connecticut Program. More detailed analysis of the ground water data is forthcoming in a Yale University Master's thesis [26]. A DOHS report to be released in June 1987 will be a synthesis of 1986 water and indoor air data.

BACKGROUND

During the summer of 1985, the Toxic Hazards Section of DOHS conducted preliminary testing of domestic well water in granitic-type bedrock wells. EPA sampling protocol was followed for all radon water tests [11]. Of the 63 wells tested, 25.4% had radon levels greater than 10,000 pCi/l (range 100 - 89,400 pCi/l) (picoCuries per liter : 1 pCi/l is approx. 2 nuclear disintegrations per minute per liter of air). Estimates of radon partitioning into air from water supplies [5] are an approx. air:water ratio of 1:10,000 pCi/l [9, 20]. A well water radon value of 10,000 pCi/l would correspond to an indoor air measurement of 1.0 pCi/l, which is within the range of the national average (0.3 - 2.2 pCi/l) and represents the lower

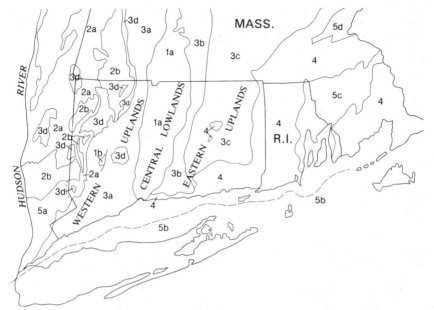

GEOLOGIC TERRANES OF CONNECTICUT

1. Newark (Rift Valley) Terrane
   a. Hartford Basin
   b. Pomperaug Basin

2. Proto-North American (Continental) Terrane
   a. Carbonate shelf
   b. Proterozoic massifs "Grenville"

3. Iapetos (Oceanic) Terrane
   a. Connecticut Valley Synclinorium
   b. Bronson Hill Anticlinorium
   c. Merrimack Synclinorium
   d. Taconic Allochthons

4. Avalonian (Continental) Terrane

5. Other geologic terranes
   a. Fordham (terrane) (= 2a + 2b + 3d)
   b. Coastal Plain
   c. Narragansett Basin
   d. Boston Basin

ROCK ASSOCIATIONS OF GEOLOGIC TERRANES

ROCKS OF THE NEWARK TERRANE — CENTRAL LOWLANDS

- Sedimentary and igneous rocks of Mesozoic age (approximately 190 million years old) which originated as sedimentary debris, lava flows, and intrusions in fault-block basins associated with continental-plate rifting during the breakup of the supercontinent Pangea and the early formation of the Atlantic ocean. (Several small areas of Mesozoic age (Newark Terrane) rocks occur in the Western and Eastern Uplands; dolerite dikes occur in both upland regions, and sedimentary basins occur in the Western Uplands near the towns of Southbury (Pomperaug Basin) and Canton.)

Sedimentary Rocks:	arkose (including brownstone)	Igneous Rocks:	basalt (traprock)
	shale		dolerite (traprock)

ROCKS OF THE PROTO-NORTH AMERICAN, IAPETOS, AND AVALONIAN TERRANES — WESTERN UPLANDS AND EASTERN UPLANDS

- Metamorphic and igneous rocks of Paleozoic and Proterozoic age (260 to 1,000 (?) million years old) which originated as oceanic and continental sediments and crust. These rocks were deformed and metamorphosed during Paleozoic plate collisions that formed the Appalachian Mountain Belt and the supercontinent Pangea. (Some of this rock existed prior to these collisions, but much was formed as sedimentary and igneous rock during and as a result of the plate collisions.)

Metamorphic Rocks:	alaskite gneiss		Igneous Rocks:	diorite
	amphibolite			gabbro
	gneiss			granite
	granite gneiss			lamprophyre
	granofels			mafic rock
	greenstone	**Figure 1.**		norite
	marble			pegmatite
	phyllite			porphyry
	quartzite	**(from Rodgers, 1985)**		syenite
	schist			ultramafic rock

limit of the range of indoor air values from some
anomalously high radon regions of the U.S. (1.0 - 1.67
pCi/l) [4]. Well water greater than 10,000 pCi/l then
represents above average indoor radon levels from a
water source alone. Other radon sources can be direct
bedrock and soil emanations [3, 30].

Indoor air sampling with Terradex Track-Etch cups
(alpha track radon detectors from the Terradex
Corporation, Walnut Creek, CA) was done in homes where
well water exceeded 10,000 pCi/l. The range of air
values measured was 0.61 to 7.80 pCi/l with a median of
2.06 pCi/l. Three of the homes (15%) exceeded 4.0
pCi/l [6], which is the EPA recommended guideline for
remedial action for homes built in the high radon
phosphate regions of Flordia [35]. The levels of air
radon found in the Connecticut study were higher than
the national average but were not unusual for regions
in the northeast [15].

DOHS concluded that because the potential for
adverse health effects exists from radon levels found
in this survey, further assessment of the distribution
of radon levels was warranted. DOHS needed information
on bedrock geology, soil surveys, ambient air
radioactivity, and hydrogeology to be considered for
additonal radon surveys. The Natural Resources Center
of DEP was asked to provide the Health Department with
guidance on the geological parameters of testing.

1986 RADON SURVEY

Meetings between the staffs of the Natural
Resources Center and the Department of Health Services
early in 1986 were used to develop a two pronged
approach for an expanded radon survey. The DOHS would
conduct more extensive water and air testing using
sites selected on the basis of geology and the Natural
Resources Center would conduct a car-borne gamma
radiation survey of the State to provide further data
on background radiation.

The expanded radon survey was conducted in 1986 to
provide additional air and water sampling in areas
previously surveyed and in new areas where little or no
information on radon levels existed. Survey
limitations included a maximum of 200 well water
samples to be collected over one summer with follow-up
indoor air analyses of the homes.

Design

A review of the Connecticut Bedrock Map [25] an
aeroradioactivity survey of the state [24] and the NURE
data (National Uranium Resource Evaluation), [33, 34]
by the Natural Resources Center (NRC) in conference

with U.S.G.S. Water Resources Division (WRD) in
Hartford lead to the selection of 20 geological areas
to be sampled. These areas were groupings of
geological units intended to cover suspected high,
uncertain, and suspected low potential radiation
sources. An attempt was made to choose wells, within
each geological unit, which represented a range of
recharge, intermediate, and discharge points within the
ground-water-flow system. Wherever possible, wells
were also selected to span a range of total depths and
thicknesses of aquifers pierced.

Each sampled well was to be located on a 1:24,000
scale U.S.G.S. 7.5' quadrangle map for comparison with
bedrock, surficial materials, and drainage basin maps
for the area. A copy of the well completion report was
to be obtained for each well, to be used for analysis
involving well parameters such as: thickness and type
of overburden, thickness of aquifer pierced, and total
depth of the well. The quality of data on the well
completion reports is highly variable and likely to
contain many inaccuracies but it is the only record of
information associated with these wells. These reports
can be used qualitatively for general consideration of
potential well parameter influences on radon levels in
bedrock wells.

Implementation

The expanded radon survey was conducted jointly by
DOHS and DEP. The selection of suggested wells for
sampling, based on bedrock geology, hydrogeology, and
the existence and quality of well logs, was the
responsibility of DEP with assistance from U.S.G.S.,
WRD, Hartford. Final well selection was done by DOHS
field personnel according to area health districts,
ability to locate wells, and homeowner approvals. Even
though confidentiality of results was not guaranteed,
homeowner participation was not a problem. As in the
1985 survey, a well water sample was collected in
accordance with EPA methodology [11] to ensure
comparability of results with data collected in other
states.   DOHS interviewed the homeowner with regard to
housing characteristics, home construction, basement
character, and water usage. The well water samples
were analyzed by DOHS Laboratory Division Radiation
Section using a Packard Tri-carb Liquid Scintillation
Spectrophotomity Model 4530. Results were sent to the
homeowner, along with literature which described the
analysis and listed recommendations.

The air testing portion of the survey employed
passive Terradex Track-Etch monitoring devices which
were placed for 3 months in the 'lowest livable area'
of each home. The air tests were conducted during the
winter of 1986/1987 when minimized air changes in the

home are expected to cause the highest radon levels
during the year [12]. A three-month measurement at
this time would represent a potential maximum average
indoor radon level.

Results

     Figure 2 illustrates the locations of all public
and private well water samples analyzed for radon
during 1985 and 1986 by DOHS. Data management and
analysis are the responsibility of DOHS, and are
ongoing throughout the duration of the survey. When
complete, these analyses will permit evaluation of the
geologic and water sampling results as predictors of
radon levels in indoor air. Additional analyses will
involve the role of various well characteristics
relative to radon levels in the 1986 ground water
survey. Analyses of the Track-Etch cups are currently
in progress, with a report from the Terradex
Corporation expected April/May 1987. Data collection
from the water sampling is complete and its synthesis
will be available as a Yale University thesis in May
1987 [26].
     Highest ground water radon was 130,240 pCi/l from
wells in the Devonian age Nonewaug Granite, a two-mica
granite (range 10,720 - 130,240 pCi/l). Elevated radon
was found in wells within several granitic gneisses:
the Glastonbury Gneiss (3070 - 80,900 pCi/l), the
Canterbury Gneiss (10,010 - 64,510 pCi/l), and the Hope
Valley Alaskite Gneiss (4060 - 59,180 pCi/l). These
Paleozoic and PreCambrian age rocks comprise approx. 5%
of the State. Intermediate radon levels were found in
water from PreCambrian and Paleozoic age stratified
metamorphic rocks (110 - 48,820 pCi/l) where radon
levels greater than 10,000 pCi/l were widely scattered.
Relatively low ground water radon values were found in
central Connecticut Mesozoic age sedimentary rock wells
(390 - 8490 pCi/l) and in Paleozoic age carbonate rock
wells (200 - 4130 pCi/l).

Discussion

With the small number of analyses provided by this
survey, it would be premature to draw conclusions
regarding a predicted range of radon levels over a
geological unit or terrane. Nevertheless, locations of
radon values in bedrock well water from this survey
show scattered areas of elevated radon and several
areas of possibly significantly elevated levels. As
expected, radon levels in well water are shown to be
variable within the same geologic unit [16].

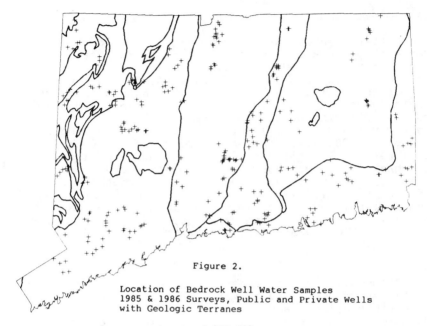

Figure 2.

Location of Bedrock Well Water Samples
1985 & 1986 Surveys, Public and Private Wells
with Geologic Terranes

1 : 1,250,000
1 inch equals approx. 20 miles

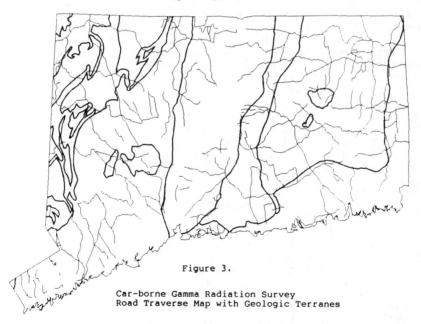

Figure 3.

Car-borne Gamma Radiation Survey
Road Traverse Map with Geologic Terranes

CAR-BORNE GAMMA RADIATION SURVEY

Better identification and characterization of
Connecticut rock bodies as radiation sources can be
provided by a ground-based profile of background levels
of gamma radiation [27]. To this end, during the late
summer and fall of 1986, staff of the Natural Resources
Center conducted a statewide car-borne gamma radiation
survey following procedures developed by the
Pennsylvania Geological Survey for use in the Redding
Prong [22, 28, 23]. The car-borne survey provides
exploration of small anomalies which were previously
integrated into gamma radiation detected over a large
area by airborne surveys [33, 34, 24]. As the nature
of high radon level occurrence is extremely variable
and seemingly sporadic, measured emanations of gamma
radiation from the ground can provide detailed mapping
of geographic areas potentially at risk [10, 27].

Methods

Reconnaissance of gamma radiation levels across the
state was accomplished by making traverses on
accessible primary and secondary roads, crossing all
major geological terranes. The vehicle was driven at a
constant speed of 20 m.p.h. on roads in good condition
and 10 m.p.h. for unpaved roads. A road log sheet was
maintained for each traverse where entries such as
weather conditions, landmarks, exceptional features of
the recording and rock outcrops were noted. Several
experimental traverses were repeated under a variety of
weather conditions, vehicle speeds, times of day, and
direction of travel in order to establish the most
reproducible results.

Like Pennsylvania, a portable Mount Sopris MS-132
Scintillation Counter containing a 1.5" x 1.5" NaI
crystal and a variable speed portable R-132A chart
recorder provided continuous measurement of gamma
radiation levels along each traverse. This equipment
was placed in a vehicle, with the scintillation counter
located on the floor of the front right side,
approximately 8" from the road surface.

Gamma radiation measurements in counts per second
(cps) on the scintillation counter (time constant = 1
sec) were taken through the body of the vehicle and
recorded on the chart paper. The chart recorder was
set at a rate proportional to the vehicle speed in
order to facilitate reconstruction of the traverse.

The chart records were interpreted into 5
categories, starting at 0-40 cps and incrementing at
intervals of 20 cps to greater than 100 cps. The road
logs were used with the interpreted chart records to
reconstruct and map the traverses onto 1:24000 scale
USGS 7.5' quadrangle base maps [28].

Data Reduction and Manipulation

The traverses were digitized from the base maps and entered into DEP's computerized geographic information system (GIS). This system utilizes ARC/INFO software, which is a state-of-the-art georelational data base management system copyrighted by Environmental Systems Research Institute of Redlands, California.

Entering the traverses into the GIS, with the associated ranges of gamma values allows somewhat unbiased data reduction and generalization at various scales. Using the GIS, background radiation, as measured along road traverses, is being compared with surficial materials, bedrock geology, and located analyses of radon in ground water and indoor air. This capability facilitates data manipulation and provides summary statistics and regional overviews of geologic materials relative to their potential for elevated radon.

Results

Car-borne gamma ray traverses covering 1367 miles of the State were completed, crossing all major geological terranes in Connecticut (Fig. 3). Inspection of preliminary results from the car-borne gamma radiation survey generally show a positive correlation with radon water analyses. Highest background gamma radiation was measured in the eastern part of the State over the Merrimack Synclinorium and Avalonian Terranes (Fig. 1) where levels of greater than 100 cps were recorded (average range 60-100 cps). This corresponds to the highest terrane average ground water analyses of approximately 14,000 and 10,000 pCi/l respectively.

The lowest gamma emanations measured during the car-borne survey were over the Newark Terrane (Fig. 1) at 20-40 cps. Bedrock well water analyzed from Newark rocks during this survey showed the lowest radon values, averaging approx. 2,000 pCi/l [7].

Discussion

Numerous factors can affect car-borne gamma radiation surveys, such as road materials and conditions [17], meteorological effects, overburden type and thickness [24, 13]. Potassium-40 and Thorium-232 and progenies are gamma emitters detectable in a car-borne survey which do not contribute to radon generation. General estimates of the magnitude of these contributions to the Connecticut car-borne gamma radiation survey can be made from inspection of the NURE separates [33, 34]. The relationship between differences in gamma emissions over various types and

thicknesses of overburden is unclear at this time.
Preliminary review of the data show measurable
differences in gamma emissions over various parts of
Connecticut. These can be generally associated with
the different geological terranes of the State, and in
some instances directly correlated with specific
geologic units. The car-borne survey confirms the
relative magnitude of radiogenic signals over various
geological terranes. It's use as a radon investigative
tool is in concert with other inventories for the study
of the relationship between earth materials and radon
levels. The car-borne survey will contribute to the
geological and geographical area characterizations for
elevated radon potential.

A STATEWIDE INDOOR AIR SCREENING PROGRAM

    Connecticut Governor William O'Neill in May 1986
directed DOHS to broaden the study of radon in
Connecticut homes and water supplies. At that time,
EPA Region 1 offered grants of air sampling devices to
New England states. Previous Connecticut studies
approached the radon issue through water testing with
follow-up air analyses. The EPA grant program
presented the opportunity to examine non-water radon
sources through air screening. It is expected that
this study will not only help examine the water/air
relationship of radon in the home, but also provide a
better estimate of Connecticut population exposure to
high radon levels. The Natural Resources Center was
asked to design a geologically stratified sampling
distribution for Connecticut's DOHS grant application
to EPA. Preliminary area characterizations using the
car-borne survey and available ground water analyses,
were used in the geological stratification of
Connecticut's statewide indoor air screening program.

Geological Stratification

    Several geologic, radiometric, and sampling
inventories were combined for the Health Department's
request that indoor air in each town of the State be
sampled. Among the inventories used were: the Bedrock
Geological Map of Connecticut [25], the Generalized
Bedrock Geologic Map of New England [21], citations of
uranium bearing minerals [29, 8], an aeroradioactivity
survey [24], an aeromagnetic survey [36], NURE
inventories [33, 34], the car-borne survey of gross
gamma radiation (NRC preliminary data), 300 located
domestic and public well water analyses [6, 7], and
miscellaneous available indoor air analyses [7]. From
the comparison of these data with population density, a
five-level sampling scheme was devised (Fig. 4). Towns

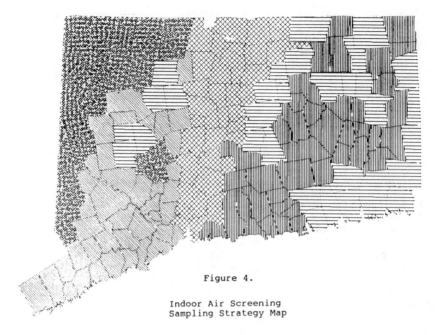

Figure 4.

Indoor Air Screening
Sampling Strategy Map

1 : 1,250,000
1 inch equals approx. 20 miles

RATING A   15 tests/town

AREAS OF MODERATE TO LOW POPULATION DENSITY WITH ROCK
UNITS WHICH ARE SUSPECTED TO HAVE A POTENTIAL FOR
LOCALLIZED ELEVATED RADON LEVELS.

RATING B   9 tests/town

AREAS OF MODERATE TO HIGH POPULATION DENSITY WITH ROCK
UNITS SUSPECTED TO HAVE SCATTERED OR LOCALLIZED
POTENTIAL FOR ELEVATED RADON.

RATING C   8 tests/town

AREAS OF MODERATE TO LOW POPULATION DENSITY WITH ROCK
UNITS SUSPECTED TO HAVE SCATTERED OR LOCALLIZED
POTENTIAL FOR ELEVATED RADON.

RATING D   7 tests/town

AREAS OF MODERATE TO HIGH POPULATION DENSITY WITH ROCK
UNITS SUSPECTED TO HAVE LESS POTENTIAL FOR ELEVATED
RADON.

RATING E   7 tests/town

AREAS OF LOW TO MODERATE POPULATION DENSITY WITH ROCK
UNITS WHICH MAY HAVE WIDELY SCATTERED OR LOCALLIZED
POTENTIAL FOR ELEVATED RADON.

of specific interest were selected for a greater number
of analyses where the geology is particularly
representative of a geological terrane or where several
of the above listed inventories showed suspected
potential for elevated radon. The highest priority
towns were to receive 15 analyses each (range 12-18)
and the lowest rated towns to receive 7 analyses each
(range 4-10), with no less than 5 analyses per town.

Program Implementation

DOHS was awarded a grant of 2000 air radon
detectors by EPA in October 1986. EPA also provided
analytical services through Eastern Environmental
Radiation Facility in Montgomery, Alabama. These air
detectors are short term, 2 day charcoal cartridges to
be placed in the lowest livable area of the homes. 500
air detectors were reserved for higher density sampling
of discovered 'hot spots' and for quality control.

Implementation of the statewide indoor air
screening program is through CONNSAVE, a private energy
auditing company supported by a consortium of public
utilities throughout Connecticut. Candidate homes for
the air radon tests are being selected from CONNSAVE's
energy audit requests in accordance with the sampling
density scheme described in Figure 4. Although
information from the energy audit is held
confidentially by CONNSAVE, confidentiality is not
being guaranteed for the radon test. The location of
each home participating in the program is being plotted
on a 7.5' USGS quadrangle base map. Complete results
are forthcoming in April/May 1987.

Analytical Plans

The located air analyses and associated attribute
information will be entered into the geographic
information system at DEP. In conjunction with mapped
ground water analyses and other inventories previously
described (see geological stratification), an in-depth
assessment and characterization of areas of the State
with elevated radon levels will be prepared by DEP and
DOHS.

DEP's geographic information system will help
examine possible relationships between the types of
earth materials present in a particular area and
associated radon levels in the ground water and indoor
air. These analyses will include: summary statistics
of source materials related to radon anomalies in
indoor air; summary of well characteristics related to
radon anomalies in ground water; characterization of
geographic and geologic regions as potential radon
source areas; and population exposure estimates in
anomalously high radon areas. Analyses of this type

will be the groundwork for a better definition of geologic source areas and possible origins of radon gas in Connecticut. Using the geographic information system, a radon potential map will be generated identifying areas of suspected low, intermediate-locally elevated, and elevated potential for radon. This will be a dynamic working map from which to assess the radon issue in Connecticut, provide public assistance, and help guide future Health Department radon programs.

REFERENCES

1. Asikainen M. "State of Disequalibrium Between U-238, U-234, Ra-226, and Rn-222 in Ground Water from Bedrock," Geochim. Cosmochim. Acta. 45:201-206 (1981).

2. Bell, C.   Personal communication (1985).

3. Breslin, A.J., and George A.C. "Radon Sources, Distribution, and Exposures in Residential Buildings," Trans Am. Nucl. Soc. (33)45 (1979).

4. Bruno, R.C. "Sources of Indoor Radon in Houses," Paper presented at Int. Symp. on Indoor Air Pollution, Health, and Energy Conservation, 13-16 Oct. 1981, Amherst, MA (Available from U.S. Environmental Protection Agency, Washington, DC) (1981).

5. Brutsaert, W. F., et al. "Geologic and Hydrologic Factors Controlling Radon-222 in Ground Water in Maine," Ground Water, 19(6):407-417 (1981).

6. Connecticut Department of Health Services. Report of a Survey of Radon Occurrence in Six Areas in Connecticut. (1985).

7. Connecticut Department of Health Services. unpublished data, 1985; 1986

8. Cooper, M. "Bibliography and index of literature on uramium and thorium and radioactive occurrences in the United States, Part 5: Connecticut, Delaware, Illinois, Ildiana, Maine, Maryland, Massachusetts, Michigan, New Hampshire, New Jersey, New York, Ohio, Pennsylvania, Rhode Island, Vermont, and Wisconsin," Spec. Paper 67, Geol. Soc. Am., p.472 (1958).

9. Cross, F.T., N. H. Harley, & W. Hofman. "Health Effects and Risks from 222Rn in Drinking Water," Health Physics 48:649-670 (1983).

10. Duval, J.   Personal communication (1985).

11. Eastern Environmental Radiation Facility. U.S. Environmental Protection Agency, "Radon in Water Sampling Program." U.S. EPA/EERF-Manual-78-1 (1978).

12. Fleischer, R.L., A. Mogro-Campero, L.G. Tuner. "Radon Levels in Homes in the Northeastern United States: Energy-Efficient Homes," in Proceedings of the Second Symposium on Natural Radiation Environment (New York: Halsted Press, 1982), pp. 497-501.

13. Goldsmith, Richard, Isidore Zietz, and H.R. Dixon. "Correlation of Aeroradioactivity and Geology in Southeastern Connecticut and Adjacent New York and Rhode Island," G.S.A. Bull. 88:925-934 (1977).

14. Hall, F. R., Donahue, P. M. and Eldridge, A.L. "Radon Gas in Ground Water of New Hampshire," Proceedings of the Association of Ground Water Scientists and Engineers, Eastern Regional Ground Water Conference, National Water Well Association, Worthington, Ohio, p. 86-100 (1985).

15. Hess, C.T. "The Occurrence of Radioactivity in the Public Water Supplies of the United States," Health Physics 48(5):553-586 (1985).

16. Hess, C.T. et al. "Radon-222 in Potable Water Supplies in Maine: The Geology, Hydrology, Physics, and Health Effects," in Proceedings of the Second Symposium on Natural Radiation Environment (New York: Halsted Press, 1982).

17. Koperski, J., T. Niewiadomski, and E. Ryba. "Indoor and Outdoor Gamma-Ray Studies in an Urban Environment in Poland," in Proceedings of the Second Symposium on Natural Radiation Environment (New York: Halsted Press, 1982).

18. Krishnaswami, S., Graustein, W.C., Turekian, K. K., and Dowd, J. F. "Radium, Thorium and Radioactive Lead Isotopes in Ground-Waters: Application to the Insitu Determination of Absorption-Desorption Rate Constants and Retardation Factors," Water Resources Research, 18 (6):1633-1675 (1982).

19. Lanctot, E. M., Tolman, A. L., and Loiselle, M. "Hydrogeochemistry of Radon in Ground Water," Proceedings of the Association of Ground Water Scientists and Engineers, Eastern Regional Ground Water Conference, National Water Well Association, Worthington, Ohio, p. 66-85 (1985).

20. Land and Water Resources Center, Division of Health Engineering, Maine Department of Human Services. "Radon in Water and Air; Health Risks and Control Measures," Land and Water Resources Center, University of Maine at Orono, 1983.

21. Olszewski, Wm. Jr., E.L. Boudette. "Generalized Bedrock Geologic Map of New England," New Hampshire Water Supply and Pollution Control Commission & U.S. Environmental Protection Agency Region 1 (1986).

22. Pennsylvania Topographic and Geologic Survey, "Map of the Reading Prong, Eastern Pennsylvania, Showing the Locations of Generalized Gamma-Ray Anomalies Detected by Carborne Survey," Commonwealth of Pennsylvania, (1985).

23. Pennsylvania Geological Survey   Written communication, 1985.

24. Poponoe, P.   "Aeroradioactivity and Generalized Geologic Maps of Parts of New York, Connecticut, Rhode Island and Massachusetts," U.S.Geological Survey. MAP GP-359 (1966).

25. Rodgers, J.   "Bedrock Geological Map of Connecticut," Connecticut Geological and Natural History Survey, (1985).

26. Rothney, L. "A Survey of Radon Levels in Private Bedrock Wells of Connecticut," Yale University School of Medicine, Departm,ent of Public Health, Division of Environmental Health STudies, New Haven, CT (in prep).

27. Rumbaugh, James. "Effect of Fracture Permeability on Rn222 Concentrations in Ground Water of the Reading Prong, Pa.," Master's Thesis, Penn State University, State College, Pa. (1983).

28. Smith, R.   Personal communications (1985;1986).

29. Sohon, Julian A.   "Connecticut Minerals Thier Properties and Occurrence," State of Connecticut, State Geological and Natural History Survey Bulletin No. 77 (1951).

30. Tanner, Allan B. "Indoor Radon and Its Sources in the Ground," U.S. Geological Survey. OFR 86-222 (1986).

31. Turner-Peterson, C.E., "Sedimentology and Uranium Mineralization in the Triassic-Jurassic Newark Basin, Pennsylvania and New Jersey," in Turner-Peterson, C.E., editor, Uranium in Sedimentary Rocks: Application of the facies concept to exploration: Society of Economic Paleontologists and Mineralogists, Rocky Mountain Section, Denver, Colorado, p. 149-175 (1980).

32. Turner-Peterson, Christine E., Olsen, P. E., and Nuccio, Vito F., "Modes of Uranium Occurrence in Black Mudstones in the Newark Basin, New Jersey and Pennsylvania," in Robinson, Gilpin R., Jr., and Froelich, Albert, eds. U.S. Geological Survey Workshop on the Early Mesozoic Basins of the eastern United States: U.S. Geological Survey Circular 946, p. 120-124 (1985).

33. U.S. Atomic Energy Commission and U.S. Geological Survey, "Preliminary Reconnaissance for Uranium in Connecticut, Maine, Massachusetts, New Jersey, New York, and Vermont, 1950 to 1959," RME-4106, TID UC-51 (1969).

34. U.S. Energy Research and Development Administration. "Airborne Geophysical Survey of a Portion of New England," GJO-1666-1 (1976).

35. U.S. Environmental Protection Agency. "Indoor Radiation Exposure Due to Radium-226 in Florida Phosphate Lands: Radiation Protection Recommendations and Request for Comment," Federal Register, 44:38664 (1979).

36. Zietz, I., J. Kirby, and F. Gilbert. "Aeromagnetic Map of Connecticut," U.S.Geological Survey. MAP GP-887 (1974).

EXTREME LEVELS OF [222]RN AND U IN A
PRIVATE WATER SUPPLY

Jerry D. Lowry, Civil Engineering
University of Maine, Orono, ME

Donald C. Hoxie and Eugene Moreau, Div. of Health
Engineering, Dept. of Human Services, Augusta, ME

INTRODUCTION

In 1985, the Maine Department of Human Services discovered a private water supply in Leeds, ME, that contains over 40,700 BqL^{-1} (1.1 x 10^{+6} pCiL^{-1}) of [222]Rn on average, and ranges between 13,300 and 59,200 BqL^{-1}. The well water also contains a gross alpha concentration of approximately 10.0 BqL^{-1} (270 pCiL^{-1}), of which more than 95 percent is U (403 ugL^{-1}). The ratio of [234]U to [238]U averages 1.17, which compares closely to sea water at 1.14. The Ra content comprises less than 2 percent of the gross alpha. The levels of [222]Rn and U are considered to be extremely high, with the [222]Rn being the highest known level the authors are aware of for a drinking water supply. This area of Maine has geologic features characteristic of those shown by others to have a high potential for elevated levels of [222]Rn and other radioisotopes [1-3].

The airborne levels of [222]Rn were also investigated prior to mitigation by water treatment, and extremely high concentrations were documented. The maximum, minimum, and average [222]Rn levels in the kitchen were 4.3, 1.1, and 0.1 BqL^{-1}, respectively. The maximum levels recorded in the home followed water use events and are among the highest levels documented in households worldwide. In the bathroom, the levels during a shower event were as high as 74 BqL^{-1} (2000 pCiL^{-1}). The [222]Rn levels representative of periods without a water use event were relatively low. Additional details of the [222]Rn and U levels at this site have been given elsewhere [4].

There are legitimate health concerns associated with internal organ cancers for waterborne [222]Rn and with U toxicity at the extreme levels existing in this water supply [5-8]. In addition, the elevated indoor air [222]Rn levels that result from waterborne [222]Rn, via water use, are a significant health concern in terms of lung cancer [9-11]. The scope of this paper is limited to the engineered solution to an environmental health problem, and does not include an in-depth consideration of the possible health effects for the occupants of the dwelling using this water supply. An extensive review of [222]Rn and U and their health effects has been published by Health Physics in an issue devoted entirely to radioactivity in drinking water [12].

The purpose of this paper is to update the information presented previously [4] about this site, in particular to the ramifications on treatment alternatives associated with the presence of both [222]Rn and U in a water supply.

WATER TREATMENT DESIGNS

Due to the obvious link between the water and air [222]Rn concentrations in this household, mitigation by water treatment appeared to be a viable solution to the health hazard posed by [222]Rn. Previous research has established that granular activated carbon (GAC) adsorption is the most cost effective method for removing [222]Rn from household water supplies [13,14]. The GAC adsorption/decay steady state allows a bed of GAC to remain effective, theoretically, for more than several decades. Radon and its first four short-lived progeny build to steady state levels in the bed during the first few weeks of operation and we assume that the fifth daughter, [210]Pb (22 y half-life), slowly accumulates with time. Field units monitored to date have shown no sign of decreased efficiency after 5 y of operation, and it is expected that other factors, unrelated to [222]Rn, will ultimately control the life of the GAC bed.

Recent research has established the effectiveness of the most popular water-grade GAC products for [222]Rn removal for both household and municipal water treatment applications [15,16]. The effectiveness of these carbon products has been defined by a steady state adsorption/decay constant, K_{ss}, which is the negative of the slope of a plot of ln Rn vs. the empty bed detention time (EBDT). The EBDT is the volume occupied by the bed of GAC divided by the average water demand. Thus, a simple first order model describes [222]Rn removal by GAC adsorption, once a steady state has been achieved. The concept described here is similar to that discussed by Evans [17] for the physical filtration of [222]Rn daughters from U mine air. The design parameters for the GAC system are given in Table 1.

The problem of U in the water supply was addressed by designing an anion exchange treatment unit. This design was based upon recent work by Hathaway and Sorg [18] at the U.S. EPA Municipal Environmental Research Laboratory (MERL) in Cincinnati, OH. The uranyl ion (UO_2^{+2}), in near neutral or alkaline pH waters, complexes with phosphate and carbonate [19]. In

Table 1. Treatment Systems Design Data

Granular Activated Carbon System:

Influent [222]Rn	55,500 BqL^{-1}
Estimated Water Demand (six-person household)	910 Ld^{-1}
GAC Bed Volume	85 L
K_{ss}, @ 10 degrees Celsius	3.2 h^{-1}
Estimated Empty Bed Detention Time (EBDT)	2.24 h
Expected Treated Water [222]Rn Concentration	42.9 BqL^{-1}
Expected Removal	99.92 %

Uranium System:

Influent [238]U + [234]U Activity	10.0 BqL^{-1}
Estimated Water Demand (6 person household)	910 Ld^{-1}
Resin Bed Volume	7.0 L
Average Empty Bed Detention Time (EBDT)	11.1 min
Expected Treated Water U Activity	not known

---

drinking water the carbonate complexes exist in various ratios of $UO_2(CO_3)_2^{2-}$ and $UO_2(CO_3)_3^{4-}$ depending upon pH and are removed by Type 1 macroporous resins.   Either hydrochloric acid/sodium hydroxide or sodium chloride can be used to regenerate, and the operating cycle is reported to be relatively long due to a high resin capacity for the concentrations of U being removed.   The parameters for the anion exchange treatment unit are given in Table 1.

A schematic for the treatment systems is shown in Figure 1. The GAC treatment unit was placed in-line immediately downstream from the hydropneumatic tank, sediment filter, and water meter. The sediment filter was installed to avoid any need to routinely backwash the GAC unit. Lowry [13] has demonstrated that scheduled routine backwashing of a GAC treatment unit for [222]Rn applications is not necessary and is detrimental to treatment efficiency.

Both treatment units were backwashed and put into service on October 11, 1985. The two treatment units were placed in series, with the GAC unit in the lead position.   This was done to investigate the potential of the GAC to adsorb U as well as [222]Rn. Previous reports have not indicated that GAC is a viable alternative for U removal from drinking water [12]; however, some related literature supports this idea [20,21].

METHODS

Waterborne [222]Rn, gross alpha, Ra, and U concentrations were determined by the Maine Public Health Laboratory, Augusta, ME. This laboratory is certified by the U.S. EPA for the gross alpha, U, and Ra analyses. Waterborne [222]Rn was analyzed by a liquid scintillation procedure after Prichard and Gesell [22], with

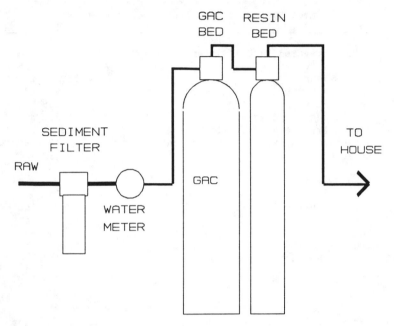

Figure 1. Treatment Systems: Flow Schematic

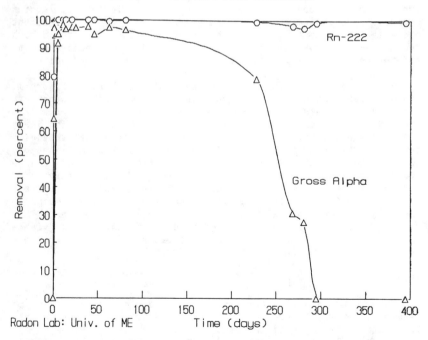

Figure 2. Radon and Gross Alpha Removal by GAC

mineral oil substituted for toluene, and radium standards used for quantification.   Two 15 mL glass vials from were set up in the laboratory using a 10 mL syringe to withdraw the field collected water samples from a single 30 mL glass septum vial. The samples were counted for 50 minutes. The airborne ^{222}Rn was determined by a portable monitor and detector. Uranium and Ra were analyzed according to previously published methods [23,24].   Uranium counting time was 24 h and the Ra samples were counted for 30 min for each of three counts (initial, one-week, and at three weeks after sample preparation).

Counting   uncertainties were 8 to 10 percent for U and Ra analyses.   Uncertainty for ^{222}Rn ranged from 20 percent at the lower values to 8 percent at the higher values. Gross alpha uncertainties ranged from 60 percent at the lowest levels to only 2 percent at the highest values.

RADON REMOVAL

The performance of the GAC system for ^{222}Rn removal from start up through 400 days is illustrated in Figure 2.   The achievement of steady state took approximately 10 days, which is typical for other field installations. The removal efficiency after this 10 day period was greater than 99.9 percent through day 228. Between day 228 and 281 the removal was lower than expected and although it returned to a very high level after that time it never reached the higher than expected removal exhibited during the first 228 days.   Based upon a measured water demand of 877 Ld^{-1} (232 gald^{-1}), the actual EBDT averaged 2.33 h and the expected treated water ^{222}Rn concentration was 24.1 BqL^{-1} (650 pCiL^{-1}). The measured steady state ^{222}Rn concentration in the treated water averaged 14.1 BqL^{-1} (380 pCiL^{-1}) prior to day 228. These results show slightly better than expected performance. The data after day 228 showed an average removal efficiency of 98.9 percent, with a low and high value of 97.5 and 99.8 percent, respectively.   The agreement of the degree of ^{222}Rn removal efficiency for the measured and calculated levels is quite good, considering the high degree of removal.   Presently, the ^{222}Rn concentrations in the treated water are approximately 5,000 to 10,000 pCiL^{-1}.   These levels are higher than expected, but appear to be related to the relatively high accumulated adsorbed mass of U on the bed.

URANIUM REMOVAL

The GAC system designed for ^{222}Rn removed essentially all the U present in the water supply until sometime between day 100 and day 200. The limited sampling makes it difficult to state exactly when a breakthrough began; however, it has been our experience that household units have a relatively sharp breakthrough of adsorbates due to a low loading rate, and we suspect that the breakthrough began closer to day 200 than shown in the simple line plot of Figure 2.   It should be noted that the gross alpha test was essentially measuring U since the Ra content was about 2 percent of the total influent gross alpha concentration. By day 295 a complete breakthrough of the influent U was measured and the GAC ceased to remove this component from the supply. During the period of progressive breakthrough (approximately day 200 to day 300) the downstream resin bed received an increasing concentration of U. The resin bed, with its relatively high capacity for the carbonate complexes of U, continues to remove essentially 100 percent (greater than 99 percent) of the U exiting the GAC unit. A summary of the removal of gross alpha (U) by the GAC and anion resin beds, as well as the influent levels treated, is presented in Figure 3.

Uranium adsorption is similar to any non-decaying adsorbate since its mass reduction by decay within the GAC bed is negligible in the time frame we are considering. Thus its breakthrough for the GAC bed was expected in a relatively short time. This is in great contrast to ^{222}Rn and its daughters. Based upon an assumption of 100 percent daughter retainment, it can be calculated that the maximum mass due to ^{222}Rn and its short-lived daughters for this particular case is 36.1 nanograms. This mass plus the maximum accumulated ^{210}Pb, thus far, is miniscule in contrast to the relatively high mass of adsorbed U. It is estimated that the total adsorbed U on the GAC bed is 100 grams, making the mass ratio of U to ^{222}Rn and daughters 2.8×10^9. This incredibly high mass of U in comparison to the adsorbed ^{222}Rn and progeny appears to have only a very slight effect on the steady state performance of the GAC bed for ^{222}Rn removal, as discussed further in the following section.

EFFECT OF ADSORBED U ON THE ^{222}RN STEADY STATE

The previous data presented in FIgure 2 suggest a lower efficiency of ^{222}Rn removal occurring simultaneously with the breakthrough, or saturation, of the GAC with respect to U. These data are amplified in Figure 4, where treated water ^{222}Rn and gross alpha removal vs. time are presented. Aside from the single questionable measurement of high ^{222}Rn on day 63, levels appeared to rise in conjunction with the onset of the U breakthrough event. After complete saturation of the GAC with U by day 300, the ^{222}Rn levels never returned to the previous levels. Radon-222 removal before and after breakthrough was 99.95+ and 98.7 percent, respectively. Given the fact that this water has an extremely high influent ^{222}Rn concentration, the apparent influence of U adsorption on the steady state was easily detected. It is speculated that this effect might go undetected or be negligible for more commonly encountered wells at a much lower ^{222}Rn concentration. In terms of removal efficiency loss the relatively high mass of adsorbed U appears to have very minimal effect on the ^{222}Rn adsorption/decay steady state.

CORRELATION OF ^{222}RN AND U

The influent ^{222}Rn and U levels for this well show a

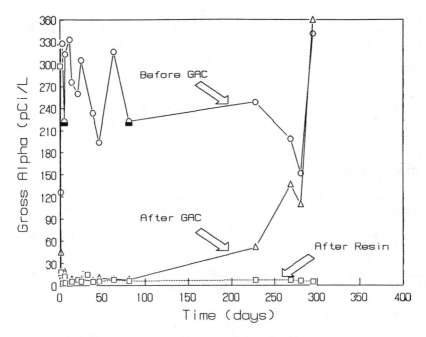

Figure 3. Gross Alpha: Levels and Removal

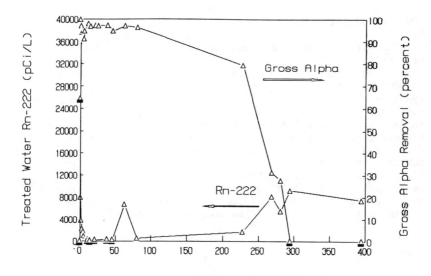

**Figure 4. Comparison of Radon and Gross Alpha
Removal on GAC**

reasonably strong positive correlation. This fact is illustrated in Figure 5 and even more clearly in Figure 6. The correlation coefficient was found to be 0.84 for the 17 samples collected. While this makes sense in terms of what is logical, it is sometimes difficult to show a similar correlation for other wells, and certainly difficult to quantify to any degree. Other water supplies in the immediate area showed no meaningful correlations or trends in terms of the occurrence of ^{222}Rn with either U or Ra. In fact, what was indicated was that there were none in this particular area.

AIRBORNE RADON

The reduction of airborne ^{222}Rn from the household water supply is evident in Fig. 7. A period of 24 to 48 hours was required for the hot water storage tank to lose its ^{222}Rn through usage and after this period the bathroom air concentrations remained below 0.006 BqL^{-1} (1.5 pCiL^{-1}). This is in contrast to levels as high as 74 BqL^{-1} (2000 pCiL^{-1}) prior to ^{222}Rn removal from the water supply. These results show that for this location the water was contributing nearly 100 percent of ^{222}Rn present in the indoor air. In general, airborne ^{222}Rn levels in this household are significantly below 0.12 BqL^{-1} (3.0 pCiL^{-1}). Thus, the significant potential health problem associated with airborne ^{222}Rn has been eliminated through the removal of ^{222}Rn from the water supply.

OTHER CONSIDERATIONS

The steady state gamma exposure rate created by the retainment of ^{214}Bi and ^{214}Pb on the GAC caused the bed to become a very significant source of radiation, and the unit has been shielded temporarily until it can be located in a small structure outside the home. Details of this aspect of the problem can be found elsewhere [4]. The GAC unit will be replaced by a fresh bed, which will be installed in series after the resin bed. In this way the permanent installation will first remove U and the GAC bed will not serve two roles. It will be interesting to see if the new GAC, without removing U, will continue to achieve the initial highest level of ^{222}Rn exhibited by the first bed prior to U breakthrough. In addition, the spent GAC vessel will be analyzed by a portable gamma detector (germanium) and colimator to determine the location of ^{222}Rn progeny with depth.

CONCLUSIONS

It has been demonstrated, in this case, that it is possible to effectively reduce elevated ^{222}Rn levels in indoor air through the removal of ^{222}Rn from the water supply. In situations where there are other significant contributors of ^{222}Rn, water treatment will only reduce the airborne ^{222}Rn in proportion to its contribution

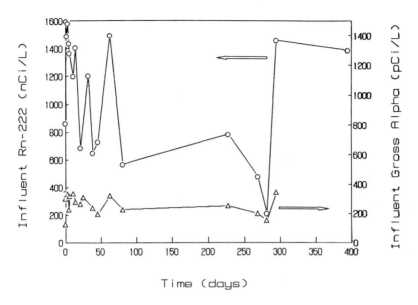

Figure 5. Influent Gross Alpha and Radon

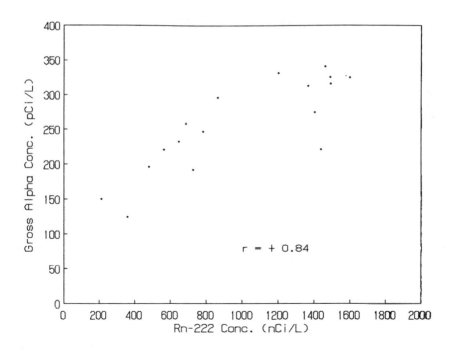

Figure 6. Correlation of Radon and Gross Alpha

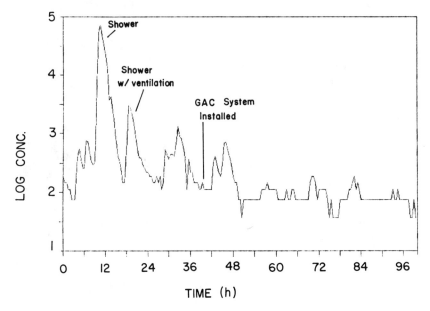

Figure 7. Airborne Radon Levels (log pCi/L)
In Bathroom

by the water supply.   This particular household had a ^{222}Rn problem almost entirely associated with its water supply.

Based upon the results of this study, the following conclusions can be made.

1. A water supply, with an extremely high ^{222}Rn level averaging 41,590 BqL^{-1} (1.124 x 10^{+6} pCiL^{-1}), was found to be contributing nearly 100 percent of the elevated ^{222}Rn measured in the indoor air.   Levels as high as 74 BqL^{-1} (2000 pCiL^{-1}) were documented in parts of the home during water use events.   In addition, a high U concentration was present in the water supply.

2.   A properly designed water treatment system is capable of bringing airborne ^{222}Rn levels down to normal background levels.

3. The GAC adsorption/decay steady state ^{222}Rn removal efficiency has ranged between 99.99 percent and 97.5 percent.

4.   GAC has a significant capacity to adsorb U and removals in this application were nearly 100 percent for 100 to 200 days (1200 to 2400 bed volumes).   Since it is expected that an ion exchange resin would be effective for at least 10,000 bed volumes, it is not likely that GAC will be cost effective for the removal of U.

5.   A saturation of the GAC with adsorbed U appeared to slightly reduce the efficiency of steady state ^{222}Rn removal, but does not appear to present a significant problem.

6.   A strong positive correlation between ^{222}Rn and gross alpha (U) was found for the influent concentraions.

ACKNOWLEDGMENTS

Funding and support for this project was from the Maine Dept. of Human Services, Division of Health Engineering.   Sample analyses and radiation measurements were made by the Public Health Laboratory and Health Engineering.   The U.S. EPA, Office of Research and Development (under Grant No. R8108290) funded the research that established the design constant, K_{ss}, used in the design of the ^{222}Rn treatment system.

REFERENCES

1. Hess C.T., Norton S.A., Brutseart W.B., Casparius R.G., Coombs E.G. and Hess A.L., 1979, Rn-222 in Potable Water Supplies in Maine: The Geology, Hydrology, Physics, and Health Effects, Completion Report, Land and Water Resources Center, University of Maine and the Office of Water Research and Technology, U.S. Dept. of the Interior, Washington, DC.

2. Lanctot E.M., Tolman A.L. and Loiselle M., 1985, "Hydrogeochemistry Of Radon In Ground Water," in: Proc. of the

Second Annual Eastern Regional Groundwater Conference, (Portland, ME: National Water Well Association).

3. Hall F.R., Donahue P.M. and Eldridge A.L., 1985, "Radon Gas in Ground Water Of New Hampshire," in: Proc. of the Second Annual Eastern Regional Groundwater Conference (Portland, ME: National Water Well Association).

4. Lowry J.D., Moreau E., and King, E., 1987 (est.), "Mitigating Extreme Levels of ^{222}Rn and Waterborne ^{222}Rn and U byWater Treatment," accepted for publishing by the Health Physics Journal.

5. National Academy of Sciences Committee on Biological Effects of Ionizing Radiation (BEIR), 1980, "The Effects on Populations of Exposure to Low Levels of Ionizing Radiation," (National Academy Press, Washington, DC)

6. Cross F.T., Harley N.H. and Hofmann W., 1985, "Health Effects and Risks From Radon in Drinking Water," Health Physics, 48, 649-670.

7. Mays C.W., Rowland R.E. and Stehney A.F., 1985, "Cancer Risks From The Lifetime Intake Of Radium and Uranium Isotopes," Health Physics, 48, 635-648.

8. Wrenn M.E., Durbin P.W., Howard B., Lipsztein J., Rundo J., Still E.D. and Willis D.L., "Metabolism of Ingested Uranium and Radium," Health Physics, 48, 601-634.

9. Harley N.H., "Editorial--Radon and Lung Cancer in Mines and Homes," 1984, New England Journal of Medicine, 310, 1525-1526.

10. Ra84   Radford E.P. and Renard K.G., 1984, "Lung Cancer in Swedish Iron Miners Exposed to Low Doses of Radon Daughters," New England Journal of Medicine, 310, 1485-1494.

11. Samet J.M, Kutrirt, D.M., Waxweiller, R.J., and Key, C.R., 1984, "Uranium Mining And Lung Cancer In Navajo Men," New England Journal of Medicine, 310, 1481-1484.

12. Health Physics Society/Pergamon Press, 1985, Health Physics, 48, 5.

13. Lowry J.D. and Brandow J., 1985, "Removal of Radon from Water Supplies," Journal of Environmental Engineering, 111, 511-527.

14. Lowry J.D., Brutseart W.F., McEnerney T., and Molk C., "Point-Of-Entry Experiments For Radon Removal," 1987, American Water Works Journal, April.

15. Lowry J.D., 1985, "Design of a GAC Water Treatment System for Radon," in: Proceedings of the Am. Soc. of Civil Engineers Conf. on Env. Engineering, (Boston, MA: Environmental Engineering Division, American Society of Civil Engineers)

16. Islam S., 1986, "Technology Development and Economics of [222]Radon Removal From Water Supplies," M.S. Thesis, Dept. of Civil Engrg., University of Maine, Orono, ME 04469.

17. Evans R.D., 1969, "Engineer's Guide to the Elementary Behavior of Radon Daughters," Health Physics, **17**, 229-252.

18. Hathaway S.W., Sorg, T.J., 1985, "Ion Exchange Technology for Removing Uranium from Drinking Water," Unpublished draft document, Drinking Water Laboratory, U.S. EPA, Cincinnati, OH 45268.

19. Hostetler P.B. and Garrels, 1962, "Transportation and Precipitation of Uranium and Vanadium at Low Temperatures, with Special Reference to Sandstone-Type Uranium Deposits," Econ. Geol., **57**, 137-167.

20. Tolmachev Y.U., 1943, "Adsorption of Uranyl Salts Onto Solid Adsorbents," IZV.ANSSSR (Bull. of Acad. Sci. USSR, Chemistry Ser., No. 1, p. 28.

21. Moore G.W., 1954, "Extraction of U from Aqueous Solution by Coal and Some Other Materials," Econ. Geol., Vol. 49, pp. 652-8.

22. Prichard H.M. and Gesell T.F., 1977, "Rapid Measurements of Rn-222 Concentrations in Water with a Commercial Liquid Scintillation Counter," Health Physics, **33**, 577-581.

23. Bodnar, L.Z., and D.R. Percival, RESL Analytical Chemistry Branch Procedures Manual, Spec. AS-5-1, US Dept. of Energy, Spec. AS-5, 1979, Revised 10/81 IDO-12096.

24. Radiological Analysis of Environmental Samples, Spec. RC-88A-1, 1959, U.S. Dept. of Health, Education, and Welfare.

RADIUM-226 AND RADON-222 IN DOMESTIC WATER
OF HOUSTON-HARRIS COUNTY, TEXAS

Irina Cech, Mengistu Lemma and Howard M. Prichard,
The University of Texas Health Science Center at Houston,
School of Public Health, Houston, Texas

Charles W. Kreitler
The University of Texas, Bureau of Economic Geology, Austin, Texas

INTRODUCTION

Ingestion or inhalation of radium (Ra) and products of its decay are considered a health risk. In the human body bones, myeloid stem cells, and lungs are particularly sensitive to radiation exposure. Epidemiologic studies have established a strong association between certain occupational and therapeutic exposures to isotopes of Ra and bone cancer [1-5]. Radon (Rn)-222, an airborne progeny of decay of Ra-226, has been associated with lung cancer in miners [6,7], and recently has received national attention because of high concentrations observed in houses in some parts of the United States [8,9]. Studies have also raised the issue of cancer risk associated with Ra and Rn in domestic water [8-12].

Regulatory standards for radioactivity in water have been established for Ra but not Rn. The combined activity of the two isotopes of Ra (226 and 228) should not exceed a maximum contaminant level (MCL) of 5 pCi/ℓ [13,14]. Rn in water is not yet regulated by Federal and State standards, although this is expected to change in the near future.

Ra regulation under the Safe Drinking Water Act of 1974 and related amendments went into effect in 1977, but because of the sheer number of public water systems in a State as large as Texas, the assessment of Ra concentrations and compliance with EPA and State regulations is a formidable task, not yet completed. In addition, approximately 4 million people in Texas use private water supplies which are not covered by current regulations and for which data on Ra are not available. No regulatory agency now routinely monitors Rn, in either public or individually-owned supplies.

Deposits of uranium, the progenitor of Ra-226 and Rn-222, occur in Texas [15]. The major deposits occur in Eocene and younger formations and it is believed to be predominantly associated with volcanic ash in the Catahoula formation of the Miocene Age [16,17]. In 1984, taking advantage of the cost-effective testing procedure available at the University of Texas at Houston, School of Public Health laboratory, a study was initiated by Cech and co-investigators [18], in cooperation with several ground water districts and river authorities in Texas, to assess the geographic distribution of Ra-226 and Rn-222 in domestic water in selected regions of this State. Anomalously high Ra and Rn concentrations were found in the Gulf Coast area including Harris County (Figure 1); Ra in excess of 20 pCi/ℓ and 3000 pCi/ℓ for Rn were observed in some public water supplies in this county.

Harris County, with a 1980 population of 2,684,000, has the largest number of people among Texas counties and contains 18 percent of the State population. The water supply and distribution system in Harris County is complex. The City of Houston alone operates over 200 wells. A water well field may contain from 1 to 11 wells, and pump water from various depths into several collecting plants. The eastern half of the City of Houston, roughly 50 percent of the population of over one million, is supplied primarily by surface water from Lake Houston (Figure 2); the western half is supplied by ground water. The boundaries of areas served by each well, well field, or by surface water cannot be defined accurately because systems are interconnected; these boundaries vary with variations in demand, pressure in water lines, season, and even time of day.

Population growth and the rapid development of new suburban subdivisions on the fringes of the City of Houston resulted in numerous smaller water utilities. Approximately 300 such municipal utility districts (MUDs) and 1700 community water systems supply ground water in Harris County besides the City of Houston. Most of the wells with Ra concentrations that exceed MCL were from these smaller utilities.

Approxmately 150,000 people in the rural parts of Harris County, and an estimated 10 percent of urban residents, are supplied by water from individually-owned wells. Northwest, northeast, and southwest Harris County contain undeveloped areas. All projections of urban growth indicate that these areas will be developed in the near future. Pockets of Ra-rich water found in MUD's in these parts of Harris County suggest that future subdivisions might encounter problems. However, no reliable guidance for well location is presently available to well developers in the County.

A more detailed sampling of ground and surface water was necessary in this area to assess patterns of distribution of Ra and Rn, so that some predictive tool could be developed to guide present and future water works. Thus, during 1985 and 1986, an intensive sampling for Ra-226 and Rn-222 was conducted in Houston-Harris County.

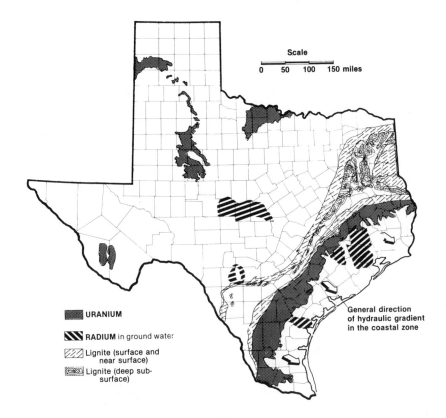

Figure 1.   Radioactive deposits in Texas.   Compiled by superim-
posing Ra-226 field data gathered by Cech et al.
[18] on a map of uranium deposits adapted from Kier
et al. [15].   Additional sources were Texas Water
Development Board [19] and Brock [20].

METHOD

One hundred sixteen samples were collected, 77 of them from
distribution systems and 39 directly from wells.   Of the samples
collected from distribution systems, 27 were taken from various
small municipal utilities districts in northwest Harris County.   Of
the 39 samples taken directly from wells, 26 were from southwest
Houston.   The rest of the samples were collected from public and
individually owned wells throughout the County.   Several samples
were collected in neighboring Fort Bend and Montgomery Counties to
estimate regional concentrations.

As a first step in this investigation, samples were gathered
from residential and commercial dwellings.   Testing of water at the
household tap involved visiting residences where permission was

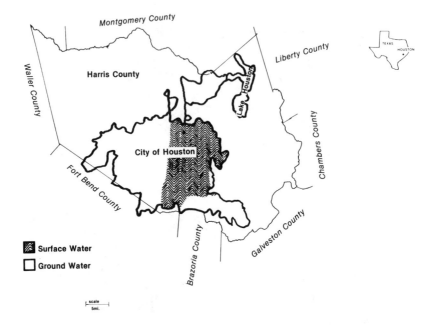

Figure 2.   Greater   Houston-Harris   County,   Texas: Approximate
            boundaries   for   surface   water-ground   water   service
            areas, 1980.

secured from heads of households to collect one-liter water samples. Duplicate samples were taken to assure the reproducibility
of the laboratory results.
        Analyses for Ra-226 and Rn-222 were conducted with modifica-
tions of liquid scintillation techniques described in [21]. Ra-226
was extracted from the 0.5 to 1 liter water samples by passage
through a 10 ml bed of cation resin (Dowex 50W-X8, 20-50 mesh).
The resin beds were rinsed with distilled water, transferred to a
20 ml liquid scintillation vials and covered with distilled water.
A 10 ml layer of toluene containing 5 g/l of 2,5-diphenyloxazole
was added, and the vials were sealed and stored for at least a week
prior to analysis. Three to twelve hours before the analysis, the
vials were vigorously shaken to ensure adequate contact between the
aqueous and the organic phases. The Rn content of the organic
phase was then assayed using a liquid scintillation counter, with
windows optimized for Rn in toluene. Because Rn originally present
in the water sample is not retained in the rinsed resin, any Rn
detected in the assay was due to ingrowth from Ra entrained in the
resin.
        Radon-222 concentrations were determined by drawing 15 ml
samples directly from a source into a clean syringe. Care was
taken to prevent aeration of the samples in the process. The
samples were then injected beneath 5 ml layers of a mineral
oil-based scintillation solution in 20 ml vials. The vials were

vigorously shaken to promote phase contact, held for at least three hours to permit Rn daughter ingrowth, and assayed with a liquid scintillation counter. The results were corrected for the amount of Rn decay between sampling and assay.

The information obtained on this phase of the investigation was mapped using the computer technique SYMAP [22], as shown on Figures 3 and 4, and this helped to identify areas where further sampling, directly from wells, could provide additional, clarifying information. The subsequent step was to identify wells in the distribution network which caused Ra and Rn to appear in the residential water. This was a straightforward task when small single-source utilities were concerned. More difficult was to pinpoint Ra and Rn sources within interconnected multiple-well municipal systems.

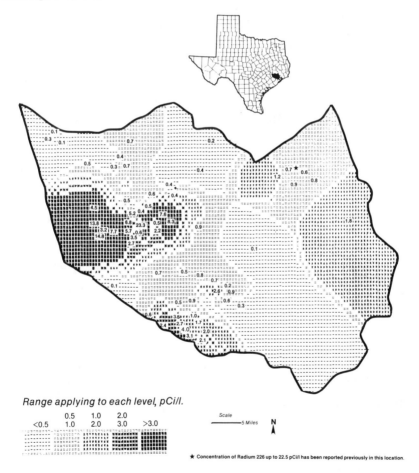

Range applying to each level, pCi/l.

	0.5	1.0	2.0	
<0.5	1.0	2.0	3.0	>3.0

Scale
———5 Miles   N

★ Concentration of Radium 226 up to 22.5 pCi/l has been reported previously in this location.

Figure 3.   Concentrations of Ra-226 observed in tap water in Harris County, Texas, 1985-1986.

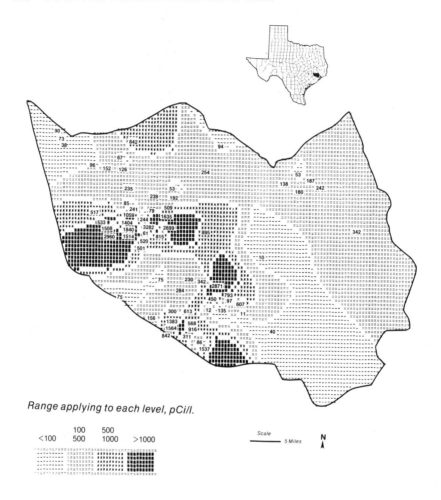

Range applying to each level, pCi/l.

Figure 4.   Concentrations of Rn-222 observed in tap water in
Harris County, Texas, 1985-1986.

In order to assess variations in Ra and Rn as related to
location and depth of wells, a subset of 64 points representing a
diverse geographic distribution of water sources was formed and
subjected to statistical analysis. This subset included 39 samples
taken directly from wells and 25 other points taken from service
areas of small utilities where the source-well definitely was
known.   The point-source locations are shown in Figures 5 and 6
(except for additional points to the north, in adjacent Montgomery
County).

The general hypothesis investigated in this study was that Ra
and Rn concentrations are a function of a) pumping depth, b)
distance from uranium deposits in sandstone aquifers, and c) some
modifying local structural features, in particular, the proximity

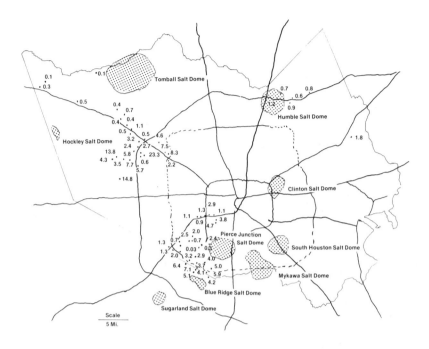

Figure 5.   Point-source measurements of Ra-226, Harris County,
            Texas,   1985-1986.   Samples   were   obtained   either
            directly at well heads or from single-source municipal
            utility districts.   Shown also are locations of salt
            domes.

of salt domes.   The variables that might have influenced the
distribution of radiochemicals were defined as follows:

Water-Bearing Formation and Pumping Depth

Depth to water sources varied from zero for surface water to
about 610 meters (m) for wells.   The data on depths of wells and
water-producing formations were derived from the materials of the
United States Geological Survey, [23].   Because many wells in the
Gulf Coast region have multiple screens, the well depth in the
statistical analyses was represented by upper and lower limits of
pumping ranges.

Location

Well location was represented by the distance (downdip) from a
reference line chosen approximately where known uranium deposits
occur [15].   On the Gulf Coast, the uranium-bearing formations crop
out parallel to the coast northwest of the study area and roughly

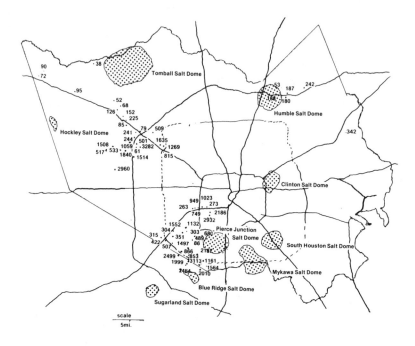

Figure 6. Point-source measurements of Rn-222, Harris County, Texas, 1985-1986.

coincide with the recharge zone of the Gulf Coast aquifers. The distance from the outcrop to the well was measured in the southeast direction to correspond with the general direction of hydraulic gradient [24-28].

Salt Domes

There are numerous salt domes in the study area (Figure 5). The proximity of salt deposits is a pertinent regional feature. Because the general trend in ground water movement may have been interrupted locally by pumping, the distance to salt domes was measured as the distance to the nearest dome, regardless of direction.

RESULTS

Figures 3-6 present Ra and Rn concentrations observed in the distribution systems and at the well heads in Greater Houston-Harris County in 1985-86. Concentrations of Ra and Rn were below limits of detection in the southeast part of the study area which is supplied by surface water, whereas detectable concentrations of these radioisotopes were present in virtually every sample of well water.   In northwest and southwest Harris County ground waters exhibited concentrations greater than trace levels of Ra and Rn; several wells in these areas had concentrations in excess of the maximum allowable for Ra.

The major "hot" spot was located in northwest Harris County. Up to 23 pCi/ℓ of Ra-226 (more than four-fold in excess of the maximum allowable) and up to 3300 pCi/ℓ of Rn-222 were observed in an area geographically adjacent to Logenbaugh and White Oak Creeks, about 13 to 16 kilometers (km) south-southeast of the salt domes near Hockley and Tomball.

The second anomaly, with concentrations up to 7 pCi/ℓ of Ra-226 at the well heads and up to 4 pCi/ℓ in the distribution system, was found on the border between Harris and Fort Bend Counties (far southwest Houston). Of the 11 wells in this section, five had concentrations of Ra-226 greater than 5 pCi/ℓ (the maximum allowed for total Ra), and the lowest concentration (one well) was about 3 pCi/ℓ.   Rn concentrations in this area were up to 3100 pCi/ℓ.   These wells were developed near the piercement-type salt dome, Blue Ridge. Some of the wells were located in the adjacent Fort Bend County but belonged to the City of Houston (Figures 5 and 6).

In the southwest-central part of Harris County, Rn concentrations up to 2900 pCi/ℓ were observed near the salt dome at Pierce Junction.   Ra-226 concentrations up to 4.7 pCi/ℓ were present at the well heads and about 3 pCi/ℓ was observed in the distribution system.   In northeast part of Harris County, a concentration of 22 pCi/ℓ of Ra-226 was encountered earlier by others in a well developed near a piercement-type salt dome at Humble [29].

Figure 7 shows variations in the concentration of Rn-222 in well water as related to the concentration of its progenitor, Ra-226.   The coefficient of correlation (R) between these two variables was 0.77 and the slope of linear regression was positive and significant (probability p of this correlation to be found by chance was less than 0.01).   The empirical equation relating waterborne concentration of Rn to Ra was

$$\hat{Rn} = 307 + 187\ Ra, \qquad\qquad /1/$$

where $\hat{Rn}$ is an estimate of Rn, in pCi/ℓ.

The proportion of variation in Rn concentrations which could be accounted for by Ra was 59 percent.   This shows that Ra and Rn measured under field conditions correlated reasonably well, although not perfectly.   The remaining scatter probably was due to differences in the underground retention opportunities for gas Rn and dissolved Ra. This, in turn, implies that our understanding of the distribution of Ra and Rn in water supplies may further profit from exploring the roles that pumping depth and well location play in this distribution.

The relationship of Ra and Rn concentrations to pumping depth was investigated first by fitting linear regression models to respective data sets. For Ra-226, the slope of regression was not significant, which indicated that the linear model was inappropriate, and this becomes immediately apparent when one inspects the scatter-plots on Figures 8a and 9a.   On each of these figures, a

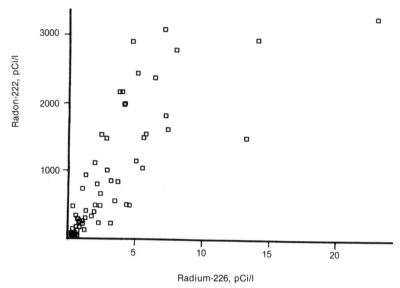

Figure 7.   Relationship between concentrations of Ra-226 and Rn-222 in well water, Harris County, Texas, 1985-1986.

non-linear distribution of Ra with depth was evident. The concentrations were observed to peak in a zone between 180 m (Figure 8a) and 320 m (Figure 9a), which corresponded to a fringe zone between Chicot and Evangeline aquifers (Figure 10a). Further increase in depth caused concentrations of Ra to decrease, although not to zero. This distribution suggests introducing a quadratic term into a previous model. If this is done, the resulting fit becomes statistically significant ($p < 0.05$).

The greatest concentrations of Rn also were observed between 180 and 320 m below the surface (Figures 8b-10b) but, differently from Ra, there was an overall underlying tendency for Rn concentrations to increase with depth. The linear model was appropriate to describe the vertical concentration profile for Rn ($p < 0.01$) and the addition of a quadratic term to the linear model was not an improvement.

The distance from the outcrop of known uranium deposits was not a statistically significant predictor for Ra, nor did the addition of a quadratic term to a linear model improve the fit. Examination of the scatter-plot on Figure 11a indicates that Ra distribution along the northwest to southeast axis downdip from uranium deposits was bi-modal, with the first increase in concentrations observed at about 90 km southeast from a reference line and the second, after 115 km. The location of the first anomaly coincided with the area of geological faulting stretching between the Hockley and Tomball salt domes, about 13 to 16 km southeast of the domes [30]. The location of a second, comparatively less severe Ra anomaly coincided with the area of the fault system associated with the Blue Ridge salt dome.

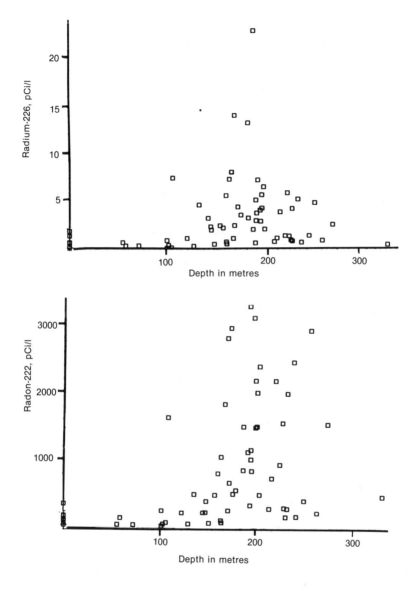

Figure 8.   Distribution of (a) Ra-226 and (b) Rn-222 with depth;
            upper limit of pumping range.

These data indicate that Ra travels in a more complex path
than could be expressed by a linear distance from known deposits of
uranium and their Ra in Houston-Harris County water wells probably
relates to local sources, instead of distant deposits in the
outcrop.

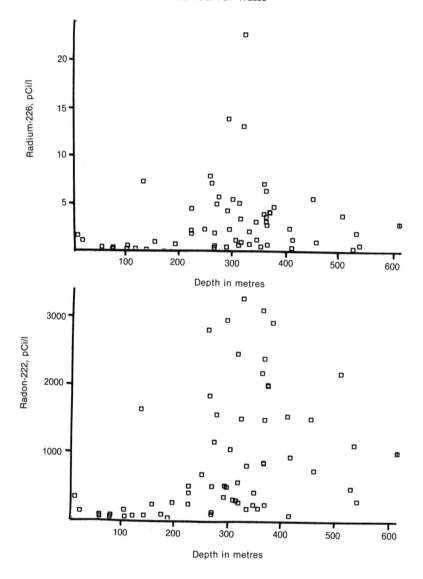

Figure 9. Distribution of (a) Ra-226 and (b) Rn-222 with depth; lower limit of pumping range.

The bi-modal distribution with distance was also noted for Rn (Figure 11b) but, again, differently from Ra, there was an underlying tendency for Rn concentrations to increase toward the coast. The linear model fitted to this distribution had a positive slope and was statistically significant ($p < 0.01$).

The proximity of a salt dome was an important predictor for the presence of both Ra and Rn in well water. The inverse rela-

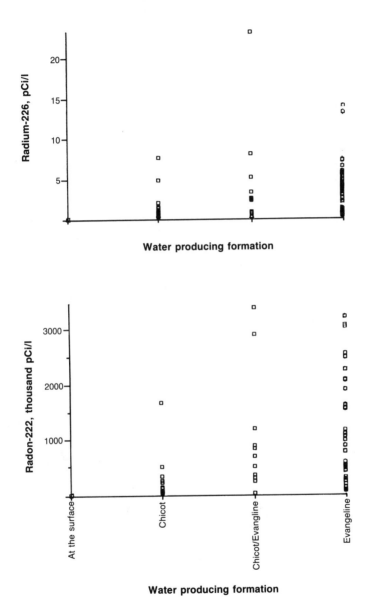

Figure 10. Distribution of (a) Ra-226 and (b) Rn-222 depending on water-producing formation.

tionship, the increase in concentrations as distances to the domes decreased, was particularly evident for wells developed near Blue Ridge and Pierce Junction salt domes (Figures 12a and b). Because of the rural location, no deep municipal wells were present in the vicinity of the Hockley and Tomball salt domes and this fact might

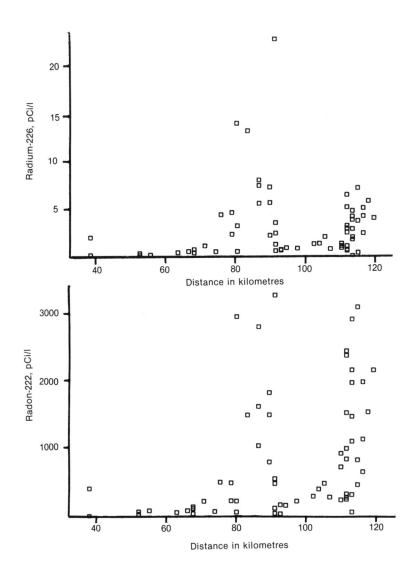

Figure 11.   Distribution of (a) Ra-226 and (b) Rn-222 with the
distance from the outcrop of known uranium deposits.

have obscured the statistical association. The correlation of Ra
with other salt domes in Harris County could not be made because
they were situated in a part of Harris County which was supplied by
surface water.

Multivariate analyses showed that a combination of factors
provided still a better fit. Where Ra is concerned, both pumping
depth and the distance from the nearest salt dome were significant

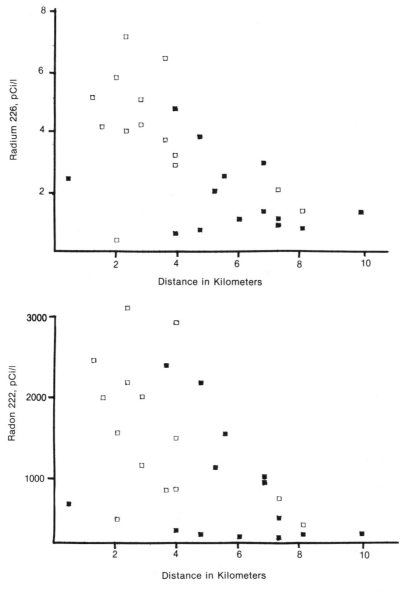

Figure 12.   Distribution of (a) Ra-226 and (b) Rn-222 with the distance to Blue Ridge and Pierce Junction salt domes.

factors (p < 0.01) and the equation relating waterborne concentrations of Ra to these factors was

$$\hat{Ra} = 0.06 + 0.04(DEPTH) - 0.09 \times 10^{-3}(DEPTH)^2 - 0.60(DNS) +$$

$$+ 0.04(DNS)^2, \qquad\qquad /2/$$

where (DEPTH) stands for pumping depth (m) measured to the top screen of pumping range and (DNS) was the distance from the nearest salt dome (km). Because the proximity of salt domes was measured as the distance (km) from the dome nearest to each well and regardless of direction, the model in equation 2 includes quadratic terms for this variable as well as for the variable of depth. The statistical analysis supporting equation 2 is illustrated in Table 1.

Table 1.   Multiple Regression Analysis Report for Ra-226 vs. Pumping Depth and Location.

Source	df	Sequential Sum-Sqr	R^2	F-Ratio	Tail Prob	Last Sum-Sqr	F-Ratio
Constant	1	632.33					
DEPTH	1	42.37	4.67	4.66	0.037	55.43	5.83
$(DEPTH)^2$	1	50.73	10.25	5.33	0.023	21.15	2.22
DNS	1	141.47	25.83	14.88	0.001	50.96	5.36
$(DNS)^2$	1	112.44	38.21	11.82	0.000	112.44	11.82
Model	4	347.0188	38.21	9.12			
Error	59	561.0715					
Total	63	908.0902					
Mean Square Regression		86.7569					
Mean Square Error		9.509686					

Parameter Estimation			
Variable	Regression Coefficient	Standard Coefficient	Standard Error
Constant	5.98E-02	0	0
DEPTH	3.91E-02	.7475541	1.62E-02
$(DEPTH)^2$	8.66E-05	-.4710227	5.81E-05
DNS	-.62	-.9650517	.27
$(DNS)^2$	4.49E-02	1.42079	1.30E-02

DEPTH - Depth of pumping, top screen, m;

DNS - Distance from salt domes, km

The R^2 equal to 38 indicated that two basic factors, pumping depth and distance from salt domes, accounted for 38 percent of the variations in Ra.

An alternative expression, for depth measured to the bottom of the pumping range, was

$$\hat{Ra} = -0.13 + 0.02(DEPTH) - 0.03 \times 10^{-3}(DEPTH)^2$$

$$- 0.61(DNS)^2 + 0.04(DNS)^2. \qquad\qquad /3/$$

For Rn, the best combination of variables was

$$\hat{Rn} = -413 + 4(DEPTH) + 11(DDR) - 197(DNS)$$

$$+ 12(DNS)^2 \qquad\qquad /4/$$

for the top of the pumping range and

$$\hat{Rn} = -190 + 2(DEPTH) + 10(DDR) - 216(DNS)$$

$$+ 12 (DNS)^2 \qquad\qquad /5/$$

for the bottom of the pumping range.

The variable (DDR) in equations 4 and 5 represents distance (km) downdip from the recharge zone, as explained in the "Method" section. The statistical analysis that led to derivation of these equations is illustrated in Table 2. The combination of three basic variables describing pumping depth and location of water source accounted for 42 percent of the variations in Rn.

If data on Ra concentrations are available, the Rn concentrations can be estimated as

$$\hat{Rn} = -367 + 171(Ra) + 2(DEPTH) + 9(DDR)$$

$$- 104(DNS) + 4(DNS)^2 \qquad\qquad /6/$$

for the top of the pumping range, and

$$\hat{Rn} = -266 + 173(Ra) + (DEPTH) + 8(DDR)$$

$$- 112(DNS) + 5(DNS)^2 \qquad\qquad /7/$$

for the bottom of the pumping range. This combination of factors described 74 percent of variations in Ra (Table 3).

DISCUSSION

The Gulf Coast is underlain by productive ground water aquifers, an important factor which has facilitated the urban and industrial development of the Greater Houston-Harris County area. For years all of the water needs in the City of Houston and Harris County have been satisfied entirely by ground water. However, years of pumping large volumes of water have caused a decline in

Table 2. Multiple Regression Analysis Report for Rn-222 vs. Pumping Depth
and Location

Source	df	Sequential				Last	
		Sum-Sqr	R^2	F-Ratio	Tail Prob	Sum-Sqr	F-Ratio
Constant	1	5.41E+07					
DEPTH	1	9408528	17.62	18.08	0.000	4851181	9.32
DDR	1	2793209	22.85	5.37	0.023	2443481	4.69
DNS	1	2863555	28.22	5.50	0.012	5003873	9.61
(DNS)^2	1	7618718	42.49	14.64	0.001	7618717	14.64
Model	4	2.27E+07	42.49	10.90	0.000		
Error	59	3.07E+07					
Total	63	5.34E+07					
Mean Square Regression		5671002					
Mean Square Error		520469					

Parameter Estimation

Variable	Regression Coefficient	Standard Coefficient	Standard Error
Constant	-413.01	0	0
DEPTH	4.37	.345	1.43
DDR	10.72	.25	4.95
DNS	-197.16	-1.27	63.58
(DNS)^2	11.79	1.54	3.08

DEPTH - Depth of pumping, top screen, m;

DDR - Distance downdip from recharge km;

DNS - Distance from salt domes, km

the potentiometric pressure and subsidence of the land, especially
in southeast Houston [31]. Since the 1950's the prevailing philos-
ophy behind water development in Houston has been that a) reorien-
tation of the water supply system is necessary from total
dependence on ground water to a combination of surface and ground
water sources; and that b) wells need to be redistributed to reduce
the rate of subsidence, phasing out most of the wells in the
southeast and developing new wells in northwest Harris County,
closer to the recharge zone.

In 1954 the first surface water from the San Jacinto River via
Lake Houston was provided to the east side of Houston. On the
other hand, the development of ground water has been actively
pursued in northwest, northeast, and southwest Harris County to
accommodate urban expansion in these directions.

Table 3.   Multiple Regression Analysis Report for Rn-222 vs. Ra-226,
           Pumping Depth, and Location

			Sequential			Last	
Source	df	Sum-Sqr	R^2	F-Ratio	Tail Prob	Sum-Sqr	F-Ratio
Constant	1	5.41E+07					
RA	1	3.19E+07	59.76	134.10	0.000	169E+07	71.06
DEPTH	1	3578880	66.47	15.04	0.002	779210	3.23
DDR	1	2591628	71.23	10.89	0.002	1614991	6.79
DNS	1	568185	72.38	2.39	0.124	1307976	5.50
(DNS)^2	1	944674	74.14	3.97	0.048	944674	3.97
Model	5	3.96E+07	74.15	33.28	0.000		
Error	58	1.38E+07					
Total	63	5.34E+07					
Mean Square Regression		7918282					
Mean Square Error		237935.6					

Parameter Estimation

Variable	Regression Coefficient	Standard Coefficient	Standard Error
Constant	-366.87	0	0
Ra	170.83	.704	20.26
DEPTH	1.82	.14	1.01
DDR	-8.73	.20	3.35
DNS	-104.07	-.67	44.39
(DNS)^2	4.49	.59	2.26

Ra - Radium 226, pCi/ℓ

DEPTH - Depth of pumping, top screen, m;

DDR - Distance downdip from recharge km;

DNS - Distance from salt domes, km

Data gathered by Brock [20] in northwest Harris County indicated that at least 12 MUD's, those serving recently developed suburban subdivisions, violated standards with respect to Ra in the public drinking water.

Our findings confirmed excess Ra in the water served by these relatively small utility districts. In addition, water from five wells belonging to the City of Houston-proper (former MUDs which were acquired by the City to accommodate newly added subdivisions on the far southwest) contained Ra at concentrations greater than 5 pCi/ℓ.

Observed excess Ra was associated with ground water. Concentrations in water from surface sources were less than 0.5 pCi/ℓ and no surface-water violations for Ra were found.

Earlier attempts to correct contaminated wells in northwest Harris County involved plugging the lower-most screens, because of an assumption that Ra concentration increased progressively with depth. The present study, however, shows that this assumption is justified for Rn, but not for Ra. The linear model that was initially applied in an attempt to describe Ra variations with depth was found to be inadequate and a quadratic term for nonlinearity was required to achieve a better fit. Concentrations of Ra tended to peak between 180 and 320 m below the surface. This depth suggests that Ra might be associated with the upper Evangeline - lower Chicot aquifers. Under such conditions plugging the deepest screens in the multiple screen wells may cause Ra concentrations to become greater instead of smaller.

The increase of Rn with depth, unlike Ra, may be related to differential retention, accumulation, and diffusion properties of these radioisotopes. Rn is a short-lived, highly volatile radioactive gas whereas Ra is a dissolved cation. For an equal amount of paternal Ra, its gaseous progeny Rn may have a better chance to accumulate in the deepest strata, while it may dissipate more readily from progressively shallower formations. Water-bearing strata in the study area dip toward the coast and, hence, water wells in the southeasterly direction tend to be progressively deeper. The trend for Rn concentrations to increase in the direction of the Gulf may relate to this factor of increasing depth.

The Ra and Rn in ground water from the Evangeline and Chicot aquifers may indicate that waters have flowed around the flanks of salt domes that pierce the aquifers. Uranium deposits have been found in the rocks that flank or overlay Gulf Coast salt domes, such as the Palangana dome [32] and the Hockley dome [33], and may be the source for the Ra and Rn. The uranium presumably is precipitated in the reducing environment surrounding the dome. As ground water flows past the dome it entrains the soluble Ra but leaves the insoluble uranium.

Natural ground-water flow in the Gulf Coast aquifer is down the stratigraphic dip toward the coast at a rate of approximately 1 m per year [34,35]. However, the extremely heavy pumpage in the Houston area has created an extensive cone of depression [36] and reversed the direction of flow. The presence of high concentrations of Ra hydrologically updip from the Pierce Junction dome (Figure 5) further suggests this reversal. The dissolved Ra has been transported up to 4 km over 30 years (the approximate age of the cone of depression); the rate of 130 m per year is significantly greater than natural ground-water flow rates of 1 m per year.

An alternative hypothesis for the source of Ra and Rn is that they originated from uranium deposits in the deeper formations and migrated up the dome flanks and associated faults into shallower formations.

The major deposits of uranium in the Gulf Coast are associated with the Catahoula formation of the Miocene age. This formation, which in South Texas is mined commercially for uranium, extends along the entire coast from Mexico to Louisiana. The Catahoula Sandstone is also the deepest fresh-water bearing unit in the Gulf

Coast aquifer complex. It crops out inland from the study area and dips toward the Gulf at a rate of about 17 m per km; near the coastline it is approximately 1.6 km below the land surface [28].

The formations which constitute the Gulf Coast aquifer range in age from the Oligocene and Miocene eras for the Catahoula sandstone to the Quarternary period for shallow alluvium. The actual production of well water involves strata of the Pliocene-Pleistocene age (Evangeline and Chicot aquifers) rather than the Miocene age. The presence of Ra and Rn in well water derived from the Evangeline and Chicot aquifers may be evidence of a cross-formational flow where deeper water with greater potentiometric head from Miocene strata leaks upward through confining strata [24;27;37;38]. Such seepage would be most prominent up the flanks of piercement-type salt domes or associated faults.

The vertical concentration profile exhibited by Rn supports the hypothesis of an upward flow from greater depths. However, the non-linear distribution with depth observed for Ra and a rather narrow range of pumping depths in which Ra concentrations occur suggest a source in the upper Evangeline aquifers. In addition, any Ra from a Catahoula source should have decayed to below detection levels considering the long flow paths downdip from a uranium deposit in the Catahoula to a dome, up the dome flanks, and then into the Evangeline and Chicot aquifers. Because the half-life of Ra-226 is 1600 years, Ra will decay to concentrations below detection limits in a ground water flow system if travel times for the Ra are more than a few thousand years. The areas of high Ra and Rn concentrations in the Evangeline aquifer are 80 to 120 km downdip from its outcrop. At a flow rate of 1 m per year, these waters are older than measurable by Ra concentrations. If a flow path through the Catahoula is envisioned, travel time will be even greater.

Ra as a ground water tracer may give conservative ground water flow velocities because of its chemically reactive nature. Dissolved Ra should function chemically similarly to dissolved calcium. Tanner [39] proposed that in saline waters the abundant cations compete with Ra for exchange sites and thus a greater portion of Ra ions remains in solution. Conversely, waters with low ionic strengths are favorable for Ra adsorption onto the aquifer matrix. Kraemer and Reid [40] studied Ra in deep saline formation waters and observed this direct correlation between salinity and Ra activity. In the Evangeline aquifer the migration of Ra may be retarded by ion exchange or conversely enhanced if dome dissolution increases the salinity around a dome.

In our study, the proximity of salt domes was found to be a strong predictor of Ra and Rn presence in well water, particularly in combination with a certain range of pumping depths. Whether Ra is related to uranium associated with salt domes or to brine leakage up the flanks of salt domes cannot be answered at this time. We advise, however, against developing domestic wells near salt domes, especially medium-to-deep wells (180 m and deeper).

The health significance of Ra and Rn-rich water in Harris County is not known. All the water wells in question were located in subdivisions that have existed for only about 10 to 15 years. In northwest Harris County the customers were notified by the respective MUD's that Ra levels in their water were in excess of MCL. In southwest Houston, where a blend of water from several

wells is distributed, the city, technically, is not under obliga-
tion to notify customers regarding Ra concentrations present in
specific component-wells, as long as the levels in the distribution
system are not in excess of the maximum allowable.

Blending water may not be the most reliable option, however,
for dealing with Ra and Rn, since resultant concentrations vary
depending on several, not entirely predictable factors, i.e., how
many wells are in operation at any given time, the pumping rate of
each contributing well, pressure in water lines, and demand.   The
boundaries of service areas for each particular component in the
distribution system are not known.   It would be advisable, instead,
for the City of Houston to take steps toward replacing, re-working,
or treating problem wells.

The EPA has published two booklets advising homeowners on
methods to reduce indoor concentrations of Rn [8,9].   In these
documents, users of private wells were informed that domestic water
might contribute Rn to the indoor air.   However, the EPA said this
usually is not a problem when large community water supplies are
concerned, since water supposedly releases most of its Rn before
reaching individual houses.   In our study in Houston-Harris County,
parallel measurements of Rn made at municipal well heads and at
consumers' taps indicated that the en route self-aeration of well
water does not occur.   There was no evidence of appreciable drop in
Rn concentrations in municipal well water upon its arrival at
household taps.

Water treatment for Ra and Rn is available.   The technology of
Rn and Ra removal does not differ much from other gases or alkaline
earth elements, i.e., well-known and common technology for removal
of calcium or magnesium hardness.   The economics of rehabilitation
by lime or lime-soda softening, weak acid ion exchange, or reverse
osmosis have been reviewed recently by Snoeyink et al. [41] and
Brock [20]).   To date, some MUD's in Harris County have been
successful in reducing levels of Ra and Rn by altering screening
depths in their wells.

SUMMARY

This study showed that anomalous concentrations of Ra-226 and
Rn-222 exist in parts of the upper Texas Gulf Coast.   While all
factors influencing the distribution and fate of pollutants in
underground porous media might not be easy to measure or even
identify, often the very expectation of this great complexity
prevents one from observing patterns that otherwise might be quite
helpful.   The findings summarized in this communication present an
encouraging picture for predicting depths and locations in the Gulf
Coast where elevated Ra and Rn concentrations may be encountered.
Two to four key variables accounted for the statistically significant variation (from 38 to 74 percent) in concentrations observed
under field conditions.

ACKNOWLEDGEMENTS

The authors are grateful to Jeffrey Strauss of the United States Geological Survey (USGS), Houston Office, for his help in data gathering and to Robert Gabrysch, District Chief of the USGS at Houston for his valuable advice during this study. Appreciation is expressed to Arlin Howles of the Houston Geological Society for peer review of this paper, and to Gay Robertson and Peggy Powell of the University of Texas at Houston, School of Public Health for their help in the preparation of the manuscript.

REFERENCES

1. Martland, M. S. "Occupational Poisoning in Manufacture of Luminous Watch Dials," JAMA 92:187-192 (1929).

2. Polednak, A. P., A. F. Stehney and R. E. Rowland. "Mortality among Women First Employed before 1930 in the United States Radium Dial-Painting Industry: A Group Ascertained from Employment Lists," Am. J. Epidemiol. 107:179-195 (1978).

3. Tountas, A. A., V. L. Fornasier, A. R. Harwood and P. M. Leung. "Postirradiation Sarcoma of Bone: A Perspective," Cancer 43:182-187, (1979).

4. Mays, C. S. and H. Spiess. "Bone Tumors in Thorotrast Patients," Environ. Res. 18:88-93 (1979).

5. Mays, C. W. and H. Spiess. "Bone Sarcomas in Patients Given Radium-224," in Radiation Carcinogenesis, Epidemiology and Biological Significance, J. D. Boice and J. F. Fraumeni, Eds. (New York: Raven Press, 1984), pp. 241-252.

6. Sevc, J. E., E. Kunz and V. Placek. "Lung Cancer in Uranium Mines and Long-Term Exposure to Radon Daughter Products," Health Phys. 30:433-437 (1970).

7. Archer, V. E., J. K. Wagoner and R. E. Lundin. "Lung Cancer among Uranium Miners in the U.S.," Health Phys. 25:351-371 (1973).

8. "A citizen's guide to radon; what it is and what to do about it," United States Environmental Protection Agency, U.S. Department of Health and Human Services, OPA-86-004 (1986).

9. "Radon Reduction Methods; A Homeowner's Guide," U.S. EPA, U.S. DHHS Report OPA-86-005 (1986).

10. Petersen, N. J., L. D. Samuels, H. F. Lucas and S. P. Abraham. "An Epidemiologic Approach to Low-Level Radium-226 Exposure," Public Health Reports 81:805-814 (1966).

11. Bean, J., P. Isacson, R. M. A. Mahne and J. Kohler. "Drinking Water and Cancer Incidence in Iowa. II. Radioactivity in Drinking Water," Amer. J. of Epidemiol. 16(6):924-932 (1982).

12. Lyman, G. H., C. G. Lyman and W. Johnson. "Association of Leukemia with Radium Ground Water Contamination," JAMA 254:621-626 (1985).

13. United States Environmental Protection Agency (1975) National Interim Primary Drinking Water Regulations. Federal Register 40(248), Part 1V.

14. "Drinking Water Standards Governing Drinking Water Quality and Reporting Requirements for Public Water Supply Systems," Texas Department of Health, Division of Water Hygiene, Austin, Texas (1977).

15. Kier, R. S., L. E. Garner and L. F. Brown. "Land Resources of Texas," The University of Texas at Austin, Bureau of Economic Geology (1977).

16. Cook, L. M. "The Uranium District of the Texas Gulf Coastal Plain," in Natural Radiation Environment III, T. F. Gessel and W. M. Lowder, Eds. (Washington, D.C.: U.S. Department of Energy, 1980), pp. 1602-1922.

17. Texas Department of Health, Bureau of Radiation Control, Environmental Assessment, Safety Evaluation Report, and Proposed License Conditions Related to Anaconda Minerals Company-Rhode Ranch Project, McMullen County, Texas, 1982.

18. Cech, I., H. M. Prichard, A. M. Mayerson and M. Lemma. Unpublished results (1984).

19. Texas Department of Health, Division of Water Hygiene, Chemical Analyses of Public Water Systems, 1983.

20. Brock, J. D. "Radioactivity of Drinking Water in Northwest Harris County," Master of Public Health Thesis, The University of Texas School of Public Health, Houston, Texas (1984).

21. Prichard, H. M., T. F. Gesell and C. R. Mewyer. "Liquid Scintillation Analysis for Radium-226 and Radon-222 in Potable Water," in Liquid Scintillation Counting - Recent Applications and Development, C. T. Peng, D. L. Horrocks and E. L. Alpen, Eds. (New York: Academic Press, Inc., 1980), Vol. 1, pp. 347-355.

22. Dougenik, J. A. "SYMAP User's Manual," Laboratory of Computer Graphics and Spatial Analysis, Harvard University, Cambridge (1975).

23. United States Geologic Survey, Houston Office. Open files and personal communication (1986).

24. Winslow, A. G. and W. W. Doyel. "Salt Water and Its Relations to Fresh Ground Water in Harris County, Texas," Texas Board of Water Engineers, Bulletin 5409, Houston, Texas, 1954.

25. Wood, L. A., R. K. Gabrysch and R. Marvin. "Reconnaissance Investigation of the Ground-Water Resources of the Gulf Coast Region, Texas," Texas Water Commission, Bulletin 6305, Austin, Texas, 1963.

26. Gabrysch, R. K. "Development of Ground Water in the Houston District, Texas, 1961-65," Texas Water Development Board, Report 63, Austin, Texas, (1967).

27. Wesselman, J. B. and S. Aronow. "Ground-Water Resources of Chambers and Jefferson Counties, Texas," Texas Water Development Board, Report 133. Austin, Texas (1971).

28. Popkin, B. P. "Ground-Water Resources of Montgomery County, Texas," Texas Water Development Board, Report 136, Austin, Texas (1971).

29. Harris County Health Department, open files and personal communication (1985).

30. Kreitler, C. W. "Fault Control of Subsidence, Houston-Galveston Area, Texas," The University of Texas at Austin, Bureau of Economic Geology, Research Note 5, Austin, Texas, 1976.

31. Turner, Collie and Braden, Consulting Engineering Co. "Comprehensive Study of Houston's Municipal Water System," Report to the City of Houston (1968).

32. Weeks, A. D. and D. H. Eargle. "Uranium at Palangana Salt Dome, Duval County, Texas." U.S. Geological Survey Prof. Paper 400-B. Austin, Texas (1960).

33. Kyle, J. R. and P. E. Price. "Metallic Sulphide Mineralization in Salt Cap Rocks, Gulf Coast, USA," Trans. Inst. Metall. 95:1310-1316, (1986).

34. Kreitler, C. W., E. Guevera, G. Granata and D. McKalips. "Hydro-Geology of Gulf Coast Aquifers, Houston-Galveston area, Texas," Transactions-Gulf Coast Association of Geological Societies, vol. XXVII, 1977.

35. Fogg, G. E., and C. W. Kreitler. "Ground-Water Hydraulics and Hydrochemical Facies of the Eocene Aquifers in East Texas Basin," The University of Texas at Austin, Bureau of Economic Geology, Report of Investigation No. 127, 1982.

36. Gabrysch, R. K. "Land-Subsidence in the Houston-Galveston Region, Texas," International Association of Hydrological Sciences, Second International Symposium on Land Subsidence Proceedings, (1972) pp. 16-24.

37. Wesselman, J. B. "Ground-Water Resources of Fort Bend County, Texas," Texas Water Development Board Report 155, Austin, Texas, 1972.

38. Kreitler, C. W. "Ground-Water Hydrology of Depositional Systems," in Depositional and Ground-Water Flow Systems in the Exploration for Uranium - A Research Colloquium, W. E. Galloway, C. W. Kreitler and J. H. McGowen, Eds. The University of Texas at Austin, Bureau of Economic Geology, 1979, pp. 118-1767.

39. Tanner, A. B. "Physical and Chemical Controls on Distribution of Radium-226 and Radon-222 in Ground Water Near Great Salt Lake, Utah," in Proceedings of International Symposium on the Natural Radiation Environment, J. C. Adams and W. M. Lowder, Eds. (Chicago, Illinois: University of Chicago Press, 1964, pp. 253-276.

40. Kraemer, T. F. and D. R. Reid. "The Occurrence and Behavior of Radium in Saline Formation Water of the U.S. Gulf Coast Region," Isotope Geoscience 2:153-174 (1984).

41. Snoeyink, V. I., J. L. Pfeffer, D. W. Snyder and C. C. Chambers. "Barium and Radium Removal from Groundwater by Ion Exchange," Environmental Protection Agency EPA-600/S2-84-093, Cincinnati, Ohio, 1984.

RADIUM, RADON AND URANIUM ISOTOPES IN GROUNDWATER FROM CAMBRIAN-ORDOVICIAN SANDSTONE AQUIFERS IN ILLINOIS

Robert H. Gilkeson, Illinois State
Geological Survey, Champaign, Illinois

James B. Cowart, Florida State University,
Tallahassee, Florida

ABSTRACT

The regional occurrences of selected radioactive isotopes in groundwater were studied in an investigation of the natural geologic sources of high concentrations of ^{226}Ra and ^{228}Ra in groundwater from wells finished in the Cambrian and Ordovician bedrock in northern Illinois. The combined dissolved concentration of the two isotopes ranges from 2.0 to greater than 50.0 pCi/L. Over 100 public water supplies in northern Illinois exceed the U.S. EPA Interim Drinking Water Standard of 5.0 pCi/L. Most supply wells are over 1000 feet deep and open to receive groundwater from sandstone, dolomite and shale lithologies. The most productive aquifer units, the sandstones, provide the high dissolved radium concentrations.

Dissolved concentrations of ^{222}Rn range from 40.0 to 1000 pCi/L. The combined dissolved concentrations of ^{238}U and ^{234}U range from less than 0.1 pCi/L to 8.0 pCi/L; concentrations greater then 1.0 pCi/L reflect marked enrichment in ^{234}U. Over large regions of northern Illinois groundwater in the Cambrian-Ordovician bedrock is uniquely enriched in ^{234}U.

Dissolved concentrations of ^{226}Ra are poorly correlated with dissolved concentrations of ^{238}U, ^{234}U, ^{222}Rn and ^{228}Ra. However, a significant source of dissolved ^{226}Ra is the chemical Precipitation adsorption of the parent ^{238}U, ^{234}U and ^{230}Th

nuclides on silica surfaces of the sandstones. A
significant source of dissolved ^{228}Ra is the
occurrence of ^{232}Th enriched accessory minerals in
the sandstone bedrock. The ionic strength of
groundwater is an important control on the
dissolution of ^{226}Ra. Mechanisms important to the
dissolution of both ^{226}Ra and ^{228}Ra are alpha recoil
and in specific localities the solubility of barite.

INTRODUCTION

Over the past decade there has developed an
increasing national concern for an assessment of the
natural radioactivity environment and its impact on
human health. Radium is one of the nuclides of
primary concern. The U.S. Environmental Protection
Agency (1) has promulgated an interim primary
drinking water standard of 5.0 pCi/L for the combined
concentration of the two radium nuclides, ^{226}Ra in
the ^{238}U decay series and ^{228}Ra in the ^{232}Th decay
series. There is also great concern for the health
impact of ^{222}Rn, the daughter nuclide formed by
disintegration of ^{226}Ra. The U.S. EPA is currently
reviewing the health impact from radium, radon and
uranium as part of the process of establishing
drinking water standards for these nuclides.
A nationwide study of the occurrence of natural
radioactivity in public water supplies (2) determined
that there are two specific geologic regions where
more than 75% of the known radium violations occur.
One region is the Piedmont and Coastal Plain
provinces in New Jersey, North Carolina, South
Carolina and Georgia. The second region is in the
north-central United States and includes parts of
Minnesota, Iowa, Illinois, Missouri, and Wisconsin.
The impacted groundwater supplies in the north-
central region are largely produced from aquifers in
bedrock of Cambrian and Ordovician age. This paper
summarizes research findings concerning the
occurrence of ^{226}Ra, ^{28}Ra, ^{222}Rn, ^{238}U and ^{234}U in
groundwater from the Cambrian and Ordovician bedrock
in the Illinois part of the north-central region.
The research is documented in a series of papers (3-
8).

HYDROGEOLOGIC SETTING AND GROUNDWATER CHEMISTRY

The study area for this paper is the 35-county
region of northern Illinois where public and
industrial groundwater supplies are produced from
aquifers in the Cambrian and Ordovician bedrock
(Figure 1). A generalized column of rock-
stratigraphic units in northen Illinois is shown in
Figure 2. The Cambrian and Ordovician bedrock is
composed of thick sections of sandstone and dolomite

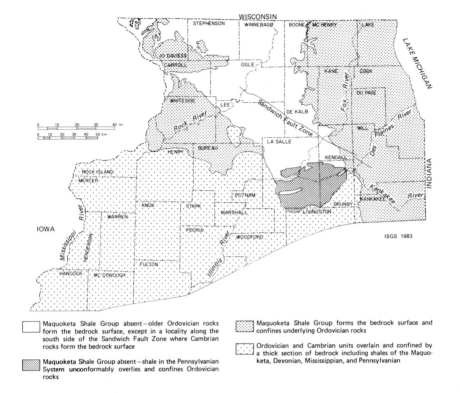

Figure 1. Geologic map of the area of northern
Illinois where groundwater supplies are produced from
the Cambrian and Ordovician bedrock.

interbedded with shale. On a regional scale the
significant aquifers are the sandstones, including
the St. Peter, the Ironton-Galesville, the Elmhurst
Member in the basal Eau Claire Formation, and the Mt.
Simon. The Ironton-Galesville Sandstone is the most
productive aquifer over a major part of the study
area.

Glacial drift deposits blanket northern Illinois
with thicknesses over bedrock varying from less than
5 meters to greater than 150 meters. Figure 1 shows
the subcrop of bedrock units beneath the glacial
drift. The stippled pattern on the figure shows that
over much of the study area aquifers in the Cambrian
and Ordovician bedrock are overlain and confined by
shale bedrock. The most significant confining bed is
the Maquoketa Shale Group. In the study area
aquifers with high dissolved radium concentrations
occur below the Maquoketa. Groundwater from poorly
productive aquifers within the Maquoketa Shale Group

SYSTEM	SERIES	GROUP OR FORMATION	LOG	THICKNESS (m)	DESCRIPTION
Quarternary	Pleistocene			0-180	Unconsolidated glacial deposits – pebbly clay (till), silt, and gravel; loess (windblown silt), and alluvial silt, sand, and gravels.
Tertiary and Cretaceous				0-30	Sand and silt.
Pennsyl-vanian				0-150	Mainly shale with thin sandstone, limestone and coal.
Mississippian	Valmeyeran	St. Louis, Salem, Warsaw, Keokuk-Burlington		0-180	Limestone; cherty limestone; green, brown and black shale; silty dolomite.
Mississippian	Kinderhookian				
Devonian				0-120	Shale, calcareous; limestone strata, thin.
Silurian	Niagaran			0-140	Dolomite, silty at base, locally cherty.
Silurian	Alexandrian	Kankakee, Edgewood			
Ordovician	Cincinnatian	Maquoketa		0-75	Shale, gray or brown; locally dolomite and/or limestone, argillaceous.
Ordovician	Champlainian	Galena		0-140	Dolomite and/or limestone, cherty; Dolomite, shale partings, speckled;
Ordovician	Champlainian	Platteville			Dolomite and/or limestone, cherty, sandy at base.
Ordovician	Champlainian	Glenwood		0-200	Sandstone, fine- and coarse-grained; little dolomite; shale at top.
Ordovician	Champlainian	St. Peter			Sandstone, fine- to medium-grained; locally cherty red shale at base.
Ordovician	Canadian	Prairie du Chien Group: Shakopee		30-400	Dolomite, sandy, cherty (oolitic), sandstone.
Ordovician	Canadian	Prairie du Chien Group: New Richmond			Sandstone, interbedded with dolomite.
Ordovician	Canadian	Prairie du Chien Group: Oneota			Dolomite, white to pink, coarse-grained, cherty (oolitic), sandy at base.
Cambrian	Croixan	Eminence and Potosi			Dolomite, white, fine-grained, geodic quartz, sandy at base.
Cambrian	Croixan	Franconia			Dolomite, sandstone, and shale, glauconitic, green to red, micaceous.
Cambrian	Croixan	Ironton		20-80	Sandstone, fine- to medium-grained, well sorted, upper part dolomitic.
Cambrian	Croixan	Galesville			
Cambrian	Croixan	Eau Claire: Proviso, Lombard, Elmhurst		75-140	Shale and siltstone; dolomite, glauconitic; sandstone, dolomitic, glauconitic.
Cambrian	Croixan	Mt. Simon		150-790	Sandstone, coarse-grained, white, red in lower half; lenses of shale and siltstone, red, micaceous.
Precambrian					

Figure 2. Generalized column of rock-stratigraphic units in northern Illinois.

and groundwater in aquifers above the Maquoketa have low radium concentrations.

Aquifers in the Cambrian and Ordovician are unconfined or poorly confined over a large area of north-central and northwestern Illinois. In this region, the Cambrian and Ordovician bedrock aquifers are recharged by local precipitation and groundwaters and are generally dilute even in the deeply buried Mt. Simon Sandstone. Total dissolved solids range from approximately 200-400 mg/L with the dominant ions being calcium, magnesium and bicarbonate. Concentrations of sodium, chloride, and sulfate are generally very low.

A portion of the recharge in the north-central region enters regional flow systems that extend to the east and south where the Cambrian and Ordovician aquifers are confined by shale bedrock. In the confined aquifers there is an increase in total dissolved solids to the east, south and southwest in the study area. Dissolved constituents that show a marked increase are sulfate, chloride and sodium. Beyond the southern boundary of Figure 1, groundwater in the Cambrian and Ordovician bedrock is not used for potable water supplies because concentrations exceed 1500 mg/L in the southern, and 3000 mg/L in the southwestern parts of the area. In the confined region of the study area, there is also an increase in groundwater mineralization with increasing depth in the Mt. Simon Sandstone. In northeastern Illinois, brackish groundwaters (>3000 mg/L) are present at moderate depths in the Mt. Simon Sandstone and dissolved mineral concentration increases to 75,000 mg/L or greater at the bottom of the sedimentary pile above the granite basement.

In unconfined and poorly confined aquifers in north-central and northwestern Illinois, the isotopic composition of the water molecule exhibits values that are comparable to modern precipitation in the area. However, in confined aquifers to the east, south and southwest, there is a shift in the isotopic composition that indicates the source of the groundwater was meteoric water formed at high latitude in a much colder thermal regime. Gilkeson et al. (4) concluded that a source of recharge related to glaciation was required for groundwater in the confined aquifers in the study area. Siegel and Mandle (9) report similar isotopic evidence for a source of recharge related to glaciation for confined Cambrian and Ordovician aquifers in Minnesota, Iowa, Missouri, Illinois, Wisconsin, and Indiana. This is the north-central region of high dissolved radium concentrations noted by Hess et al. (2).

The stable isotope composition of dissolved

sulfate was determined to investigate the source of
the high concentrations that occur in confined
aquifers in the study area. The isotopic data were
used to show that marine evaporites were the source,
but that the source beds were of a different age than
Cambrian or Ordovician (4,8). In the eastern part of
the study area the dissolved sulfates have an
isotopic composition that is comparable to the range
shown for evaporites of Silurian and Devonian Age
(10). Extensive sequences of evaporites occur east
of the study area in Silurian and Devonian rocks that
flank and underlie Lake Michigan. During the recent
glacial stages in Illinois (approximately 13,000
years before the present) continental glaciers
occupied Lake Michigan with ice boundaries extending
across the eastern part of the study area. Gilkeson
et al. (4) postulated that pressure from the glacial
ice overburden caused basal meltwater containing
sulfate and other salts to recharge into the Cambrian
and Ordovician bedrock through the floor of Lake
Michigan. Subglacial recharge to the confined
aquifer may also have occurred: 1) where glacial ice
advanced over the Des Plaines Disturbance, an
anomalous region of deeply fractured rock and complex
structure in the northeastern part of DuPage County
and western Cook County that is suspected to be a
meteorite impact feature and, 2) where glaciers
advanced over locally unconfined regions in Grundy
and LaSalle Counties (Figure 1).

RADIOACTIVE ISOTOPES IN GROUNDWATER

Radium-226 and Radium-228

    The present research program on the occurrence of
226Ra and 228Ra in groundwater from the Cambrian and
Ordovician bedrock greatly benefitted from earlier
research (11-16), and unpublished analyses from the
files of Henry Lucas and Richard Holtzman.
    Gilkeson et al. (5) compared new analyses to the
historical data and concluded that the concentration
of dissolved radium in discrete aquifer units of the
Cambrian and Ordovician has not changed over time.
Also, the concentration of 226Ra and 228Ra in
groundwater produced from most public supply wells
has not changed over time. An important finding of
Gilkeson et al. (5) was that high dissolved
concentrations of 226Ra and 228Ra are present in
groundwater from sandstones-the productive aquifers
in the Cambrian and Ordovician bedrock. Groundwater
produced from dolomites has lower concentrations.
Radium concentrations in groundwater from the
Maquoketa Shale is very low (less than 1 pCi/L).

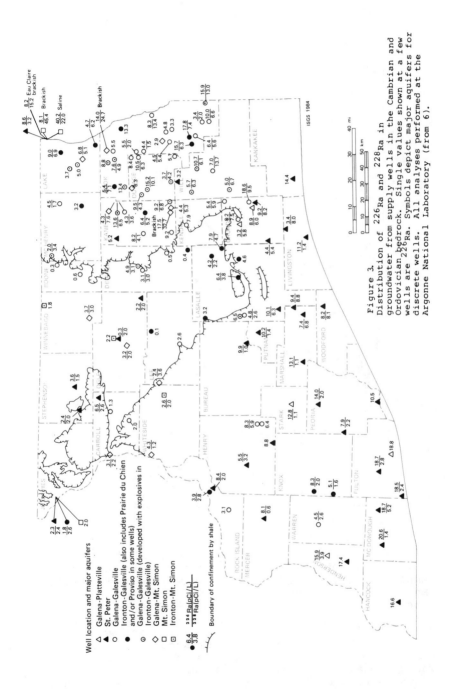

Figure 3.
Distribution of 226Ra and 228Ra in groundwater from supply wells in the Cambrian and Ordovician bedrock. Single values shown at a few wells are 226Ra. Symbols depict major aquifers for discrete wells. All analyses performed at the Argonne National Laboratory (from 6).

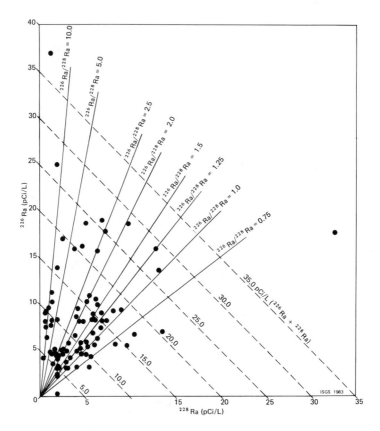

Figure 4. The low correlation coefficient between the concentration of radium-226 and radium-228 in groundwater from the Cambrian and Ordovician bedrock. For a least square line fit to the analyses, the correlation coefficient, r, = 0.275. Isolines show the combined concentration of the two nuclides and the ^{226}Ra ^{228}Ra activity ratio (from 5).

The data set from the study area includes analyses of both ^{226}Ra and ^{228}Ra for groundwater samples from 129 wells that produce from the Cambrian and Ordovician. The range in the combined concentrations is from 2.3 to 53.5 pCi/L, with only nine of the samples in compliance with the maximum contaminant level of the U.S. EPA drinking water standard. Over 100 public water supplies in the study area are not in compliance with the standard. The regional variation in ^{226}Ra and ^{228}Ra in groundwater from the Cambrian and Ordovician is shown in Figure 3. On this figure, symbols are used to identify the aquifers that provide groundwater to supply wells or were sampled by packer tests. The lowest combined concentrations occur in north-central and northwestern Illinois where aquifers in the Cambrian and Ordovician bedrock are confined. Combined concentrations greater then 20 pCi/L occur in brackish groundwater in the southwestern and eastern part of the study area. Figure 4 shows the poor correlation between ^{226}Ra and ^{228}Ra concentrations. For groundwater samples from 90 wells at widely scattered locations throughout the study area the ^{226}Ra ^{228}Ra activity ratio varies from less than 0.75 to greater than 10.

Geochemical and radiological mechanisms that control the occurrence of ^{226}Ra and ^{228}Ra in groundwater are presented in (5). The mechanisms are discussed briefly here.

Radium-228 concentrations in the study area are poorly correlated with the concentration of total dissolved solids; low ^{228}Ra concentrations occur in dilute groundwater in the northern part of the study area and also in brackish groundwater in the southwestern area. Radium-228 concentrations are generally less than 3.0 pCi/L over the western half of the study area. In the eastern part of the study area ^{228}Ra concentrations are often greater than 5.0 pCi/L with the highest concentrations in groundwater from the Mt. Simon Sandstone.

Radium-228 is the first daughter nuclide in the ^{232}Th decay series. Because of its short half-life (5.75 years), ^{228}Ra will not migrate significant distances in confined aquifers from its site of generation. Thorium ions have very slight solubilities in groundwater, therefore the distribution of the ^{232}Th in the bedrock is an important control on the occurrence of ^{228}Th in groundwater. An important source of ^{228}Ra is thought to be thorium-rich accessory minerals in the sandstone bedrock. The variation of ^{228}Ra concentrations that occurs in the study area is compatible with source rocks as a primary control.

The mechanisms that control the occurrence of ^{226}Ra in groundwater are more complex than for ^{228}Ra because of the position of the nuclide in the ^{238}U decay series. The precursor nuclides to ^{226}Ra, in order of occurrence, are ^{238}U, ^{234}Th, ^{234}Pa, ^{234}U and ^{230}Th. Significant disequilibrium between the nuclides may occur in groundwater flow systems because of differences in the chemical and radiological properties of the nuclides. The extreme disequilibrium between ^{238}U and ^{234}U in groundwater of the study area is discussed in the uranium section of this paper. An important source of ^{226}Ra in groundwater in confined aquifers in the Cambrian and Ordovician is the accumulation of the precursor nuclides ^{238}U, ^{234}U and ^{230}Th on the sandstone matrix due to mobilization mechanisms and to the groundwater transport of uranyl ions. Silica surfaces in the sandstone matrix have a greater affinity for uranyl and thorium ions than for radium (17). Radium-226 formed from the disintegration of ^{230}Th present on the sandstone matrix may be placed in solution by alpha recoil or mobilized from the matrix as a function of the groundwater chemistry.

Figure 5 presents the correlation between ^{226}Ra concentrations and the total dissolved solids concentrations for 157 groundwater samples collected at widely scattered locations in the study area. The regression analysis indicates a direct relation between the ionic strength of groundwater and the concentration of dissolved ^{226}Ra. This relation was reported earlier (18). The effect of ionic strength on desorption of ^{226}Ra would be most significant for the sandstone bedrock. Unfortunately, most groundwater samples in Figure 5 are from wells that produce groundwater from both sandstone and dolomite bedrock.

Radon-222

The range in ^{222}Rn concentrations for 80 groundwater samples collected from widely scattered wells in the study area is from 40 to 1000 pCi/L. Most values are less than 300 pCi/L. Figure 6 presents the relation between the ^{222}Rn concentration and the concentration of the parent ^{226}Ra isotope for 70 groundwater samples from the Cambrian and Ordovician bedrock. Little correlation was found between the concentrations of ^{222}Rn and ^{226}Ra in the groundwater samples. Some of the highest ^{222}Rn concentrations were measured in groundwater with the lowest ^{226}Ra concentrations. However, for all of the analyses the ^{222}Rn concentrations were at least one order of magnitude greater than the ^{226}Ra concentrations.

The measured ^{222}Rn concentrations in groundwater are evidence that ^{226}Ra is ubiquitous on the matrix of the Cambrian and Ordovician bedrock, even in regions where the concentrations of ^{226}Ra in groundwater are low. Geochemical mechanisms that may partition ^{226}Ra onto the rock matrix include cation-exchange and coprecipitation with iron, calcium, magnesium, or barium to form mineral phases. Radon-222 generated by disintegration of ^{226}Ra present on the matrix would be efficiently released by these mechanisms to groundwater, with one exception.

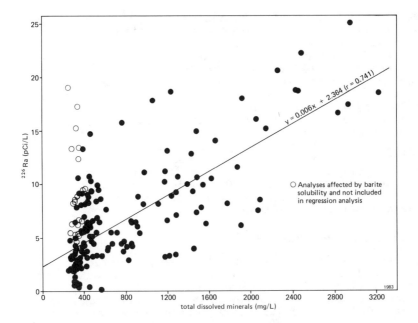

Figure 5. The relation between the concentration of radium-226 and total dissolved minerals in ground-water from the Cambrian and Ordovician bedrock (from 5).

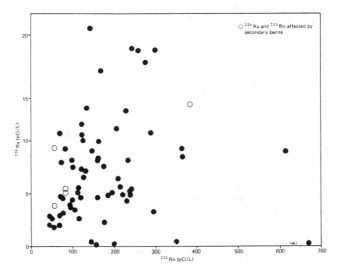

Figure 6. The low correlation coefficient between the concentrations of radium-226 and radon-222 in groundwater from Cambrian and Ordovician bedrock. For a least square line fit to the analyses, the correlation coefficient, r, = 0.164 (from 5).

Precipitates of barite do not release ^{222}Rn generated by disintegration of coprecipitated ^{226}Ra (19).   Some anomalously low and anomalously high ^{222}Rn concentrations measured in groundwater from wells in northeastern Illinois are attributed to the secondary accumulation of barite on the rock matrix of the well bore.   At these wells low concentrations of ^{222}Rn and ^{226}Ra in groundwater were measured during periods when extreme supersaturation of barium and sulfate ions in solution resulted in secondary precipitation of barite (4,5); high concentrations of ^{226}Ra and ^{222}Rn in groundwater were measured at these wells during periods of dissolution or corrosion of the barite.   For these wells the temporal change from the condition of precipitation to dissolution of barite was a function of pumpage rate.

Uranium-234 and Uranium-238

Analyses for ^{234}U and ^{238}U were performed on fresh and brackish groundwater samples collected from the Cambrian and Ordovician bedrock.   The mass concentration of dissolved uranium ranged from less than 0.002 ug/L to 1.10 ug/L; the combined activity concentration of ^{234}U and ^{238}U in the samples ranged from less than 0.1 pCi/L to 7.9 pCi/L. All samples with activity concentrations greater than 1.0 pCi/L were greatly enriched in ^{234}U.   It is obvious that a measurement of the mass concentration of uranium would not accurately represent the uranium activity concentration in these enriched groundwaters.

Figure 7 presents the distribution of the ^{234}U ^{238}U activity ratios for groundwater samples collected from the Cambrian and Ordovician aquifers. The ratios range from 2.0 to greater than 40.   The lowest ratios occur in primary recharge zones in regions where the Cambrian and Ordovician are unconfined.   In this hydrogeologic setting the activity ratio increases with an increase in flow path and residence time of groundwater in the aquifers.

The greatest dissolved uranium concentration measured in the study area was a mass concentration of 15.3 ug/L and a combined activity concentration of 92.2 pCi/L in a saline groundwater sample collected from the basal section of the Mt. Simon at a United States Geological Survey research borehole in the northeastern corner of the study area.   An air rotary method was used to drill this test well.   The oxidation of the zone around the borehole may have increased the concentrations or uranium in groundwater smaples collected from this site and perhaps also altered the activity ratios.

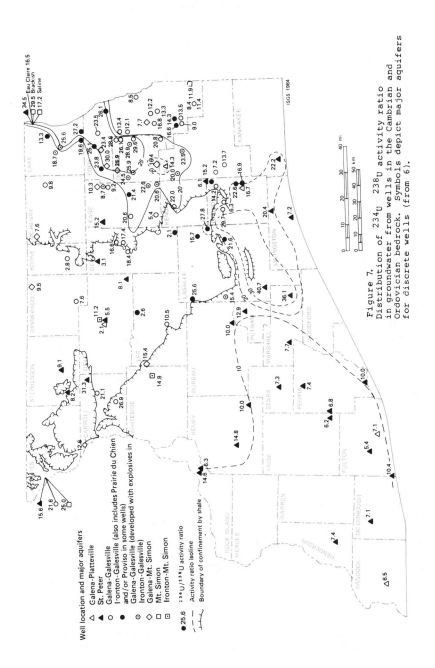

Figure 7.
Distribution of 234U 238U activity ratio
in groundwater from wells in the Cambrian and
Ordovician bedrock. Symbols depict major aquifers
for discrete wells (from 6).

Figure 8 presents the relation between Eh and uranium concentration in groundwater for 77 wells in the study area that produce water from the Cambrian and Ordovician bedrock and for 5 wells that produce groundwater from shallower aquifers. The data confirms that the oxidation-reduction potential is an important control on the uranium concentration in groundwater. Dissolved uranium concentrations were less that 0.08 ug/L in all groundwater samples in which Eh values were -50 millivolts or lower. The highest uranium concentrations are present where Eh values were +100 millivolts or greater. Many groundwater samples with Eh values between -50 to +50 millivolts have appreciable uranium concentrations. It is important to note that the wells have boreholes that are open to several stratigraphic units; groundwater from discrete units may differ markedly in dissolved sulfide concentrations, oxidation-reduction potential, and uranium concentration. The sensitivity of Eh values measured with the platinum electrode method to the presence of sulfide in groundwater is discussed in (5).

Figure 8 presents the values for the ^{234}U ^{238}U activity ratio with the analyses. The ratios range from 2.4 to 40.7 for the analyses on groundwater from the Cambrian and Ordovician bedrock and from 1.3 to 2.5 for the analyses on groundwater from aquifers within and above the Maquoketa Shale Group.

The analyses presented in Figure 7 illustrate the marked increase in ^{234}U ^{238}U activity ratios that occurs where the Cambrian and Ordovician are overlain and confined by shale bedrock. Also note that the activity ratios in the confined aquifers exhibit regional trends. In the southwestern region of the study area the activity ratios range from 5.4 to 14.8; over a large part of this region the ratios are less than 8.

The extreme disequilibrium between ^{234}U and ^{238}U that occurs in groundwater from the Cambrian and Ordovician bedrock of the study area is unique. Such high activity ratios have not been found in other flow systems. Other researchers have reported values for groundwater that range from 0.5 to 12.3 (20). Only a few investigators have reported ratios greater than 5. Kronfeld (21) reported ratios as high as 12.3 for groundwater in a confined sandstone aquifer in Texas. Activity ratios ranging from 5.5 to 10.5 were measured in confined aquifers in the Cambrian and Ordovician bedrock in the tri-state region of Missouri, Kansas and Oklahoma (22). These aquifer units (sandstones and dolomites) have lithologies that are similar to the Cambrian and Ordovician aquifers in northern Illinois.

The comparison of data from adjacent wells in the northeastern part of the study area shows that the ^{234}U ^{238}U activity ratio is higher in wells that are constructed to produce a major part of their supply from the Ironton-Galesville Sandstone. Lower ratios are present in wells that produce a significant component of water from the dolomite units. Radium-226 values are also comparatively higher in wells that produce more water from the sandstones. However, note that wells in the southwestern area with low ^{234}U ^{238}U activity ratios have high dissolved ^{226}Ra concentrations.

A significant feature on figure 7 is the zone of high ratio groundwater that extends from the northeastern part of the study area southwestward through DuPage County into central Kane County, northeastern Kendall County and northwestern Will County. Groundwater samples from within this zone have activity ratios greater than 20;in the northeastern part of the zone ratios range from 25 to 35. An exception are two wells near the boundary between DuPage and Will County with ratios around 16. These wells produce a greater component of groundwater from dolomite units and also have lower ^{226}Ra values. Uranium-234 concentrations in the zone range from 0.4 pCi/L to greater than 4.0 pCi/L with the highest values present on the eastern margin of the zone and in the vicinity of the Des Plaines Disturbance.

A second zone of high ratio groundwater is present in Grundy County, southern LaSalle County, the northern margin of Livingston County, and eastern Marshall County. Activity ratios in this zone range from 20 to 40.7; Uranium-234 activity concentrations range from 0.2 pCi/L to 7.9 pCi/L. It is interesting to note that both high activity zones exist in areas where there is speculation from stable isotope data (6) that basal meltwater beneath continental ice sheets was recharged to the Cambrian and Ordovician bedrock. Gilkeson et al. (6) also speculated that the recharging water contained dissolved uranium, enriched in ^{234}U that formed coatings on the aquifer matrix. Data from a limited investigation into uranium activity ratios for uranium coatings on the sandstone bedrock supports this speculation. The ^{234}U ^{238}U activity ratios for 0.1 N hydrochloric acid extracts of uncrushed samples of sandstone bedrock were greater than 10 for a core sample of St. Peter Sandstone from the southern part of LaSalle County (the region of the groundwater activity ratio of 40.7) and greater than 4 for a core sample of Ironton Sandstone from the western part of Kankakee County (the region of the groundwater activity ratio of 22.2). Data are too limited at the present time to confirm the relationship of glacial recharge to the extreme activity ratios that occur in groundwater.

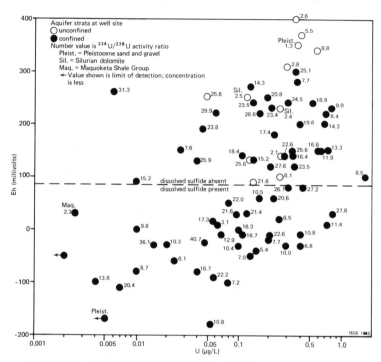

Figure 8.   The relation between Eh (platinum electrode) and the concentration of uranium in groundwater from the Cambrian and Ordovician bedrock. The ^{234}U ^{238}U activity ration is shown for each analysis.   A few analyses are listed for groundwater from other aquifers.

SUMMARY

Stable and radioactive isotopes in groundwater were studied in an investigation of the natural geologic sources of high concentrations of ^{226}Ra and ^{228}Ra in groundwater from the Cambrian and Ordovician bedrock in northern Illinois.   The combined concentrations of the two nuclides ranged from 2.3 to 53.5 pCi/L; the majority of analyses exceed the maximum contaminant level of 5.0 pCi/L. The lowest concentrations are in primary recharge zones in north-central Illinois where the Cambrian and Ordovician aquifers are unconfined.   The highest combined concentrations are in mineralized groundwater in regions where the Cambrian and

Ordovician bedrock are overlain and confined by shale bedrock. The secondary accumulation of ^{238}U, ^{234}U and ^{230}Th as diffuse coatings on the sandstone matrix is am important source of ^{226}Ra. Alpha-recoil mechanisms are important to the distribution of nuclides on the bedrock matrix and in groundwater. The mobilization of ^{226}Ra from the matrix is related to alpha recoil mechanisms and to the chemistry of the groundwater. The distribution of thorium rich accessory minerals in the sandstones is thought to be important control on dissolved ^{228}Ra concentrations.

Stable isotope studies of the composition of the water molecule and dissolved sulfate molecule established that the chemical evolution of groundwater in the study area is complex and that glacial processes are important to the recharge of groundwater in confined aquifers. Groundwater from the Cambrian and Ordovician exhibits a unique enrichment in ^{234}U. The activity ratio of ^{234}U to ^{238}U ranges from 2.0 to 40.7. The lowest ratios are present in unconfined aquifers in primary recharge zones. Ratios greater than 20 occur over large areas where the Cambrian and Ordovician are confined. Alpha recoil mechanisms are important to the disequilibrium, however, the regional distribution of activity ratios and ^{234}U concentrations suggest that glacial recharge may have contributed to the high ratios present in the eastern part of the study area. Additional studies are necessary to accurately define the mechanisms responsible for the high ^{234}U ^{238}U disequilibrium that is present in groundwater in the study area.

ACKNOWLEDGMENTS

The research program was funded in part by the Bureau of Reclamation, U.S. Department of the Interior. Glenn E. Stout, Director of the Water Resources Center, University of Illinois, was program administrator.

REFERENCES

1. <u>National</u> <u>Primary</u> <u>Drinking</u> <u>Water</u> <u>Regulations</u>:
Federal Register, Dec. 24, 1975. (Washington:
U.S. Environmental Protection Agency, 1975) v. 40
p 248

2. Hess, C.T., J. Michel, T.R. Horton, H.M. Prichard
and W.A. Conglio. "The Occurrence of
Radioactivity in Public Water Supplies in The
United States,"<u>Health</u> <u>Physics</u> 48(5):553-586

3. Gilkeson, R.H., S.A. Specht, K. Cartwright, R.A.
Griffin and T.E. Larson. "Geologic Studies to
Identify The Source for High Levels of Radium and
Barium in Illinois Groundwater Supplies: A
Preliminary Report," University of Illinois at
Champaign-Urbana Water Resources Center Research
Report 135 p 27 (1978).

4. Gilkeson, R.H., E.C. Perry and K. Cartwright.
"Isotopic and Geologic Studies to identify the
Sources of Sulfate in Groundwater Containing High
Barium Concentrations," University of Illinois at
Champaign-Urbana Water Resources Center Research
Report 81-0165. p 39 (1981).

5. Gilkeson, R.H., K. Cartwright, J.B. Cowart and
R.B. Holtzman. "Hydrogeologic and Geochemical
Studies of Selected Natural Radioisotopes and
Barium in Groundwater in Illinois," University
of Illinois at Champaign-Urbana Water Resources
Center Report No. 83-0180, p 93 (1983).

6. Gilkeson, R.H., E.C. Perry, J.B. Cowart and R.B.
Holtzman. "Isotopic Studies of the Natural
Sources of Radium in Groundwater in Illinois,"
University of Illinois at Champaign-Urbana Water
Resources Center Report No. 187, p 50   (1984).

7. Gilkeson, R.H. and J.B. Cowart. "A Preliminary
Report on Uranium-238 Series Disequilibrium in
Groundwater of the Cambrian-Ordovician Aquifer
System of Northeastern Illinois," in <u>Isotope</u>
<u>Studies</u> <u>of</u> <u>Hydrologic</u> <u>Processes</u>, E.C. Perry and
C. Montgomery, Eds. (DeKalb, IL: Northern
Illinois University Press, 1982), p 109-118.

8. Perry, E.C., T. Grundt and R.H. Gilkeson. "H, O, and S Isotope Study of Groundwater in the Cambrian- Ordovician Aquifer System of Northern Illinois," in *Isotope* Studies of Hydrologic Processes, E.C. Perry and C. Montgomery, Eds. (Dekalb, IL: Northern Illinois University Press, 1982), p 35-43.

9. Siegel, D.I. and R.J. Mandle. "Isotopic Evidence for Glacial Melt-water Recharge to the Cambrian-Ordovician Aquifer, North Central United States," *Quat. Res.* 22(3): 328-335 (1984).

10. Claypool, G.R., W.T. Holser, I.R. Kaplan, H. Sakai and I. Zak. "The Age Curves of Sulfur and Oxygen Isotopes in Marine Sulfate and Their Mutual Interpretation," *Chem. Geol.* 28: 199-260 (1980)

11. Stehney, A.F. "Radium and Thorium X in Some Potable Waters, "*Acta Radio.* 43:43-51(1955).

12. Lucas, H.F. and F.H. Ilcewicz. "Natural Radium-226 Content of Illinois Water Supplies," *Jour. Am. Water Works Assoc.* 50: 1523 (1958).

13. Krause, D.P. "Radium-228 (Mesothorium I) in Illinois Well Waters," Argonne National Laboratory Radiological Physics Division Semiannual Report No. ANL 6049, p 52-54 (1959).

14. Krause, D.P. "Radium-228 (Mesothorium I) in Midwest Well Waters, "Argonne National Laboratory Radiological Physics Division Semiannual Report ANL 6199. P 85-87 (1960).

15. Holtzman, R.B. "Lead-210 (RaD) and Polonium-210 (RaF) in Potable Waters in Illinois," in *The Natural Radiation Environment* J.A.S. Adams and W.M. Lowder, Eds. (Chicago: University of Chicago Press, 1964) p 227-237.

16. Bennett, D.L., C.R. Bell and I.M. Markwood. "Determination of Radium Removal Efficiencies in Illinois Water Supply Treatment Processes," U.S. Environmental Protection Agency, Office of Radiation Programs, Technical Note GRP/TAD-76-1, p 108 (1976).

17. Stumm, W. and J.J. Morgan. *Aquatic Chemistry* (New York: Wiley-Interscience, Inc., 1970) p 583.

18. Emrich, G.H. and H.F. Lucas. "Geologic Occurrence of Natural Radium-226 in Ground Water in Illinois, " <u>Bull.</u> <u>IASH</u> 8(3): 5-19 (1963).

19. Hahn, O. <u>Applied</u> <u>Radiochemistry</u> (Ithaca: Cornell University Press, 1936) p 278.

20. Osmond, J.K. and J.B. Cowart. "The Theory and Uses of Natural Uranium Isotopic Variations in Hydrology," <u>Atomic</u> <u>Energy</u> <u>Rev.</u> 14:621-679 (1976).

21. Kronfeld, J. "Uranium Deposition and Th-234 Alpha- Recoil: An Explanation for Extreme U-234 U-238 Fractionation Within the Trinity Aquifer," <u>Earth</u> <u>Planet.</u> <u>Sci.</u> <u>Lett.</u> 21:327-330.

22. Cowart, J.B. "Uranium Isotopes and Ra-226 Content in the Deep Groundwater, Tri-State Region, USA," <u>Jour.</u> <u>Hyd</u>. 54: 185-193 (1981).

RADON PRODUCTION IN PUMPING WELLS

Chia-Shyun Chen and John L. Wilson,
New Mexico Institute of Mining and Technology, Socorro, New Mexico

INTRODUCTION

Uranium is widely distributed in crustal rocks. The decay of
^{238}U gives rise to the uranium series which includes ^{222}Rn (radon)
as one of the radioactive daughters. Radon is a chemically inert
gas with a short half-life (3.82 days). Because of its
instability, radon can rapidly decay into a number of nonvolatile
radionuclides that can be adsorbed onto human lungs once inhaled,
thus causing lung cancer. In the ground, radon emanated from
rocks can encroach houses via water-supply wells drilled through
radon-rich fractured formation [e.g., Brutsaert et al., 1982; Hess
et al., 1985; Sasser and Watson, 1978], or via cracks in concrete
slab foundations laid above radon-contaminated aquifers or rocks
[e.g., LeGrand, 1987]. Numerous geological and hydrogeologic
studies have been conducted in investigating the radon-
encroachment related problems [e.g., Andrews and Wood, 1972;
Brutsaert et al., 1981; LeGrand, 1987; Tanner, 1964 and 1980].
Theoretical analyses of radon gas diffusion through homogeneous
soils or cracks have been done by Clements and Wilkening [1974],
Edwards and Bates [1980], Holford [1986], Landman [1982], Landman
and Cohen [1983], Landman and Rosenblat [1985], and Schery et al.
[1982, 1984 and 1986]. Nelson et al. [1983] gave some simple
analytical solutions for radon movement into flowing boreholes.
To the authors knowledge, however, there is no analytical solution
describing radon transport toward water supply wells. Therefore,
as a first attempt to fill this gap this paper is aimed at
determining analytical solutions to the practical problem
illustrated in Figure 1a, where a water supply well penetrates a
layer of saprolite overlying a fractured crystalline rock such as
granite. The low-permeability saprolite contains the groundwater
for use, and the higher-permeability fractures serve as effective
pathways conveying water from the saprolite to the pumping well.
Since most granites contain and release radon, the groundwater
originally existing in saprolite becomes radon-contaminated when
it flows through these fractures into the water supply well
connected to the houses. Hence, water-borne radon can enter the
houses and imposes a health threat.

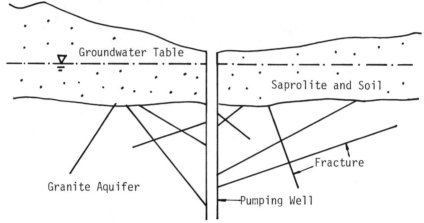

Figure 1a, Diagram for a pumping well withdrawing groundwater from saprolite through fractured granite aquifer (after LeGrande, 1987)

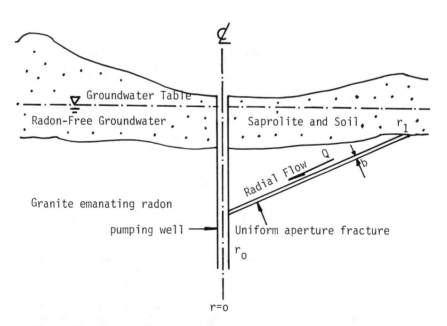

Figure 1b. The studied condition of Figure 1a; a single fracture of a uniform aperture is situated in granite emanating radon.

MODELS AND RESULTS

The practical problem described by Figure 1a is mathematically simulated herein by the idealized radial flow system depicted in Figure 1b. Only a single fracture with a uniform aperture, b, is used to analyze how the radon concentration entering the well bore would change with pumping rate, radon emission rate from granite, the fracture length, the fracture aperture, and hydrodynamic dispersion phenomena in the fracture. The groundwater involved in saprolite is assumed to be free of radon; that is, "clean" groundwater can continuously feed the fracture from the end of the fracture at the radial distance r_1. The focus of this study is on the maximum radon concentration expected at the well bore, which occurs under steady state conditions. Other assumptions include:

1. The fracture is simulated by two smooth, parallel plates separated by a constant aperture distance, b (m).

2. The granite has a constant radon production rate per unit fracture surface area, E ($Bq/d/m^2$), and thus each of the two fracture surface supplies radon to the groundwater at this constant E.

3. The tilting of the fracture is neglected, and thereby gravity effects are ignored. The steady state groundwater velocity in the fracture is described by

$$V(r) = \frac{A}{r} \qquad (1)$$

    where r is the radial distance, A is the advection parameter defined by $Q/2\pi b$ (m^2/d), and Q (m^3/d) is the pumping rate from the single fracture.

4. The fracture is always saturated with groundwater.

5. Radon is transported through the fracture by advection and longitudinal dispersion. Vertical and lateral dispersion as well as the molecular diffusion are neglected.

By making use of the mass balance principle and the above assumptions, the steady state transport equation can be formulated in a similar way used by Chen [1986] as

$$\frac{\alpha A}{r} \frac{d^2C}{dr^2} + \frac{A}{r} \frac{dC}{dr} - \lambda C + \frac{2E}{b} = 0 \qquad (2)$$

where α (m) is the constant longitudinal dispersivity, C is the radon concentration (Bq/m^3), and λ is the decay constant of radon. A value of 0.1814 d^{-1} is used for λ by considering the radon half-life as 3.82 days. The four terms included in equation (2) from left to right respectively represent the longitudinal dispersion effect, the advection effect, the radioactive decay process, and the radon production from both fracture surfaces.

The water supply well is taken into account in the model as a boundary condition; namely

$$\frac{dC}{dr}\bigg|_{r_0} = 0 \qquad (3)$$

where r_0 is the well radius. This indicates that the radon concentration entering the well bore remains unchanged inside the well bore; i.e., a continuity of radon concentration is imposed across the well bore.

Another boundary condition for equation (2) is prescribed at the end of the fracture denoted by r_1. Considering the saprolite contains radon-free groundwater, it is appropriate to state that at r_1

$$\frac{A}{r_1} C + \frac{\alpha A}{r_1} \frac{dC}{dr} = 0 \qquad (4)$$

which indicates that there is no radon mass flux transported from the saprolite into the fracture at r_1.

Analytical solutions to equation (2), subject to the two boundary conditions (3) and (4), are presented for two cases. The first case refers to a long fracture such that r_1 is mathematically set as infinity. In the second case, the fracture length r_1 is finite. The derivations for the solutions are not included in this paper but will be discussed elsewhere.

Solution for an Infinite Fracture

If the fracture length r_1 is set to infinity, a uniform, steady state concentration distribution is obtained,

$$C(r) = 2E/\lambda b \qquad\qquad r_0 \leq r \leq \infty \qquad (5)$$

that is identical to the concentration before pumping is initiated. Thus, the radon concentration entering the well bore is also $2E/\lambda b$. This constant concentration is proportional to the radon production rate E, and inversely related to the fracture aperture. A larger aperture means that there is more groundwater in the fracture. Accordingly the radon concentration becomes smaller provided that a constant amount of radon atoms are generated, as assumed herein. A higher production rate E results in higher radon concentration in the fracture when the fracture aperture is held constant.

Absence of the advection parameter A and the longitudinal dispersivity α in the solution given by equation (5) indicates that the radon concentration is independent of the pumping rate and the dispersion process. This is explainable as follows. Although the boundary condition of equation (4) implies that "clean" groundwater is introduced to the fracture at r_1, yet this effect can never be realized when r_1 is infinite. The granite produces a constant amount of radon atoms at any specific location within the fracture, resulting in no concentration gradient in the fracture. Under this circumstance, hydrodynamic dispersion and diffusion processes become unimportant. The advection process moves radon atoms but it alone can not change the concentration gradient. Therefore, the result is a constant concentration everywhere in the fracture and in the well, no matter what the pumping rate.

Solution for a Finite Fracture

Considering a finite fracture of length r_1 the analytical solution to equations (2), (3) and (4) can be shown to be

$$C(r) = \frac{2E}{\lambda b} \left\{ 1 - \exp\left[\frac{r_1 - r}{2d}\right] \frac{Bi(Y)H_0 - Ai(Y)L_0}{L_1 H_0 - H_1 L_0} \right\}, \quad r_0 \leq r \leq r_1 \quad (6)$$

where $H_0 = \alpha^{1/3} Ai'(Y_0) - \frac{1}{2} Ai(Y_0)$

$$H_1 = \alpha^{1/3} Ai'(Y_1) + \frac{1}{2} Ai(Y_1)$$

$$L_0 = \alpha^{1/3} Bi'(Y_0) - \frac{1}{2} Bi(Y_0)$$

$$L_1 = \alpha^{1/3} Bi'(Y_1) + \frac{1}{2} Bi(Y_1)$$

$$Y_0 = \alpha^{1/3} \left[\frac{r_0}{d} + \frac{1}{4\alpha}\right]$$

$$Y_1 = \alpha^{1/3} \left[\frac{r_1}{d} + \frac{1}{4\alpha}\right]$$

$$Y = \alpha^{1/3} \left[\frac{r}{d} + \frac{1}{4\alpha}\right]$$

$$\alpha = \lambda d^2 / A$$

In the above equations, $Ai'(x)$ and $Bi(x)$ are the two linearly independent Airy functions, and $Ai'(x)$ and $Bi'(x)$ denote their first derivative, respectively.

Results of equation (6) are plotted in Figures 2 and 3. The symbol $C*$ denotes the dimensionless concentration as

$$C* = \frac{C(r)}{2E/\lambda b} \quad (7)$$

which can be considered as the dimensional concentration normalized with respect to the limiting concentration $2E/\lambda b$. The dimensionless parameter β_1 can be interpreted as the dimensionless pumping rate, and

$$\beta_1 = \frac{Q}{\pi b \lambda r_1^2} \quad (8)$$

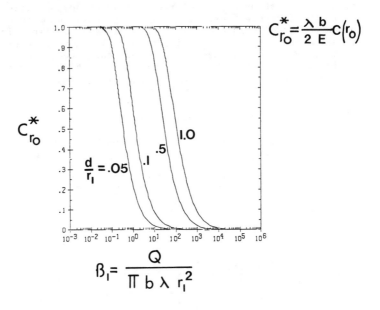

$$C_{r_o}^* = \frac{\lambda}{2} \frac{b}{E} C(r_o)$$

$$\beta_1 = \frac{Q}{\pi b \lambda r_1^2}$$

Figure 2.   Concentration at the well bore as a function of β_1 for various d/r_1 determined by equation (6).

Figure 2 presents radon concentrations in the well bore as a function of pumping rate and other pertinent parameters. In this figure the well radius, r_o, is taken as zero. Increasing pumping

rate reduces water residence time in the fracture and causes radon concentrations to drop. Well concentrations increase with longer fracture lengths, r_1, and larger residence times. These

theoretical predictions have been substantiated in detailed statistical analyses of Maine wells by Brutsaert et al. (1981). They found that wells with higher yields tend to have lower radon concentrations. Deeper wells, implying longer equivalent fracture

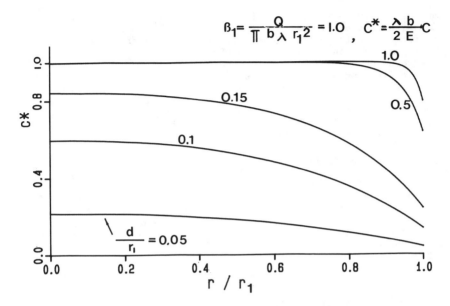

Figure 3b   Concentration distributions in the fracture from equation (6) for β_1 = 1 and various d/r_1.

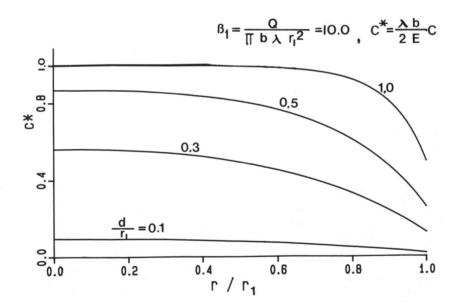

Figure 3a Concentration distributions in the fracture from equation (6) for β_1 = 10 and various d/r_1

lengths, had higher concentrations, at least up to a depth of 50 to 75 meters and were essentially constant at greater depths. Figure 2 demonstrates that for sufficiently long fractures (or small pumping rates) the radon in the well stabilizes at the background concentration, $2E/\lambda b$. These fractures are essentially infinite in length. The figures also indicate that increasing the dispersivity, and thus the mixing in the fracture, results in increased radon concentrations. Figure 3 illustrates how the radon concentration varies with distance along the fracture. The larger concentration gradients occur near the boundary of the fracture with the saprolite.

Solution for No Dispersion

As equation (1) is used to describe the groundwater velocity, one can argue that since V is generally large due to the smallness of b, the advection may dominate the transport process in the fracture. As a result of this argument, the longitudinal dispersion may be neglected and the governing equation (2) can be rewritten as

$$\frac{A}{r}\frac{dC}{dr} - \lambda C + \frac{2E}{b} = 0 , \qquad r_0 \leq r \leq r_1 \qquad (9)$$

where r_1 is taken finite. Only one boundary condition is needed for equation (9). This boundary condition is appropriate to be imposed at r_1 to ensure that the saprolite feature is properly incorporated in the problem. That is, the boundary condition for equation (9) is

$$C(r_1) = 0 \qquad (10)$$

which is an adaptation of equation (4) by neglecting the dispersion.
The analytical solution to equation (9) and (10) is

$$C(r) = \frac{2E}{\lambda b}\left\{1 - \exp\left[\frac{\lambda}{2A}(r_1^2 - r^2)\right]\right\} \qquad (11)$$

The radon concentration at the well bore is

$$C(r_0) = \frac{2E}{\lambda b}\left\{1 - \exp\left[-\frac{\lambda}{2A}(r_1^2 - r_0^2)\right]\right\} \qquad (12)$$

Equation (12) also was obtained by Nelson [1983] while using a different approach on a similar problem. It is noted that if r_1 is set to infinity, equation (11) and (12) yield a constant concentration of $2E/\lambda b$ as determined earlier for the case of an infinite fracture with dispersion. This coincidence supports the earlier discussion that if the fracture length is effectively set to infinity, the hydrodynamic dispersion and the advection are not important, and the radon concentration is independent of the pumping rate. Figure 4 is another plot of radon concentration in the well, and is based on equation (12) with $r_0 = 0$. Its behavior is similar to Figure 2, in that it shows lower radon

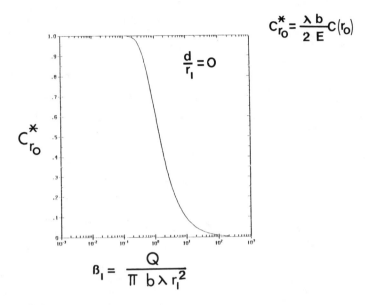

Figure 4. Concentration at the well bore determined by equation (12).

concentrations with higher pumping rates or smaller fracture
lengths. Radon concentration as a function of distance along the
fracture is pictured in Figure 5a. At higher pumping rates the
entire radon concentration profile is reduced. The radon
concentration profile for a finite radius well, $r_0 = 0.1\ r_1$, is

given in Figure 5b. The profile is identical to Figure 5a, except
that concentrations within the well are fixed at the value found
at well radius r_0. In practice the well radius is usually smaller

than fracture length, and should not significantly affect the well
concentration. Figure 4 can be used to determine the fracture
radius or pumping rate, such that the fracture appears to be
effectively infinite. This occurs for the value of β_1, such that

the normalized concentration $C*$ begins to significantly depart
from $C* = 1$. This value of β_1 is roughly 0.1. Thus whenever the

fracture length r_1 is roughly greater than

$$r_1 > \left(\frac{10\ Q}{\pi b \lambda}\right)^{1/2} \simeq 4\left(\frac{Q}{b}\right)^{1/2} \tag{13}$$

the fracture appears to be effectively infinite and the radon
concentration in the well stabilizes at the background level.

DISCUSSION

    The two solutions, with and without dispersion, have the same
character. For example, radon concentrations increase with longer
fractures and smaller pumping rates. The solutions are, however,
inconsistent. As the dispersivity d goes to zero for the
advective-dispersion solution of Figures 2 and 3, the radon
concentration drops to zero in the well and the fracture. But the
solution specifically derived for zero dispersivity, shown in
Figures 4 and 5, reveals a concentration profile that only goes to
zero for infinite pumping or zero fracture length (or zero
emission rate). The inconsistency stems in part from the
different mathematics of equations (2) and (4), and the choice of
boundary conditions.
    We've tried a variety of boundary conditions for both
problems. Using $C = 0$ at $r = r_1$ for the advection-dispersion

problem yields a result very similar to (6) and Figures 2 and 3.
It differs only for very small d, but is still inconsistent with
the no-dispersion case. Applying both boundary conditions for
equation (2) at $r = r_1$, say $C = 0$ and $\partial c/\partial r = 0$, yields an

unrealistic solution. Revising the condition for equation (4) to
$\partial c/\partial r = 0$ at $r = r_0$, leads to the trivial result of equation (5),

even for short fractures. Other attempts ended similarly. We see
no acceptable physial explanation for the inconsistency. More
work on the mathematics is needed. Never-the-less, the character
of the radon production in wells is captured by both solutions.
Only their numerical predictive capability is questionable.
    There are several other reasons to question the numerical
predictive capability of these solutions. They assume radial flow
through a single fracture, when obviously a more complex flow is

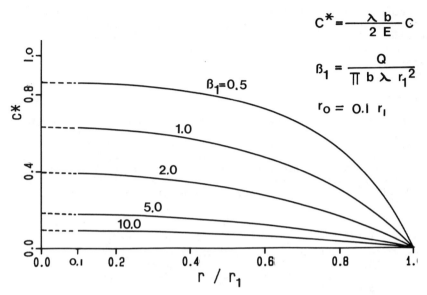

Figure 5b Concentration distributions in the fracture from equation (11) for various β_1 and $r_o = 0.1\ r_1$

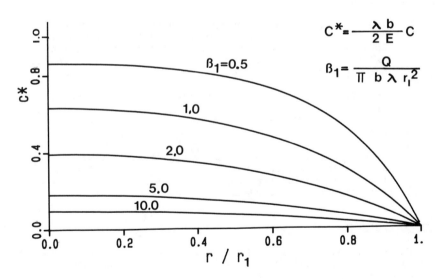

Figure 5a Concentration distributions in the fracture from equation (11) for various β_1 and $r_o = 0$

occuring in an unknown fracture network. The amount of flow in
any one fracture will depend on its length, aperture and
interconnection to other fractures and the overlying saprolite. A
true predictive model of radon concentrations in fractured
crystalline rock obviously would have to deal with this problem.

CONCLUSIONS

A simple conceptual model of radon production in wells pumping
water from fractured crystalline bedrock has been examined
analytically. The fracture serves as a radon exposed conduit for
'clean' water originally derived from an overlying layer of
saprolite. The analytical solutions provide insight into the
nature of radon production, but are not yet adequate for
predictive use. In particular, we discovered the following
specifics:

1. If the fracture is finite (i.e., $r_1 < 4(Q/b)^{1/2}$) the radon
    concentration decreases as the pumping rate increases.
    This is in agreement with the field observation that a
    lower-yielding well tends to give lower radon
    concentration.

2. If the fracture is finite, the radon concentration
    increases with the fracture length. This is also in
    agreement with field observations that deeper wells tend
    to give higher radon concentrations.

3. If the fracture is effectively infinite (i.e., $r_1 >$
    $4(Q/b)^{1/2}$), the radon concentration is independent of the
    pumping rate and of the hydrodynamic dispersion process.
    In this event, a constant radon concentration of $2E/\lambda b$
    exists in the fracture and in the well bore. This too has
    been confirmed by field data.

4. In general, the radon concentration increases as the radon
    production rate of the crystalline rock increases, as the
    dispersivity increases, or as the fracture aperture
    decreases.

5. More research is needed to investigate the influence of
    hydrodynamic dispersion on the radon transport through
    fractures, and on the complications introduced by
    accounting for more realistic fracture networks.

ACKNOWLEDGEMENTS

Useful discussions with F.M. Phillips are acknowledged. The
numerical calculations made by L.-W. Chiang and G. Woodside are
appreciated.

REFERENCES

Andrews, J.N., and D.F. Wood. Mechanism of radon release in rock
    matrices and entry into groundwaters, Trans. Inst. Min. Met.,
    81:B197-B209 (1972).

Brutsaert, W.F., S.A. Norton, C.T. Hess, and J.S. Williams.
    Geologic and hydrologic factors controlling radon-222 in
    groundwater in Maine, Ground Water, 19(4):407-417 (1981).

Chen, C.-S.. Solutions for radionuclide transport from an
    injection well into a single fracture in a porous formation,
    Water Resour. Res., 22(4):508-518 (1986).

Clements, W.E., and M.H. Wilkening. Atmospheric pressure effects
    on ^{222}Rn Transport across the earth-air interface, J. Geophys.
    Res., 79(33):5025-5029 (1984).

Edwards, J.C., and R.C. Bates. Theoretical evaluation of radon
    emanation under a variety of conditions, Health Physics,
    39(4):263-274 (1980).

Hess, C.T., J. Michel, T.R. Horton, H.M. Pinchard, and W.A.
    Coniglio. The occurence of radioactivity in public water
    supplies in the United States, Health Physics, 48(5):553-586
    (1985).

Holford, D.J. Finite element model of two-dimensional gas flow
    and transport of radonn in cracked soil, Hydrology M.S.
    Thesis, New Mexico Institute of Mining and Technology,
    Socorro, New Mexico (1986).

Landman, K.A. Diffusion of radon through cracks in a concrete
    slab, Health Physics, 43(1):65-71 (1982).

Landman, K.A., and D.S. Cohen. Transport of radon through cracks
    in a concrete slab, Health Physics, 44(3):249-257 (1983).

Landman, K.A., and S. Rosenblat. Diffusion of a contaminant in
    fractured porous media, Health Physics, 48(1):19-28 (1985).

LeGrand, H.E. Radon and radium emanations from fracture
    crystalline rocks - A conceptual hydrogeological model, Ground
    Water, 25(1):59-69 (1987).

Nelson, P.H., R. Rachiele, and A. Smith. Transport of radon in
    flowing boreholes at Stripa, Sweden, J. Geophys. Res.,
    88(B3):2395-2405 (1983).

Sasser, M.K., and J.E. Watson. An evaluation of the radon
    concentration in North Carolina groundwater supplies, Health
    Physics, 34(June):667-671 (1978).

Schery, S.D., D.H. Gaeddert, and M.H. Wilkening. Transport of
    radon from fractured rock, J. Geophys. Res., 87(B4):2969-2976
    (1982).

Schery, S.D., D.H. Gaeddert, and M.H. Wilkening.   Factors
    affecting exhalation of radon from a gravelly sandy loam, J.
    Geophys. Res., 89(D5):7299-7309 (1984).

Schery, S.D., and D. Siegel.   The role of channels in the
    transport of radon from the soil, J. Geophys. Res.,
    91(B12):12,366-12,374 (1986).

Tanner, A.B.   Radon migration in the ground: A review, in The
    Natural Radiation Environment III, Vol. 1, T.F. Gessell and
    W.M. Lowder (eds.), Tech. Inform. Center, U.S.D.O.E.,
    Washington, DC (1964).

Tanner, A.B.   Radon migration in the ground: A supplementary
    review, in The Natural Radiation Environment III, pp. 5-56,
    N.T.I.S., Springfield, VA (1980).

RADIUM-228 AND RADIUM-226 IN GROUND WATER OF
THE CHICKIES FORMATION, SOUTHEASTERN PENNSYLVANIA

L. DeWayne Cecil,
U.S. Geological Survey, Water Resources Division,
Malvern, Pennsylvania

Robert C. Smith II and Margaret A. Reilly,
Pennsylvania Department of Environmental Resources,
Harrisburg, Pennsylvania

Arthur W. Rose,
Pennsylvania State University, University Park, Pennsylvania

ABSTRACT

    Routine sampling of public water supplies by the Pennsylvania
Department of Environmental Resources (PaDER) and the U.S.
Environmental Protection Agency (USEPA) revealed three supplies in
the Chickies Formation that exceeded the USEPA drinking-water
maximum contaminant level (MCL) of 5 pCi/L (picoCuries per liter)
dissolved radium.
    The Chickies Formation, a quartzite, typically forms
ridges and borders uplands over an area of 112 square miles in the
Piedmont physiographic province. Thickness ranges from 500 feet
near the Delaware River to 950 feet in York County. Reported
yields of 100 domestic wells range from 2 to 73 gal/min (gallons
per minute) with a median yield of 12.5 gal/min. Total dissolved-
solids concentration of the ground water tends to be low (specific
conductance as low as 10 microsiemens per centimeter at 25°
Celsius) and pH is acidic (as low as 4.4). The ground waters are
slightly reducing and contain detectable Fe and Mn concentrations.
    Data from PaDER suggest that the radium anomalies are
limited to the Chickies Formation. In these acidic waters, Ra2+
is probably mobilized and remains in solution rather than being
adsorbed as in most ground-water systems. A study being conducted
by the U.S. Geological Survey, in cooperation with the Pennsylvania Topographic and Geologic Survey (PaT&GS), which includes the
sampling of 180 wells completed in or near the Chickies Formation,
is expected to characterize the hydrogeology and geochemical

environment associated with elevated radium concentrations. An
understanding of the mechanisms controlling the occurence of
radium in solution could facilitate the location of alternative
water supplies and help to anticipate radium concentrations in
similar lithologies.

Radium-228 concentrations for 48 wells ranged from less than
0.5 to 50 pCi/L, and radium-226 concentrations ranged from less
than 0.1 to 9.5 pCi/L. Forty-three of the 48 analyses had
radium-228 concentrations in excess of those for radium-226
suggesting that gross alpha screening should be used with caution
because radium-228 is not an alpha emitter.

INTRODUCTION

The Chickies Formation is a Precambrian quartzite that contains quartzose schist and ranges from massive to thin bedded. It
crops out in the Piedmont physiographic province of southeastern
Pennsylvania, generally forming narrow ridges. The Chickies
Formation covers an area of 112 square miles in parts of Adams,
Berks, Bucks, Chester, Lancaster, Montgomery, and York Counties.
The Chickies is estimated to be 500-950 feet thick, the lower portion of which is the Hellam Conglomerate member [1]. Reported
yields of 100 domestic wells range from 2 to 72 gal/min (gallons
per minute) with a median yield of 12.5 gal/min. The Chickies is
a minor aquifer, but is the only source of water for thousands of
people served by private wells.

Routine sampling by the Pennsylvania Department of Environmental Resources (PaDER) at three trailer parks that derive their
water supply from the Chickies Formation showed concentrations of
radium in the water (table 1) were in excess of the 5 pCi/L maximum contaminant level (MCL) in the U.S. Environmental Protection
Agency's (USEPA) drinking water standards [2]. Radium concentrations in ground water are normally less than 1 pCi/L [3]. The
anomalous radium concentrations do not appear to be limited to any
particular location in the Chickies Formation, and therefore, may
be related to unusual ground water geochemical conditions in the
formation.

GEOCHEMISTRY OF HIGH-RADIUM WATERS

In July 1985, field measurements were made of pH, Eh, alkalinity, dissolved O_2 and H_2S, and specific conductivity. In addition, filtered samples were collected for determination of radium
and other chemical constituents. Table 1 lists available data
from these samples plus data from earlier samples taken from the
same wells.

The low specific conductance (19 to 130 microsiemens per
centimeter) indicates relatively low dissolved solids. The water
is relatively acid (pH 4.6 to 5.3), and the major dissolved species are Na, Cl, and SO_4. Detectable amounts of dissolved Fe and
Mn, in combination with the measured Eh and pH, indicate that the
waters have been slightly reducing, although there is also an
appreciable amount of dissolved O_2, possibly acquired in the process of pumping and sampling. Visible Fe-oxide was recovered on

Table 1.   Chemical analyses of ground waters from three trailer parks in Chester and Lancaster Counties, Pennslyvania.

Parameter or Constituent	Trailer Park 1 Well 1	1 Well 2	2 Well 1	2 Well 2	2 Well 3	3 Well 1
Sp.C. (uS/cm)	75	94	130	92	19	100
TDS (mg/L)	–	–	–	–	–	116
Temp. (°C)	12	12	13	12.5	13	12
pH (field)	5.27	4.76	4.71	4.64	4.92	4.68
Eh (mv)	+467	+476		+516	+489	+492
O_2 (mg/L)	6.7	3.7	4.2	8.6	6.0	8.0
Alk. (meq/L)	0.16	0.097	0.043	0.025	0.11	0.038
H_2S (mg/L)	n.d.	–	–	–	n.d.	n.d.
Ca (mg/L)	–	–	3.4	3.5	–	0.5
Mg (mg/L)	–	–	4.1	4.8	–	1.0
Na (mg/L)	–	–	–	–	–	10.1
K (mg/L)	–	–	–	–	–	2.3
Fe (mg/L)	0.24	0.10	0.10	0.10	0.10	0.17
Mn (mg/L)	0.05	0.05	0.05	0.07	0.05	0.10
Cl (mg/L)	22	21	20	10	18	10
SO_4 (mg/L)	10	14	21	24	14	46
NO_3 (mg/L)	0.5	3.7	5.6	2.8	11	1.1
PO_4 (mg/L)	0.03	0.02	0.02	0.04	0.05	0.03
Ba (mg/L)	–	–	–	–	–	0.15

Radionuclides reported in pCi/L:

Ra-226	7.4	7.0	7 to 8.7			9.8
Ra-228	29.2	2.9	8.8 to 55.1			76
Gross alpha	62	33	14	41	12	110
Gross beta	71	<50	–	70	–	120
Total U	14.9	14.9	10	9.6	8.5	12.7
Radon-222	1060	2950	6860	1800	8620	3030

Abbreviations: Sp.C. = specific conductance (in microSiemens); TDS = total dissolved solids.

NOTE: Specific conductance through H_2S measured in field by P.E. Dresel and A.W. Rose, July 10-11, 1985;   dissolved Ca through Ba analyzed by PaDER laboratories, for samples collected earlier in 1985, except for trailer park 3 which was sampled July 11, 1985; radionuclides analyzed by EPA laboratory, Montgomery, Alabama, for samples collected 1985.

the filter at one site.

These chemical characteristics suggest that the waters originated as rainwater that has been somewhat concentrated by evapotranspiration to enrich the original Na and Cl, plus possible contamination with road salt. The SO_4, acidity, Fe concentration, and slightly reducing character appear to reflect the oxidation of pyrite, which is known to occur in the Chickies Formation. Low concentrations of Ca and Mg and low pH may be accounted for by the relative sparsity of easily dissolved carbonate, feldspar, and other silicates. The pH of these waters is distinctly lower than most ground waters.

DETAILED STUDY

A more detailed investigation of the ground water in the Chickies Formation is being conducted by the U.S. Geological Survey in cooperation with the Pennsylvania Topographic and Geologic Survey (PaT&GS). The primary objective is to identify the extent and magnitude of radium anomalies in the Chickies Formation. Public water supplies in the Chickies will be identified and sampled. The geochemical environment associated with radium concentrations will be characterized. An understanding of the conditions and processes that favor the dissolution and retention of radium in solution may help predict anomalous radiation concentrations in the Chickies Formation and similar geologic units. The project is organized into three program elements beginning with the collection of hydrologic and chemical data. A field sampling program is being designed as the data become available. Data analysis and report writing constitute the final element. The hydrology and general water quality of the study area is summarized in regional water-resource appraisal reports [4-7].

The field sampling program, developed from well records, is being designed to augment the available data. Concentrated data-collection efforts are underway in Lancaster, Chester, and York Counties, but a few sites will be sampled in Adams, Berks, Bucks, and Montgomery Counties. In general, sampling sites will be distributed throughout the Chickies, but detailed sampling in one or more small areas where the hydrogeology is well known will be performed to characterize the geochemical and hydrologic conditions indicative of elevated radium concentrations.

Phase one of the sampling program was a reconnaissance to determine the extent and magnitude of radium anomalies as rapidly as possible, and was performed between October 1986 and January 1987. Fifty-one samples were collected to supplement available radiochemical data and provide evenly distributed areal coverage in the seven counties where the Chickies crops out. Results of the reconnaissance sampling to date are summarized in table 2 and discussed below. Phase two of the sampling program will occur during the period between March and December 1987. One hundred and thirty samples will be collected for detailed chemical analyses.

The chemical constituents and properties being analyzed are listed in table 3. About 140 samples will be collected from wells in the Chickies Formation. An additional 40 samples will be

Table 2.   Concentration of Ra-226 and Ra-228 in ground water in
the Chickies Formation, southeastern Pennsylvania

Local well number	Ra-226 (pCi/L)	Ra-228 (pCi/L)	Total Radium (pCi/L)	Field pH (units)
BE-1440	1.0	2.2	3.2	6.3
BK-396	0.34	<0.9	<1.2	6.3
BK-1200	0.42	<1.0	<1.4	5.8
BK-1201	0.26	<0.5	<0.8	5.6
BK-1202	3.8	50	54	5.4
CH-293	1.6	3.5	5.1	6.4
CH-333	9.5	22	32	4.7
CH-417	0.48	<1.0	<1.5	5.2
CH-418	2.9	7.6	10.5	4.9
CH-427	0.78	1.8	2.6	5.2
CH-945	0.82	<1.0	<1.8	5.5
CH-1265	2.7	12	15	4.8
CH-1286	0.4	12	12	4.6
CH-1616	2.5	26	29	5.3
CH-2113	1.7	12	14	5.7
CH-2115	0.6	<1.0	<1.6	6.6
CH-2410	2.3	4.9	7.2	5.8
CH-2418	<0.7	2.6	<3.3	5.4
LN-1684	1.9	5.3	7.2	5.6
LN-1685	3.9	5.0	8.9	6.2
LN-1686	1.0	1.6	2.6	5.2
LN-1687	2.7	5.3	8.0	4.8
LN-1688	7.4	34	41	4.4
LN-1689	1.4	6.8	8.2	5.7
LN-1690	1.1	1.1	2.2	5.1
LN-1691	0.56	<0.5	<1.1	5.7
LN-1692	0.84	1.1	1.9	5.8
LN-1693	0.24	<0.5	<0.7	5.6
LN-1694	1.6	5.2	6.8	5.1
LN-1695	2.3	4.0	6.3	4.9
MG-1000	<0.2	<1.0	<1.2	6.3
MG-1001	<0.2	0.83	<1.0	6.3
YO-371	0.34	<0.6	<0.9	5.2
YO-792	0.85	1.5	2.4	5.5
YO-1148	5.4	20	25	5.0
YO-1149	0.62	<0.7	<1.3	6.1
YO-1150	<0.2	<1.0	<1.2	6.2
YO-1151	0.58	<1.0	<1.6	6.7
YO-1152	<0.2	<1.0	<1.2	6.4
YO-1153	1.0	<0.7	<1.7	6.5
YO-1154	1.0	<0.9	<1.9	5.4
YO-1155	<0.2	<0.9	<1.1	6.1
YO-1156	<0.2	5.6	<5.8	5.5
YO-1157	4.7	6.4	11.1	4.7
YO-1158	<0.2	<0.9	<1.1	4.9
YO-1159	<0.2	<1.0	<1.2	6.4
YO-1160	<0.1	<0.6	<0.7	6.2
YO-1161	<1.0	<1.0	<2.0	6.9

Table 3.   Dissolved constituents and chemical properties analyzed
in ground water from the Chickies Formation

Chemical Constituent	Chemical Property
Barium	Alkalinity (field)
Calcium	Eh (field)
Chloride	Dissolved organic carbon
Iron	Gross alpha radiation
Magnesium	Gross beta radiation
Manganese	pH (field)
Potassium	Specific conductance (field)
Radium-226	Total dissolved solids
Radium-228	
Silica	
Sodium	
Sulfate	
Uranium	

collected from wells adjacent to the Chickies Formation to
establish the lack of anomalous radium concentrations in other
rock units and to help determine the transport and fate of radium.
Samples collected adjacent to the Chickies will be analyzed for
all constituents and properties in table 3 so that the chemistry
can be compared.

In addition to laboratory analyses, ground-based scintillation counter surveys will be made across the contact between the
Chickies Formation and adjacent units. Radium may be deposited
along the contact as ground water moves from the Chickies into
adjacent units. Detectable anomalies in gamma radiation are most
likely in high water-table areas.

More detailed investigation of the radionuclides present in
the ground water and determination of whether the radionuclides
are dissolved or colloidal will be performed on samples from
selected wells. A few thin-section analyses may be performed to
determine whether the mineralogy of the Chickies or other formations influences radium concentrations in ground water.

Descriptive statistics will indicate the magnitude, variance,
and frequency distribution of radium concentrations in ground
water within or close to the Chickies Formation. Analysis of
variance will determine the significance of differences between
concentrations in the Chickies Formation and other rock types
(principally quartz schist, gneiss, and limestone) in contact with
the Chickies. Transport of radium across contacts and the extent
of the radium fringe into other units will be tested.

Physiochemical factors that affect the source and mobility of
radium will be examined. Geology, mineralogy, ground-water flow
paths, well depth and yield, weathering processes, chemistry of
the aqueous phase, and the distribution of parent isotopes may all
influence the concentration of radium in solution. Correlations
between dissolved radium and other radionuclide concentrations and

various physiochemical factors are intended to provide the basis of study concerning the processes contributing to the anomalies.

A preliminary map showing the extent and magnitude of radium concentrations in southeastern Pennsylvania will be prepared as soon as analytical results of the reconnaissance phase (51 samples) are received. The map will be released by the PaT&GS and will incorporate all available radium data.

A report discussing the occurrence of radium anomalies in ground water in the Chickies Formation, and the conditions under which anomalies are likely to be found is planned. It will allow the public to determine the possibility of elevated concentrations of radium in a particular area. It will also be used by public water purveyors and municipal planners to ensure that public water supplies are not developed in areas with elevated radium concentrations. The report will be published by the PaT&GS in the Water Resource Report series.

DISCUSSION OF RADIUM DATA

The elevated concentrations of dissolved radium seem to be confined to the Chickies Formation. For 25 well-water samples collected by PaDER in the vicinity of one of the three trailer parks, values for gross alpha and gross beta radiation in the Chickies Formation are distinctly higher than in water samples from adjacent rock units. Gross alpha radiation for water from nine wells in the Chickies averaged 40 pCi/L, but was consistently less than 3 pCi/L for water from 16 nearby wells outside the Chickies Formation. Similarly, gross beta radiation averaged 46 pCi/L for samples from the Chickies and was consistently less than 4 pCi/L beyond the contact.

Results of the radiochemical sampling to date show that most of the anomalous concentrations of radium in ground water of the Chickies Formation occur in Chester and Lancaster Counties (table 2). Water from eight of 13 wells in Chester County and seven of 12 wells in Lancaster County had radium concentrations that exceed the allowable MCL. In Chester County, radium-228 (Ra-228) concentrations ranged from less than 1.0 to 26 pCi/L, and radium-226 (Ra-226) concentrations ranged from 0.4 to 9.5 pCi/L. In Lancaster County, Ra-228 concentrations ranged from less than 0.5 to 34 pCi/L, and Ra-226 concentrations ranged from 0.24 to 7.4 pCi/L.

In York County, water from three of 16 wells sampled exceeded the MCL. Ra-228 concentrations ranged from less than 0.6 to 20 pCi/L, and Ra-226 concentrations ranged from less than 0.1 to 5.4 pCi/L.

No water samples from wells in Berks and Montgomery Counties exceeded the MCL for radium. Water from one well in Berks County had a Ra-228 concentration of 2.2 pCi/L and a Ra-226 concentration of 1.0 pCi/L. Water from two wells in Montgomery County had Ra-228 concentrations of less than 0.2 and less than 1.0 pCi/L and Ra-226 concentrations of less than 0.2 pCi/L.

In Bucks County, water from one of four wells sampled exceeded the MCL. Ra-228 concentrations ranged from less than 0.5 to 50 pCi/L, and Ra-226 concentrations ranged from 0.26 to 3.8 pCi/L.

Fourty-three of the 48 analyses received have Ra-228 concentrations greater than the Ra-226 concentrations (table 2). Each of the fifteen samples with Ra-228 concentrations greater than 5 pCi/L have Ra-228/Ra-226 ratios greater than unity, with a median ratio of 4.4. This Ra-228/Ra-226 ratio is much higher than usually found, and for many samples appears to exceed ratios measured elsewhere for ground water samples with appreciable radium concentrations (greater than 5 pCi/L). Some ratios greater than unity are reported by King and others [8] in South Carolina, mainly for Ra-228 concentrations less than 5 pCi/L, and by Lucas [9] for the mid-continent region, also for concentrations less than 5 pCi/L. For rocks with mean crustal activity ratio of Th-232/U-238, the ratio of activities of Th-232 daughter nuclides to U-238 daughter nuclides should be in the range 1.2 to 1.5 at equilibrium. The elevated Ra-228/Ra-226 ratios in these waters suggests that an anomalously high Th-232/U-238 ratio may exist in the Chickies Formation, possibly because of previous thorough leaching of uranium.

Current USEPA guidelines suggest that it is not necessary to analyze for Ra-228 if the concentration of Ra-226 in the sample does not exceed 3 pCi/L. However, in the Chickies Formation, significant Ra-228 anomalies exist with corresponding Ra-226 concentrations less than 3 pCi/L. These results suggest the need to use gross alpha screening with caution because Ra-228 is not an alpha emitter. Similar recommendations have been made by Hess and others [10] and Lucas [9].

Studies by Krishnaswami and others [11] suggest that radium is rapidly adsorbed by aquifer materials. However, in some ground waters of the Chickies Formation, the data presented here show that radium is not being adsorbed or otherwise immobilized. A possible reason for this behavior is the low pH. Langmuir and Melchior [12] cite experiments by Riese [13] showing that adsorption of radium on quartz and kaolinite is inhibited by low pH and elevated concentrations of $Ca2+$ because of competition for adsorption sites. The waters in the Chickies Formation may be acid enough to inhibit adsorption. In addition, lack of Fe-oxides at the Eh-pH conditions in the aquifer may also contribute to radium mobility, because the Fe-oxides commonly have large surface areas and are strong adsorbents for heavy metals. In the reconnaissance samples reported to date, radium concentration shows a significant negative correlation with pH, although a considerable scatter of the relation indicates that other factors are also involved (figure 1). Maximum values for total radium concentrations reported in table 2 are plotted with field pH in figure 1. All samples that have a field pH below 4.9 exceed the USEPA maximum contaminent level of 5 pCi/L combined Ra-228 and Ra-226 concentrations.

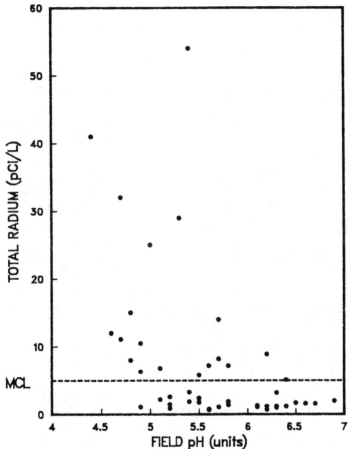

Figure 1.--Plot of field pH and total radium in
the Chickies Formation, Pennsylvania

REFERENCES CITED

1. Stose, G.W. and Jonas, A.J. "Geology and mineral resources of York County, Pennsylvania", Pennsylvania Geological Survey, 4th ser., County Report 67 (1939), pp. 199.

2. "National Interim Primary Drinking Water Regulations, U.S. Environmental Protection Agency, Publication CFR 40, Parts 100-149 (1983), pp.399.

3. Hem, J. D. "Study and interpretation of the chemical characteristics of natural water", U.S. Geological Survey Water Supply Paper 2254 (1985), pp. 363.

4. Poth, C. W. "Hydrology of central Chester County", Pennsylvania Geologic Survey, 4th ser., Water Resources Report 25, (1968), pp. 84.

5. McGreevy, L. J. and Sloto, R. A., 1977, Ground-water resources of Chester County, Pennsylvania", U.S. Geological Survey Water Resources Investigation 77-67 (1977), pp. 76.

6. Lloyd, O. B., Jr., and Growitz, D. B. "Ground-water resources of central and southern York County, Pennsylvania", Pennsylvania Geologic Survey, 4th ser., Water Reources Report 42 (1977), pp. 93.

7. Poth, C.W. "Summary ground-water resurces of Lancaster County, Pennsylvania", Pennsylvania Geologic Survey, 4th ser., Water Reources Report 43 (1977), pp. 80.

8. King, P. T., Michel, J., and Moore, W. S. "Ground water geochemistry of Ra-228, Ra-226, and Rn-222", Geochem. et Cosm Acta 46:1173-1182 (1982).

9. Lucas, H. F. "Ra-226 and Ra-228 in water supplies", J Am Water Works Assoc 77:56-77 (1985).

10. Hess, C. T., Michel, J., Horton, T. R., Prichard, H. M., and Coniglio, W. A. "The occurrence of radioactivity in public water supplies in the United States", Health Physics 48:553-586 (1985).

11. Krishnaswami, S., Graustein, W. C., and Tarkian, K. K., "Radium, thorium, and radioactive lead isotopes in groundwaters: Application to the in situd determination of adsorption-desorption rate constants and retardation factors" Water Resources Research 18:1663-1675 (1982).

12. Langmuir, D. and Melchior, D. "The geochemistry of G, Sr, Ba, and Ra sulfates in some deep water brines from the Palo Duro Basin, Texas" Geochem. et Cosm. Acta 49: 2423-2432 (1985).

13. Riese, A. C. "Adsorption of radium and thorium onto quartz and kaolinite: A comparison of solution surface equilibria models", Ph.D. dissertation, Colorado School of Mines (1982).

HYDROGEOLOGIC CONTROLS ON THE
OCCURRENCE OF RADIONUCLIDES IN
GROUNDWATER OF SOUTHERN ONTARIO

Paul J. Beck,
Ontario Ministry of the Environment, Toronto, Ontario,
Canada

Daniel R. Brown,
Ontario Ministry of the Environment, London, Ontario,
Canada

ABSTRACT

Groundwater samples from 29 municipal drinking
water sources and 137 private supplies were sampled
across Southern Ontario and analysed for the
radionuclides; radon (Rn), radium - 226 (Ra-226) and
uranium (U). Considering yearly average
concentrations, results from all locations sampled fall
within Ontario drinking water objectives for Ra-226 and
U. Currently there is no Provincial objective for Rn
levels in drinking water.

In Southeastern Ontario, private well sampling was
focussed in two areas in the Grenville Province of the
Precambrian Shield:
  (i) Kennebec - Sharbot Lake, an area containing
      anomalous airborne gamma-ray spectral
      signatures in Lennox, Addington and Frontenac
      Counties, and
  (ii) Bancroft area in northern Hastings County which
      contains economic deposits of uranium.
Radionuclide concentrations in Southeastern Ontario
ranged accordingly: Rn, 0.037-1650 Bq/L
(1-44,600pCi/L), Ra-226, <37-925 mBq/L (<1-25 pCi/L)
and U, <3-110 µg/L.

In Southwestern Ontario, sampling was focussed on municipal and private wells which were finished in the Middle Devonian limestone and dolostone of the Detroit River Group which consists of a lower carbonaceous reef facies carbonate overlain by an evaporitic carbonate sequence containing anhydrite. Groundwaters, which recharge through up to 30m of glacial till with variable amounts of kame moraine and spillway sands, are predominantly calcium-bicarbonate to calcium-sulphate in character. Rn levels ranged from 5.18-407 Bq/L (140-11,000 pCi/L) Ra-226 ranged from <37-680 mBq/L (<1-18.4 pCi/L) and U concentrations ranged from <3-39μg/L.

Additional samples from 24 domestic wells from elsewhere across the Province were sampled. Wells finished in overburden and Middle Ordovician limestone generally contained low radionuclide levels.

Samples taken from discrete water bearing horizons within a municipal well show variations in radionuclide concentrations vertically within the stratigraphy. A conceptual model is presented for radionuclide distribution in the Detroit River Group in Southwestern Ontario whereby uranium is mobilized in groundwater under oxidizing conditions. At discrete sites containing carbonaceous material, uranium is reduced and precipitated with the eventual release of Ra and Rn to the groundwater.

INTRODUCTION

Groundwater sampling for radionuclides has been conducted in Ontario on a non-routine basis to determine concentrations of Rn, Ra and U in drinking water supplies.

In 1981 the Joint Federal-Provincial Working Group on Drinking Water proposed an Interim Maximum Acceptable Concentration (I.M.A.C.) of 20 μg/L for uranium. Several private wells in South- eastern Ontario were found to contain uranium concentrations in excess of the I.M.A.C. As a result, a detailed sampling program of municipal and private drinking water supplies was conducted by the Ontario Ministry of the Environment (MOE) in 1981. [1] In Southeastern Ontario, wells are finished in a variety of geological environments ranging from metamorphosed igneous and sedimentary rock of the Precambrian Shield, Lower Paleozoic sandstones and carbonates, to unconsolidated Pleistocene sands and gravels. Radionuclide sampling of private drinking water supplies was carried out in the Kennebec-Sharbot Lake area. This area of the Precambrian Shield, containing granitic and syenitic pegmatite deposits was the subject of uranium exploration activity in the 1970's. [2] Additional sampling of private wells was conducted in the Bancroft area, where uranium mineralization was first discovered in 1922. [3] The Bancroft area is known for its diversity of uraniferous minerals and complex metamorphic geology.

In the early 1980's routine monitoring of groundwater supplies in the vicinity of a nuclear generating station in Southwestern Ontario by MOE led to the discovery of elevated Ra-226 at concentrations below the Provincial Maximum Acceptable Concentration (M.A.C). of 1 Bq/L (27pCi/L). [4] An extended sampling program indicated that elevated radionuclides were related to wells finished in limestone and dolostone of the Middle Devonian Detroit River Group.

In 1986 a Province-wide survey of domestic wells was initiated by MOE to establish background levels of radionuclides. This study will continue through 1987. A location map showing sample sites is presented in Figure 1.

The purpose of this paper is to:
  (i)   present the distribution of Rn, Ra and U concentrations in groundwater samples across Southern Ontario, and
  (ii) identify hydrogeologic controls for radionuclide occurrence in groundwaters of Southwestern Ontario.

GEOLOGY

The bedrock geology in Southern Ontario consists of crystalline rock of the Precambrian Shield which is overlain by marine sandstones, dolostones, limestones and shales ranging in age from Cambro-Ordovician to Upper Devonian. [5] The simplified bedrock geology is depicted in Figure 2. Much of the Paleozoic bedrock is covered by variable thicknesses of glacial overburden and is described elsewhere [6].

In Southeastern Ontario uranium in groundwaters appears to be derived from two main sources:
  (i)   weathering of primary uranium associated with metamorphosed basement rock;
  (ii) weathering of secondary remobilized uranium in Lower Paleozoic sedimentary rock.

In the Bancroft and Kennebec-Sharbot Lake areas uranium minerals consist mostly of uraninite, uranophane, and allanite. They occur as accessories with pegmatites, calc-silicates, and hydrothermal vein deposits, associated with regional metamorphism of the Grenville Orogeny dated at 960-975 m.y. [2] In portions of Southeastern Ontario where the Precambrian basement is overlain by Paleozoic sedimentary rock, local concentrations of uranium have been reported in Cambro-Ordovician sandstones and dolostones. [7] Secondary uranium is found associated with bituminous sequences near the Precambrian-Phanerozoic unconformity. The

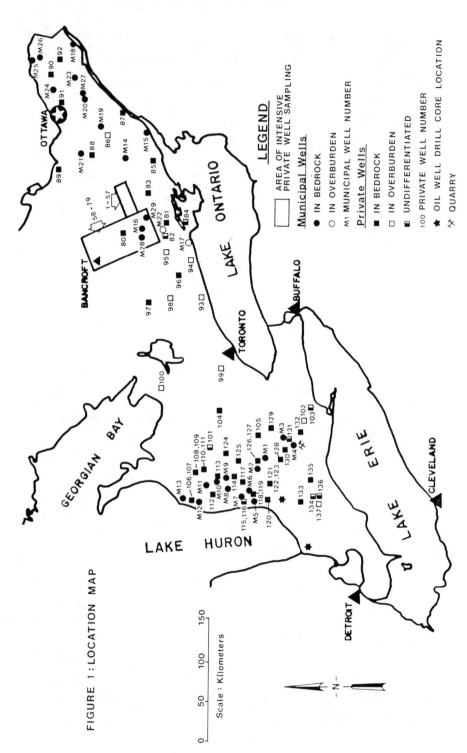

FIGURE 1: LOCATION MAP

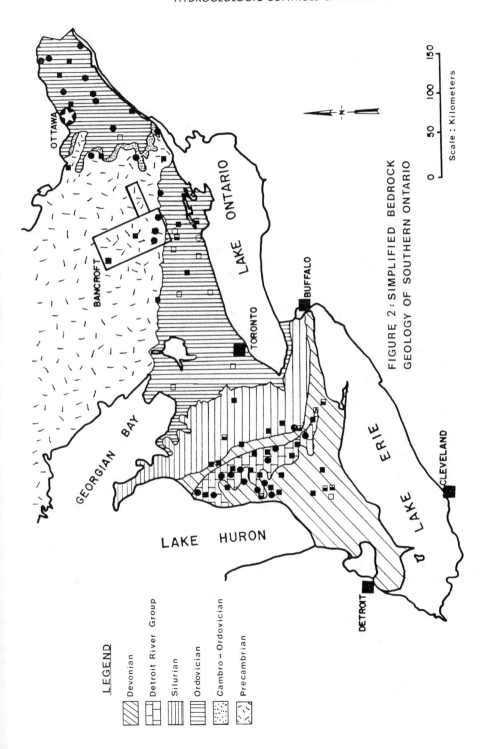

FIGURE 2 : SIMPLIFIED BEDROCK
GEOLOGY OF SOUTHERN ONTARIO

LEGEND

Devonian
Detroit River Group
Silurian
Ordovician
Cambro – Ordovician
Precambrian

Scale : Kilometers

LAKE HURON

GEORGIAN BAY

LAKE ERIE

LAKE ONTARIO

OTTAWA

BANCROFT

TORONTO

BUFFALO

DETROIT

CLEVELAND

model for this mineralization derives uranium from the
Precambrian basement by oxidation by shallow
circulating groundwater. Dissolved uranium is
transported by groundwater through bedrock fractures
and along the unconformity and is fixed by reduction in
bituminous-rich units. [8,9]
In Southwestern Ontario, the formations of interest lie
within the Middle Devonian Detroit River Group, and
consist of the lower Amherstburg Formation and the
overlying Lucas Formation. The Amherstburg which
reaches a thickness of approximately 46 m is made up of
grey brown or dark brown, fine to coarse grained, coral
stromatoporoid bioclastic limestone, locally black,
bituminous and cherty [10, 11]. The Lucas Formation
varies from light tan and brown aphanitic to
microsucrosic dolostones with minor anhydrite beds to
grey-brown, finely-crystalline to aphanitic or
sublithographic limestones of exceptional purity.

In the Amherstburg Formation, bioherms are common.
The largest of these, the Formosa Reef, covers an area
of 39,000 hectares (96,000 acres) in Bruce and Huron
Counties and is up to 27 m in thickness. The reef core
is composed of high calcium limestone with a framework
of stromatoporoids and crinoid debris. This core is
surrounded by dark brown cherty and bituminous
limestone [11]. The presence of the reef is believed
to have been critical in the deposition of the
organic-rich, bituminous strata by creating low-energy
conditions in the adjacent waters. Bituminous rich
zones within the Amherstburg have been observed in a
quarry rock face at site M4 and in core from oil
exploration holes located 60 km west northwest and 112
km west of the quarry (see Figure 1). At the quarry,
the lowermost 30 m of the Amherstburg is exposed.
Bituminous limestone lenses approximately 1 metre thick
occur at the quarry floor and approximately 12 m above
the quarry floor. This rock emits a strong odour of
hydrocarbon when freshly broken. Samples were taken
from the two bituminous lenses and from relatively
clean looking limestone about 3 m above each bituminous
zone. Samples were sent to Barringer Magenta in
Toronto for uranium analysis by fluorometry using two
different digestions. First the amount of readily
leachable uranium i.e. that which is loosely held by
the rock was determined using a 4 normal nitric acid
cold extraction while the total uranium was determined
using a hydrofluoric acid hot extraction. The results
of the analyses are given in Table 1. The bituminous
layers contain between 1.5 to 3 times the uranium
concentration of cleaner layers. Cold extractable
uranium constitutes only about 20-50% of the total
uranium in the rock.

Table 1.   Uranium Concentrations for Limestone Samples
from Amherstburg Formation

Sample No.		U (ppm) (Cold Ext.)	U (ppm) (Hot Ext.)
87 R-1	Bituminous limestone at quarry floor	2.2	6.1
87 R-2	Non-bituminous ls.3 m above quarry floor	1.4	3.0
87 R-3	Bituminous ls.10 m above quarry floor	2.4	8.8
87 R-4	Non-bituminous ls.15 m above quarry floor	0.8	4.7

DISTRIBUTION OF RADIONUCLIDES

Radon, radium-226 and uranium were analysed at the
Ministry of Labour's Radiation Protection Laboratory in
Toronto. Radon was measured directly by alpha particle
counting. Radium-226 was determined by counting Rn
daughters after Ra-226 was coprecipitated with $BaSO4$.
Detection limit for this technique is 37 mBq/L
(1 pCi/L). Uranium was determined by fluorometry on a
LiF fused pellet. Detection limit is 3 µg/L.
Figures 3,4, and 5 present the distribution of Rn,
Ra-226, and U for wells in Southern Ontario. Tables
2, 3, 4 summarize the frequency distribution of those
same parameters. Uranium concentrations in private
wells ranged from <3 µg/L to 91 µg/L. The frequency
distributions show that for private wells, higher
concentrations of Rn, Ra-226 and U occurred more
commonly in wells in Southeastern Ontario compared to
Southwestern Ontario and reflects the sampling bias
toward areas in the Precambrian Shield known to contain
primary uranium mineralization. However, municipal
supply wells in Southeastern Ontario, which are mostly
finished in carbonate rock and overburden away from
mineralized areas, contained fewer elevated
concentrations of Ra-226 and U compared to municipal
supplies in Southwestern Ontario which were finished in
the Detroit River Group.

Table 2. Frequency Distribution For Radon

Radon Concentration (Bq/L)	No. Priv. Wells		No. Municipal Wells
0 - 18.5	23 SE*	2 SC	2 SE, 19 SW
18.6 - 37.0	12 SE	-	3 SE, 5 SW
37.1 - 74.0	3 SE,	1 SC	- 3 SW
74.1 - 185	9 SE	-	- -
186 - 370	4 SE	-	- 1 SW
371 - 740	1 SE	-	- -
741 - 1110	2 SE	-	- -
1111 - 1480	-	-	- -
1481 - 1850	1 SE	-	- -

* SE - Southeastern Ontario
  SW - Southwestern Ontario
  SC - South Central Ontario

Table 3. Frequency Distribution for Radium - 226

Radium - 226 (mBq/L)	No. Private Wells			No. Municipal Wells
0 - 37	80 SE,	13 SC,	27 SW	26 SE, 25 SW
38 - 74	-	-	4 SW	2 SE -
74 - 111	5 SE,	-	1 SW	- 5 SW
112 - 148	-	-	-	- -
149 - 185	-	-	-	1 SE -
186 - 222	-	-	-	- 3 SW
223 - 259	1 SE	-	-	- -
260 - 297	-	-	-	- -
298 - 333	-	-	-	- -
334 - 370	1 SE	-	-	- -
>370	1 SE	-	-	- 1 SW

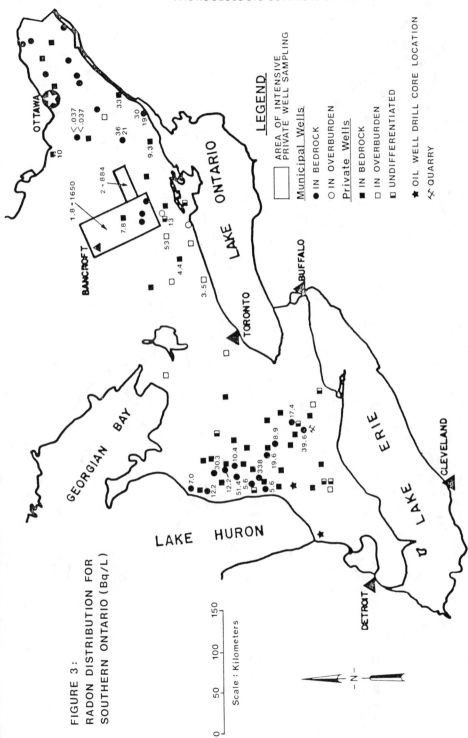

FIGURE 3:
RADON DISTRIBUTION FOR
SOUTHERN ONTARIO (Bq/L)

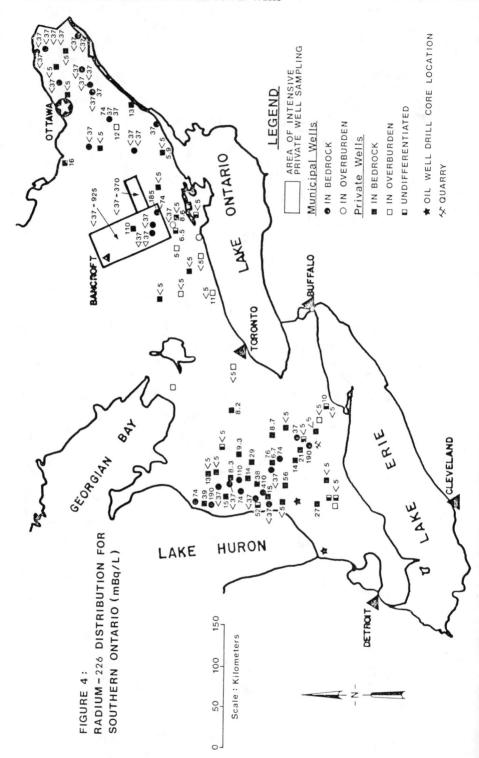

FIGURE 4:
RADIUM-226 DISTRIBUTION FOR
SOUTHERN ONTARIO (mBq/L)

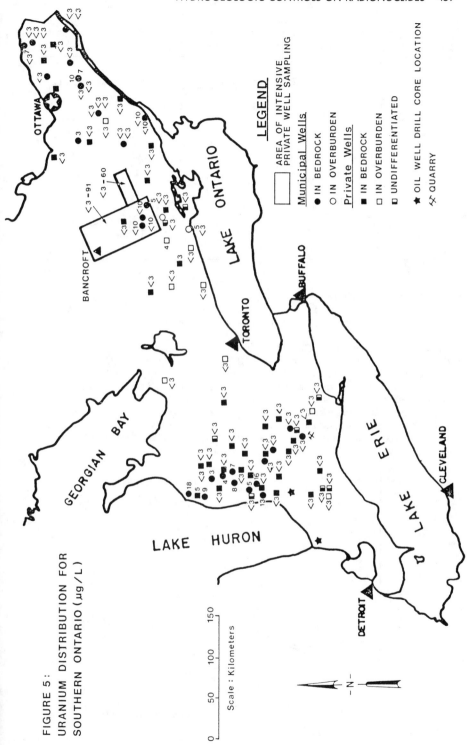

FIGURE 5:
URANIUM DISTRIBUTION FOR
SOUTHERN ONTARIO (µg/L)

Table 4. Frequency Distribution for Uranium

Uranium Concent'n (µg/L)	No. Private Wells			No. Municipal Wells	
0 - 10	75 SE,	13 SC,	32 SW	28 SE,	30 SW
11 - 20	8 SE	–	–	–	4 SW
21 - 30	–	–	–	–	1 SW
31 - 40	–	–	–	–	–
41 - 50	2 SE	–	–	–	–
51 - 60	1 SE	–	–	–	–
61 - 70	–	–	–	–	–
71 - 80	–	–	–	–	–
81 - 90	–	–	–	–	–
91 - 100	1 SE	–	–	–	–
>100	1 SE	–	–	1 SE	–

Figure 6 plots log Ra-226 vs log U concentrations for samples exceeding 37 mBq/L Ra-226. Samples from Southeastern Ontario show a strong correlation (r = 0.73) which is significant at greater than the 99% confidence level. Samples from Southwestern Ontario show a poor correlation (r= 0.35) and for Ra-226 concentrations greater than about 63 m Bq/l corresponding U concentrations are considerably lower compared to Southeastern Ontario. The data indicate that both U and Ra-226 are mobile in groundwaters in Southeastern Ontario, while in Southwestern Ontario Ra-226 is more mobile relative to U. This implies that

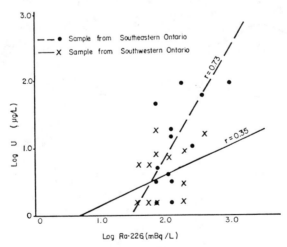

FIGURE 6. Correlation of log Radium concentration vs log Uranium concentration for samples from Southeastern and Southwestern Ontario

in Southwestern Ontario, there is either a solubility
control on uranium, or uranium is depleted at source
relative to Ra-226.

Figures 7a and 7b plot log concentrations of Rn vs
Ra-226 and U respectively. Samples from Southeastern
Ontario show positive correlations in both cases.
Samples from Southwestern Ontario show a weak
correlation between Ra-226 and Rn (r = 0.39). There
does not appear to be any correlation between U and
Rn.

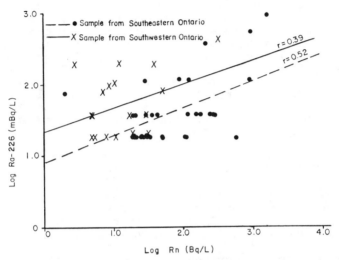

FIGURE 7A   Correlation of Log Rn concentration vs Log Ra-226 concentration

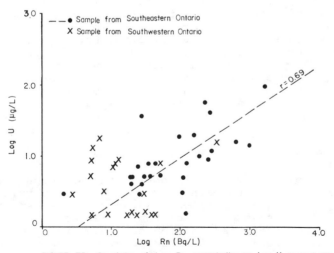

FIGURE 7B   Correlation of Log Rn concentration vs Log U concentration

Implications to Drinking Water Objectives

The M.A.C. for Ra-226 in Ontario drinking water is 1 Bq/L. All samples fall within the objective.

The current I.M.A.C. for U is 20 µg/L. This is an interim number only and was based on an incomplete series of tests taken from the Russian literature with few confirmatory data. This value is considered to be very conservative and the Federal-Provincial Working Subcommittee on Drinking Water is proposing a new M.A.C. for uranium of 100 µg/L, based on a lifetime of consumption. All sample results with the exception of some results from site M29 were within the proposed new M.A.C. for U. At M29, a maximum U value of 110 ug/L was recorded, however, uranium concentration varies seasonally in this well from 70-110 µg/L. The yearly average lies below the proposed M.A.C. for uranium.

At the present time, there is no Provincial objective for Rn in drinking water, since the major health concern is with the inhalation of radon gas, rather than its ingestion.

HYDROGEOLOGY OF SOUTHWESTERN ONTARIO

The hydrogeology of the portion of Southwestern Ontario containing the subcropping Detroit River Group was evaluated to help explain the distribution of radionuclides in groundwaters in this area.

Groundwater Flow Directions

The potentiometric surface for wells completed in Middle Devonian bedrock was plotted after a review of MOE water well records, and is depicted in Figure 8. Also shown is the subcrop edge of the Detroit River Group and the location of municipal supply wells M1-M13. The potentiometric surface contours indicate generally westward groundwater flow toward Lake Huron except in the southern portion of the map area where flow becomes predominantly southwestward toward the Michigan Basin.

Major Ion Chemistry

Major ion chemistry for some of the municipal water supplies collected between 1972 and 1986 was obtained from MOE water quality files. Water chemistry at each site remained essentially constant. Representative analyses are listed in Table 5 and are plotted on a Durov diagram in Figure 9. The Durov diagram has been found to be a useful plot for depicting chemical evolution in groundwaters. [12]

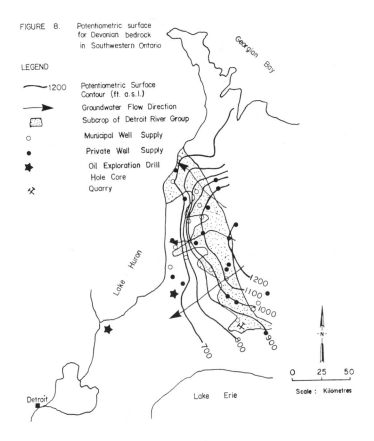

FIGURE 8. Potentiometric surface for Devonian bedrock in Southwestern Ontario

LEGEND

—1200   Potentiometric Surface Contour (ft. a.s.l.)

——→   Groundwater Flow Direction

▨   Subcrop of Detroit River Group

○   Municipal Well Supply

●   Private Well Supply

★   Oil Exploration Drill Hole Core

⚒   Quarry

Groundwater chemistry in this area is controlled by carbonate and sulphate mineralogy of calcium and magnesium. Approximately half of the groundwaters display a Ca-HCO$_3$ character and samples from sites M1 and M5 are Ca-SO4 type. Chloride reaches a maximum of only 15 meq/L % (meq/L = milliequivalents/litre = concentration x charge ÷ at.wt.). Anions are therefore dominated by bicarbonate and sulphate. Cations are dominated by Ca with Mg ranging from 20-40 meq/L %. Samples from sites M1, M2, M4, and M12 show increases in Na to between 20 and 27 meq/L %. The excess of sodium over chloride is likely due to ion exchange between Na and Ca and to a lesser extent Mg described by the following reaction:

$$(Ca^{2+}, Mg^{2+}) + 2 \ Na \ (ad) = + Ca, \ Mg \ (ad) + 2 \ Na^+ \ (1)$$

Table 5. Representative Major Ion Chemistry from
         Municipal Wells in Southwestern Ontario

Loc.	Well	Ca*	Mg	Na	K	Alk.	SO4	Cl
M1	Lorne	52	18.2	31	2.1	225	19.0	21.5
	#2	124	39	17	2.2	216	28.0	3
	#3	302	52	23	2.6	216	782	5
	#4	73	22	27	2.0	250	77	3
	#6	93	26	28	2.1	204	185	3
M2	#2	55.3	27.2	32.8	1.4	202	95	5.4
M3	#6		NO DATA					
M4	#6	66	38	46.5	2.3	221	165	21.0
	#7	92	48	76	3.3	222	270	70
M5	#2	155	34.6	46	1.6	169	420	18
	#3	114	28.8	38	1.4	164	300	14
M6	#1	74.5	27.6	11.2	1.6	238	61	18.5
	#2	115	29.0	23.2	1.8	283	116	41.0
M7	#1	82	26.6	16.2	1.65	274	30.0	34.0
	#2	77	21.6	11.8	1.50	258	22.0	17.0
	#3	70	24.8	7.8	1.10	250	41.0	9.0
M8			NO DATA					
M9			NO DATA					
M10	#1	89.0	29.4	13.5	0.65	225	125	6.0
	#3	58.0	20.2	5.1	0.50	224	18.5	1.0
M11		75.5	25.6	1.2	0.23	240	41	2.0
M12		43.5	18.6	28.0	2.52	178	91	11.5
M13			NO DATA					

* All concentrations in mg/L.

FIGURE 9. Durov plot for municipal
well waters in Southwestern
Ontario

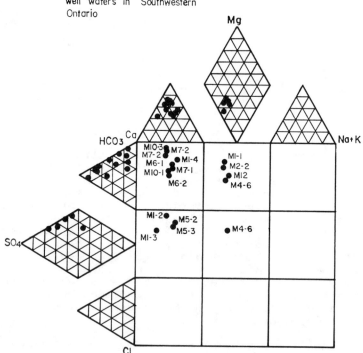

Measurements of Eh and pH

Field measurements of Eh and pH of selected municipal wells was conducted in the field in 1987 using an Orion Model SA 210 digital pH and millivolt meter. A combination platinum electrode calibrated with quinhydrone solution was used to measure Eh. Eh and pH measurements were carried out using an airtight cell attached to a sample tap at the wellhead, which provided a continuous sample at a low flow rate while the well pump was operating. Readings were taken every 5 minutes until stabilization was reached, usually within 5-15 minutes for pH and 15-45 minutes for Eh. Groundwater temperature ranged from 7.5 - 9.5 °C. Eh and pH values are listed in descending order of Eh readings in Table 6 along with Rn, Ra-226 and U concentrations.

Table 6. Eh-pH Measurements and Radionuclide Concentrations for Selected Municipal Groundwater Supplies in Southwestern Ontario

Loc.	Well	T(°C)	Eh(mV)	pH	Rn (Bq/L)	Ra-226 (mBq/L)	U(µg/L)
M7	#1	7.5	+112	7.21	–	<37	5
	#3	7.5	+101	7.36	5.18	<37	9
M1	#6	8.0	+ 86	7.53	5.18	37	5
M11		8.5	+ 75	7.35	30.3	<37	3
M6	#2	9.0	+ 45	7.45	11.1	<37	8
	#1	9.0	+ 29	7.39	338	410	16
M12	#1	8.0	- 40	7.53	12.2	190	9
M10	#3	–	- 55	7.51	7.77	<37	3
M8	#2	–	- 68	7.60	2.66	190	3
M9	#2	8.5	- 84	7.48	10.4	110	7
M13		9.0	-118	7.59	7.03	74	18
M1	Lorne	8.0	-123	7.67	8.88	74	<3
M5	#2	9.0	-126	7.62	5.55	<37	13
M8	#1	8.5	-133	7.62	51.4	74	8
M2	#2	8.0	-148	7.69	17.0	<37	<3
M3	#6	9.5	-178	7.29	17.4	37	<3
M4	#7	9.0	-255	7.59	29.6	37	<3
	#3	9.0	-279	7.54	5.18	37	<3
	#8	9.0	-293	7.29	39.6	190	<3

Eh-pH values are plotted on Figure 10. Sample pH ranged from 7.29 to 7.69. This resulted in a very narrow field on the Eh-pH diagram over the range of Eh values from +112 mV to -293 mV. An inspection of Table 6 shows that all samples which had Eh values ≤ -148 mV contain U values below detection, while all those sites,

with the exception of the Lorne Ave. well at M1, which
had Eh values > -148 mV contain concentrations of U
above detection. The correlation between log U
expressed in μg/L and Eh in mV, is positive with a
correlation coefficient r = 0.53 which is significant at
between the 95-99% confidence levels.

The low temperature aqueous geochemistry of uranium
is described elsewhere. [13, 14] Under oxidizing
conditions U tends to mobilize via its hexavalent state
and its solubility is enhanced by the formation of
carbonate complexes. [15]

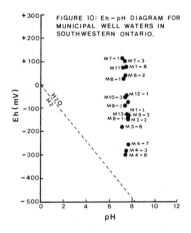

FIGURE 10: Eh-pH DIAGRAM FOR
MUNICIPAL WELL WATERS IN
SOUTHWESTERN ONTARIO.

Uranium is readily precipitated under reducing conditions at sites of sulphate reduction or carbonaceous
rich horizons within the aquifer. Figure 11 illustrates the stability field for uraninite (UO_2) as
described by Hostetler and Garrels [14] and Langmuir
[13] for the system $U-O_2-CO_2-H_2O$ at $25^\circ C$, $PCO_2 = 10^{-2}$
atm, and $\Sigma U = 10^{-6}$ M. The diagram shows that at
equilibrium, under oxidizing conditions soluble uranyl
carbonate complexes are stable in groundwater. As the
oxidation potential is reduced to a point on the
boundary of the uraninite stability field, dissolved
uranium becomes unstable and the system favours
uraninite precipitation. With further reduction,
uraninite will remain the stable form of uranium.
Recognizing the differences between field system
conditions and conditions at equilibrium, the data from
Southwestern Ontario have been compared with this
diagram. The Eh reading of -148 mV at a measured pH of
7.69 from site M2 lies between the two stability fields
for uraninite calculated at equilibrium. The Eh value
of -148 mV may represent the lower limit of stability

for soluble uranium species in groundwaters in Southwestern Ontario. Under more positive Eh values, dissolved U will be stable while at Eh values ≤ -148 mV, dissolved U will be unstable and tend to precipitate. This would explain the lack of detectable uranium at oxidation potentials below -148 mV.

Both Rn and Ra-226 failed to show a statistically significant correlation with Eh, however four of six sites which measured positive Eh values contain Ra-226 concentrations below detectable levels, while ten of thirteen sites which measured negative Eh values contain detectable Ra-226. Radium has been reported to be mobile under reducing conditions,

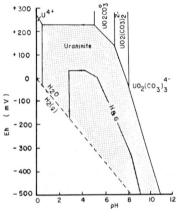

FIGURE II. Eh-pH diagram for system $U-O_2-CO_2-H_2O$ described by Langmuir (1978) and Hostetler and Garrels (1962). $T = 25°C$, $PCO_2 = 10^{-2}$ atm, $\leq U = 10^{-6}$ M

but its mobility is limited by adsorption onto clays and organic matter. [17, 18,] Its mobility under oxidizing conditions is further reduced by adsorption onto iron and manganese hydroxides, as well as coprecipitation with barite and gypsum. [19]

It is worth noting that those waters containing greater than 20 meq/L % Na (eg M1, M2, M4, M12) fall within the range of negative Eh readings. If these Na-enriched groundwaters represent some degree of ion exchange, then the associated low Eh is consistent with observations by Edmunds who noted that Eh decreased and ion exchange increased along the groundwater flow path in the Lincolnshire Limestone [18].

Spatial Variation in Radionuclide Concentration

Well #1 at sample site M6 showed highest concentrations of Ra-226 and Rn in Southwestern Ontario.

This well is completed as an open hole in bedrock.   The
MOE wished to determine whether radionuclides were
present throughout the entire stratigraphic sequence or
originated from discrete water-bearing lenses within the
Detroit River Group.   The driller's log was first consulted to identify shaley layer aquitards.   These were
identified on the caliper log as zones having a larger
diameter.   A packer system was then installed down the
hole to isolate each water bearing section, and allow
discrete sampling for major ion chemistry, Ra-226 and U.
The results are given in Table 7 for Ra-226 and U.

Table 7.   Depth Sampling for Ra-226 and U in Well #1 at
           Municipal Well Site M6.

Sample No.	Depth (m)	Ra-226 (mBq/L)	U (μg/L)
S-1	58.2 +	440	12
S-2	45.4 – 51.5	450	13
S-3	50.3 – 54.6	350	13
S-4	39.0 – 43.0	510	11
S-5	34.8 – 38.7	620	10
S-6	30.5 – 34.8	680	12
S-7	27.4 – 31.7	500	16

Major ion chemistry remained essentially constant, but
uranium displays a 60% variation in concentration while
radium - 226 shows a 94% variation in concentration.
These values are of the same order of magnitude as the
variation in uranium concentration from the Detroit
River Group limestone listed in Table 1.

CONCEPTUAL MODEL FOR RADIONUCLIDE DISTRIBUTION IN
SOUTHWESTERN ONTARIO

    A conceptual model which accounts for the
distribution of radionuclides in Southwestern Ontario
must be consistent with the following observations in
the data:

    (1)   At low Eh readings, concentrations of uranium
          in municipal wells are not detectable, while at
          higher Eh readings, detectable uranium was
          measured.

    (2)   At highest positive Eh readings, Ra-226 is at
          or below detectable concentrations in municipal
          groundwaters.   As oxidation potential is
          reduced Ra-226 concentrations become variable
          ranging from below detection to 190 mBq/L.

(3)   Radon occurs in concentrations considerably
      above detection in all municipal wells where it
      was measured. Its distribution does not appear
      related to Eh-pH conditions. There does not
      appear to be any correlation between Rn and U,
      while Rn and Ra-226 show a weak positive
      correlation.

Additional considerations include:

(4)   A natural source for low levels of uranium
      occurs within the Amherstburg Formation of the
      Detroit River Group as evidenced by uranium
      concentrations in rock (see Table 1).

(5)   Sampling of isolated water bearing horizons
      within the Detroit River Group indicates that
      zones of slightly elevated U and Ra-226 exist.
      Analyses of rock samples indicate that these
      zones of preferential U accummulation are
      bituminous and observation of oil well core
      indicate that these zones are extensive in the
      Detroit River Group in and beyond the study
      area.

The conceptual model is depicted in Figure 12.
Uranium, which was fixed during diagenesis or by
reduction at some later time, is dissolved by
circulating groundwaters which have a positive oxidation
potential. Uranium solubility is further enhanced by
the formation of carbonate complexes. Uranium is
transported by groundwater and under reducing oxidation
potentials, is reduced and fixed at bituminous rich
sites in the Detroit River Group. At any point along
the uranium pathway, from its source in the oxidized
zone to its fixing in the reduced zone, U-238 can
disintegrate to produce Ra-226. Ra-226 in the oxidized
zone can either be transported by groundwater or fixed
through adsorption onto iron hydroxides, clay minerals
and carbonaceous material. As well, Ra-226 can be fixed
by coprecipitation with calcium, strontium and barium
sulphates. Under reducing conditions, there are fewer
mechanisms for fixing radium, since iron hydroxides are
unstable, and sulphate reduction precludes the
precipitation of sulphate minerals. At any point
(soluble or precipitated) along the radium pathway Rn
can be produced by disintegration, which explains its
rather widespread occurrence.

FIGURE 12:
CONCEPTUAL MODEL FOR RADIONUCLIDE DISTRIBUTION IN SOUTHWESTERN ONTARIO.

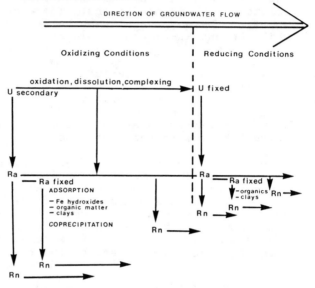

CONCLUSIONS

(1) Hydrogeologic conditions exert a major control or the distribution of uranium in groundwaters Uranium from primary or secondary sources are solubilized under conditions of positive oxidation potential and are fixed by reduction at appropriate sites such as carbonaceous rich zones of chemically reducing potential.

(2) Radium appears to show a variable range of concentrations in groundwaters except under conditions of high oxidation potential where it is removed from solution by adsorption and/ or coprecipitation with Ca and Ba sulphates. Under more reducing conditions, there appears to be fewer mechanisms restricting radium solubility; but it is still subject to adsorption by clays and carbonaceous material. Its dissolved concentration will depend on the distribution of these materials within the aquifer.

(3) Radon concentrations in groundwater indicate proximity to a radium source since the half life for Rn is only 3.8 days, and groundwater flow velocities are commonly low. Its distribution must be directly tied to the distribution of radium in either the fixed or dissolved state.

REFERENCES

1. Tooley, J. W. "An Assessment of Uranium in Drinking Water Supplies - Southeastern Region," MOE Draft Report (1982), pp. 29.

2. Robertson, J.A. "Uranium Deposits in Ontario," in Mineralogical Association of Canada Short Course in Uranium Deposits. Edited by M. M. Kimberley (Toronto: University of Toronto Press, 1978), pp. 229-280.

3. Ellesworth, H. V. "Radium-bearing Pegmatites of Ontario," Geological Survey of Canada, Summary Report, Part D (1921).

4. "Ontario Drinking Water Objectives," Ontario Ministry of the Environment. (1983), pp. 56.

5. Ontario Division of Mines. "Paleozoic Geology of Southern Ontario," Map 2254 (1972).

6. Chapman, L. J. and D. F. Putnam. "The Physiography of Southern Ontario," Ontario Geological Survey, Special Volume 2 (1984), pp. 270.

7. Grasty, R. L., G. Charbonneau, and H. Steacy. "A Uranium Occurrence in Paleozoic Rocks West of Ottawa," Geological Survey of Canada, Paper 73-1A (1973), pp. 286-289.

8. Steacy, H. R., R. Boyle, B. Charbonneau, and R. Grasty. "Mineralogical Notes on the Uranium Occurrences at South March and Eldorado, Ontario," Geological Survey of Canada, Paper 73-1B (1973), pp. 103-105.

9. Ruzicka, V. and H. R. Steacy. "New Radioactive Occurrences in the Precambrian West of Ottawa," Geological Survey of Canada, Paper 75-1A (1975), p. 234.

10. Uyeno, T. T., P. Telford and B. Sanford. "Devonian Conodonts and Stratigraphy of Southwestern Ontario," Geological Survey of Canada, Bulletin 332 (1982), pp. 11-15.

11. Sanford, B. V. "Devonian of Ontario and Michigan," in Proceedings of the International Symposium on the Devonian System, Edited by D. H. Oswald (Calgary: Alberta Society of Petroleum Geologists, 1968), pp. 973-985.

12. Howard, K. W. F., and P. Beck. "Hydrochemical Interpretation of Groundwater Flow Systems in

Quaternary Sediments of Southern Ontario," Canadian Journal of Earth Sciences. 23(7): 938-947 (1986).

13. Langmuir, D. "Uranium Solution-Mineral Equilibria at Low Temperatures with Applications to Sedimentary Ore Deposits," Geochimica et Cosmochimica Acta. 42: 547-569 (1978).

14. Hostetler, P. B., and R. M. Garrels. "Transportation and Precipitation of Uranium and Vanadium at Low Temperatures, with Special Reference to Sandstone-Type Uranium Deposits," Economic Geology 57(2): 137-167 (1962).

15. Hagmaier, J.L. "Groundwater Flow, Hydrochemistry and Uranium Deposition in the Powder River Basin, Wyoming," PhD Thesis, University of North Dakota, Grand Forks, N.D. (1971).

16. Felmlee, J. K., and R. A. Cadigan. "Radium and Uranium Concentrations and Associated Hydrochemistry in Ground Water in Southwestern Pueblo County, Colorado," United States Geological Survey, Open-File Report 79-974 (1979), pp. 54.

17. Bloch, S., and R.M. Key. "Modes of Formation of Anomalously High Radioactivity in Oil-Field Brines," Bulletin of American Association of Petroleum Geology, Jan. (1981), pp. 154-159.

18. Langmuir, D., and J. R. Chatman. "Groundwater Prospecting for Sandstone-Type Uranium Deposits: A Preliminary Comparison of the Merits of Mineral-Solution Equilibria, and Single-Element Tracer Methods," Journal of Geochemical Exploration 13: pp. 201-209 (1980).

19. Edmunds, W. M. "Trace Element Variations Across an Oxidation-Reduction Barrier in a Limestone Aquifer," in Proceedings of IAGC Symposium on Hydrogeochemistry and Biogeochemistry, Tokyo. (Washington, D.C: The Clarke Co., 1970), pp. 500-526.

Acknowledgements

The authors are indebted to Anita Foley for typing the
manuscript, Peter Yu and Mike Scafe for drafting the
figures and to Jim Dart who provided useful criticism
and constructive comment.

Session VI: Predictive Models for the Occurrence of Radon and Other Radioactivity

Moderator: Henry F. Lucas, Argonne National Laboratory

A PREDICTIVE MODEL
FOR INDOOR RADON OCCURRENCES --
A FIRST APPROXIMATION

Harry E. LeGrand, Independent Hydrogeologist, Raleigh,
North Carolina

INTRODUCTION

Knowledge of how radon gas is transmitted in the shallow
ground environment and how it emanates into buildings is grossly
incomplete. Admittedly, some excellent research studies have
been made (1, 2), and some general associations between certain
aspects of the environment and radon occurrences in buildings
are recognized. Yet, a technique for precisely predicting the
radon concentrations indoors is not likely to be developed soon.
As knowledge increases, successive approximations toward a final
predictive model may be required (3). An early approximation of
a predictive model for indoor radon is presented in this paper.
It applies specifically to the crystalline rock region of the
eastern United States, but it should have some application on a
broader basis.

The predictive model described focuses on understanding the
wide-ranging permeability characteristics in the soil and rock
fracture system. Radon is thought to accrete in confined subsurface air and moves under ground to low-pressure places, such as
houses niched in hill slopes. Driving forces for the air-laden
and entrapped radon gas are considered to be a rising water table
and infiltrating moisture from the land surface (4).

A crude direct relation between the amount of uranium in the
rock and its radon progeny in overlying buildings has long been
recognized (5). Also recognized is the fact that buildings in
which air from soil or rock can enter readily but not leave readily tend to have higher radon concentrations than those constructed in other ways (6). These relationships, concerning uranium
content of the rock and construction of buildings, have been
noted by many workers and now provide collectively a partial
framework that can be used in developing the best screening

mechanism for solving indoor radon problems. The hydrogeologic framework from which radon emanates has been described (4), providing a basis for the preparation of this predictive model.

JUSTIFICATION FOR A PREDICTIVE MODEL

A predictive model faces several problems. These include (a) the imprecise results of a predictive model, and (b) the relative inexpensiveness of radon measurements in buildings, which can negate the need for a predictive system in some cases.

Sampling and analysis techniques are available that allow indoor radon to be measured for prices ranging from about $20 to $100 (7). Where measurements are adequately made there would be no need to use a predictive model, of course. Yet, measurement of radon in the millions of buildings is not feasible, variable concentrations of radon may exist within space and time conditions in a room, and questions of reliability of sampling and analytical procedures arise in many cases.

A predictive model must be regarded with modest pretensions. Predicting the precise concentration of radon will never be attainable. Attempting to measure at specific sites the changing complex variables would result in costs that would outweigh those of direct measurements of radon, and the results would be less precise than those of direct measurements. To be cost effective, a predictive model would have to be expedient. The expediency is possible by making useful approximations of values of some variables; it would be partly subjective and would produce results that can be cast only in categories of probable approximate degree of seriousness.

In spite of the shortcomings cited above, a predictive model is needed. Although radon measurements are relatively inexpensive, only a small percent of buildings that need to be examined are likely to be sampled and analyzed. A technique that can be expedient, that can be used for particular situations in broad regions, and that can categorize each situation as to probable degree of seriousness is clearly needed. The predictive model proposed here is intended to fill these requirements, even though it is likely to be modified as additional information is analyzed.

The prevailing approach now undertaken by most workers toward a desired predictive model is directed toward statistical studies, involving radon measurements and compilation of data. This inductive approach can eventually be productive. Yet, unless key factors are early recognized and properly weighted, misleading interpretations and slowness in getting best answers will result. The approach taken here is largely deductive, directing efforts to understanding the hydrogeologic framework in which radon acts and making inferences about the data available. It is realized that deductively producing a predictive model at this early stage runs the risk of imperfect reliability. However, it can be vindicated by its current usefulness and by the display of key hydrogeological factors that must be considered.

THE PREDICTIVE SYSTEM

The predictive model that follows (Figure 1) provides a standardized system for describing and rating buildings and sites in terms of potential harmful concentrations of indoor radon. It offers a screening method that can allow a rough categorizing of hazard potential. Follow-up sampling and analysis would be suggested in some cases.

Evaluating the hazard potential of indoor radon in buildings or prospective buildings is done in four stages. STAGE 1 is directed to estimating approximate or relative values of four geologic and hydrogeologic factors and the extent of vertical emplacement of a building into the ground; these factors were shown to be important for the igneous and metamorphic terranes of eastern North America (4). The estimated values are identified and displayed. In STAGE 2 the coded values are entered in a simple equation to produce a numerical rating. The rating at the end of STAGE 2 is for unbuilt houses or for buildings of unknown construction. STAGE 3 allows an adjustment of the rating according to the degree to which construction of a building will let radon enter and leave. The numerical rating of a particular situation is applied (STAGE 4) to a hazard potential scale of five categories ranging between "almost certain" to "very unlikely".

STAGE 1. Four factors are considered in STAGE 1:
A. Rock type -- uranium concentration
B. Topographic slope and vertical emplacement of building
C. Hydrogeologic setting and average water-table position
D. Water-table behavior and characteristics.

Each factor has ten categories, ranging from 1 to 10; the higher the number, the higher is the likelihood of that particular category contributing to indoor radon concentration. It is emphasized that in all cases, estimates rather than precise measurements will suffice.

The "Rock Type" factor (A) includes most of the terranes in eastern United States underlain by igneous and metamorphic rocks. Other types of rocks can be interpolated as needed. A user of the system can make modifications, as required, without completely revamping the categories. Use of geologic maps and personal experience may be needed to arrive at the approximate numbered category. The presence of uranium, and its radon progeny, is more complex than the rating system implies. For example, uranium "hot spots" may occur locally in sedimentary Triassic rocks.

The categories of "Topographic Slope and Vertical Emplacement of Buildings" (B) indicate that buildings niched greatly into steep hill slopes have a higher likelihood of providing indoor radon than buildings on flat upland surfaces. Reasonable interpolations allow the best numerical value to be assigned.

The categories of "Hydrogeologic Setting and Average Water-table Position" (C) are difficult to distinguish. In many cases, only an experienced hydrogeologist can make a close approximation. If conditions are completely unknown, or if a better estimate cannot be made, it is suggested that a "6" be recorded.

STAGE 1. For each situation, identify the numerical value for each
of the four factors.

A. ROCK TYPE -- URANIUM CONCENTRATION

10. Recognized uranium hot spot area

9. Presumed uranium content slightly higher than that of normal granite

8. Average granite or granite gneiss

7. Rhyolitic tuff or some mica schist

6. Mixed rock complex with some granite or pegmatite stringers
or beds being common

5. Mixed rock complex with some granite or pegmatite stringers
not noticeably common

4. Sedimentary Triassic rocks

3. Quartzite or schists without granite stringers

2. Diorite or horblende gneiss

1. Gabbro or ultra-mafic rocks

B. TOPOGRAPHIC SLOPE AND VERTICAL EMPLACEMENT OF BUILDING

10. 35-45% slope; 10 to 12 feet of vertical emplacement

9. 25-35% slope; 8 to 12 feet of vertical emplacement

8. 15-25% slope; 8 to 12 feet of vertical emplacement

7. 10-15% slope; 7 to 12 feet of vertical emplacement

6. 5-10% slope; 5 to 8 feet of vertical emplacement

5. 2-5% slope; 3 to 6 feet of vertical emplacement, or
flat land surface with 8 to 12 feet of vertical emplacement

4. 2-5% slope; 1 to 3 feet of vertical emplacement, or flat
land surface with as much as 8 feet of vertical emplacement

3. Relatively flat land surface; no more than 4 feet of
vertical emplacement

2. Flat land surface; concrete slab or other ground contact

1. Flat land surface; all floors of building above land surface

Figure 1.   A predictive system for indoor radon occurrences.

C. HYDROGEOLOGICAL SETTING AND AVERAGE WATER-TABLE POSITION

   10. A few feet of sandy clay underlain by a few feet of saprolite that has both intergranular and fracture permeability; water table near base of saprolite

    9. A few feet of sandy clay soil underlain by a few feet of intergranular saprolite; water table in underlying fractured rock

    8. About 10-15 feet of sandy clay overlying bedrock; water table in bedrock

    7. Saprolite at land surface; water table in saprolite

* 6. Soil at land surface; nature of saprolite and bedrock not easily determined; position of water table also not known

    5. Rock at land surface; water table deeper than 20 feet

    4. Rock at land surface; water table near land surface

    3. 30-40 feet of even-textured clay; water table in clay

    2. 40-60 feet of sandy clay; water table near land surface

    1. 60 or more feet of sandy clay; water table near land surface

* If conditions are completely unknown or if estimate cannot be made better, use "6"

D. WATER-TABLE BEHAVIOR AND CHARACTERISTICS

   10. Site over intermittent pumped well, in which water table fluctuates continually in bedrock, as much as 15 feet per day near site

    9. Site near edge of cone of pumping depression; water table fluctuates in bedrock according to pumping characteristics

    8. Site on steep slope where water table is in bedrock and makes an angular air pocket with overlying clay

    7. Site on steep slope where water table is in bedrock and seasonal fluctuation is as much as 4 feet

* 6. Site on moderate slope; position of water table and its fluctuation not easily estimated

    5. Site on moderate slope where water table is in bedrock and seasonal fluctuation is about 3 feet

    4. Site on gentle slope where water table is at land surface sometimes and where seasonal fluctuation is less than 2 feet

    3. Slow fluctuation of less than 4 feet annually in sandy clay on nearly flat upland

    2. Slow fluctuation of less than 3 feet annually in sandy clay on nearly flat upland

    1. Slow fluctuation of less than 3 feet annually in sandy clay on flat upland

* If conditions are completely unknown or if estimate cannot be made better, use "6"

Figure 1. (continued)

STAGE 2.   To calculate the rating, multiply the value of the Rock Type (A) by the sum of the other 3 factors:

Rating = A(B+C+D)

STAGE 3.   The predictive model based on hydrogeology needs an additional step -- that of considering the degree to which buildings allow radon to enter and to leave. There are many building characteristics involved, but for general purposes the three categories shown below may be used to provide final projections. After inspection of basement or lower floor, use judgment in assigning the most reasonable value below.

Above average construction -- thick concrete slab; concrete or other tight wall features; few or no cracks or openings	Average or conventional basement construction -- plain concrete block walls, or a few cracks or some obvious opening to the outside at a level near the floor	Below average construction -- thin concrete slab; cracks in walls; obvious openings
	Use values as in STAGE 2	
May need to SUBTRACT 10 to 80 points		May need to ADD 10 to 80 points

Figure 1.   (continued)

STAGE 4.   Locate where calculated rating value falls
          on the HAZARD POTENTIAL SCALE

HAZARD POTENTIAL SCALE

Likelihood of Worrisome Concentrations of Radon *

Total Points	Likelihood
More than 200	Almost certain
160-200	Very likely
80-160	Questionable
60-80	Not likely
Less than 60	Very unlikely

* Worrisome concentrations, as here defined, represent
  as much as 4 pCi/l at least 30% of the time

Examples of final descriptions and ratings:

Site 1.   8(7+9+6) = 176 (very likely)

Site 2.   3(7+7+4) = 54 + 50 = 104 (questionable)

Site 3.   5(6+6+6) = 90 - 30 = 60 (not likely)

(See text for full explanation of these ratings)

Figure 1.   (continued)

The categories of "Water-table Behavior and Characteristics" (D) are also difficult to distinguish; a selected category needs to be assigned, nevertheless. To some extent, there is some overlap with factors B and C, but this overlap is needed to provide the proper weighting in the final calculated rating. If conditions are completely unknown, or if a better selection cannot be made, it is suggested that a "6" be recorded.

STAGE 2. A preliminary rating is made by multiplying the value of Rock Type by the sum of the values of the other three factors. This rating can apply to buildings that are considered to be of average construction and to have average ventilation features. It applies also to unbuilt houses and to houses in which the construction and ventilation features have not been observed.

The factors (A, B, C, and D) are not of equal rank, and therefore an attempt is made to weight them in a simple manner. Rock Type (A) is considered the most important factor; its value is multiplied by the sum of the other assigned values. The other three factors (B, C, and D) are given equal weight in rank. Future studies may show that the factors deserve special weighting. An attempt to make the weighting more precise would make the rating procedure cumbersome. It is important to display all separate values of the factors with the rating because the description of the hydrogeologic conditions and of the building setting is very important for purposes of overview and concensus checking.

STAGE 3. The preliminary rating displayed in STAGE 2 must be adjusted to account for the ability of radon to enter a building and to be retained in it. As shown in Figure 1, this adjustment is made by considering whether the indoor radon is affected by building construction that can be categorized as average, below average, or above average. The points to be subtracted for above average construction or to be added for below average construction represent crude estimates that are arbitrary. Nevertheless, an experienced evaluator can make reasonable and useful estimates. The final coded description and rating should be expressed, as in the examples shown.

STAGE 4. The final rating (STAGE 3) is directed to the Hazard Potential Scale, as shown in Figure 1. The Hazard Potential Scale allows the evaluator to predict the probability of a building to have worrisome concentrations of radon. Worrisome concentrations, as described here, are those exceeding 4 pCi/l, recommended by the U.S. Environmental Protection Agency (7) as the level above which corrective action should be taken. The categories of ranges of probability are shown on Figure 1 with their ranges of point values. Many situations will fall within the "questionable" range. Point values in the "questionable" range and those above that range should be directed to actual sampling and analysis for indoor radon. The final rating applies to concentrations in a basement or lowest level of a building.

EXAMPLES OF DESCRIPTIONS AND RATINGS

The following three examples of descriptions and ratings of sites and situations correspond to those identified on Figure 1. Letters and numbers in parentheses below are the factors and values, respectively, identified or estimated for that site.

Site 1. A planned house is to be built over granite (A, 8). The split-level house will be on a topographic slope of about 15 percent; the ground elevation will be about 10 feet higher than the basement floor on the uphill side (B, 7). The foundation will be dug into saprolite, and the water table presumably is in the bedrock (C, 9). Since the site is on a moderate slope and the position of the water table is not known, a "6" rating is shown for D. Because the house is unbuilt and the construction is unknown, STAGE 3 adjustment is not needed.

Site 1: 8(7+9+6)=176 (very likely).

Site 2. The house is built on schists that have no granite veins or layers (A, 3). The house is on a 12 percent slope; the basement extends into the hillside where the vertical emplacement is about 12 feet (B, 7). Saprolite is at the land surface and the water table is thought to be in the saprolite, about 2 feet below the basement (C, 7). The water-table fluctuation annually may be about 2 feet, and it may be above the basement floor at times (D, 4). There are many cracks in the walls and obvious openings into the saprolite (STAGE 3, +50).

Site 2: 3(7+7+4)=54 + 50 = 104 (questionable).

Site 3. The house is built on mixed rocks with a few scattered granite veins or layers (A, 5). The house is on a slope of about 4 percent; the basement floor is about 7 feet below land surface on the uphill side (B, 6). No rock or saprolite is visible and position of the water table is difficult to estimate (C, 6). The water-table behavior is not known (D, 6). The house foundation and walls are tight, with only a few visible cracks or openings on the uphill side (STAGE 3, -30).

Site 3: 5(6+6+6)=90 - 30 = 60 (not likely).

SUMMARY AND CONCLUSIONS

Efforts to develop a model that can allow one to predict the degree of health risks from radon gas emanations in buildings must face intangible factors. Yet, a useful predictive model is needed because (a) measurements of radon in millions of buildings are not feasible, (b) variable concentrations of radon may exist within space and time conditions in a room, and (c) questions of reliability of sampling and analytical procedures arise in many cases.

A first approximation toward a predictive model is proposed that applies to the crystalline rock terrane of eastern North America. The worrisome problems of indoor radon are thought to be caused by reciprocal air and water interchange in near-surface rock materials, such as granites, that have at least moderate amounts of uranium. Thus, it is essential that a predictive model be based on fundamental hydrogeologic factors rather than solely on "hot-spot" uranium areas.

The preliminary model proposed represents an early approximation of some future improved model that could be more acceptable. In spite of its imperfections, the model is presented now to provide a framework to which future selective sampling and analyses can be associated. Present sampling is either helter-skelter or based on incomplete knowledge of the factors involved.

The predictive model described focuses on understanding the wide-ranging permeability characteristics in the soil and rock fracture system. Radon accretes in confined subsurface air and moves under ground to low-pressure places, such as houses niched in hill slopes. Driving forces for the air-laden and entrapped radon gas are thought to be a rising water table and infiltrating moisture from the land surface.

The predictive model is based on estimating values (on a 10-point numerical scale) of four factors, as follows: (a) Rock type -- uranium content, (b) Topographic slope and vertical emplacement of building, (c) Hydrogeologic setting and average water-table position, and (d) Water-table behavior and characteristics.

The values are compiled and then adjusted more specifically to conditions of indoor construction and ventilation. The final total value is directed to a rating scale that indicates a relative degree of seriousness; each rated situation is classified as: "almost certain", "very likely", "questionable", "not likely", or "very unlikely".

Test results are congruent with understanding the principles of radon behavior and with interpretation of documented measurements. Persons with a few days training can use the predictive system effectively and expediently.

In addition to the merits the model may have as a rating system, it also serves a useful purpose by displaying a coded numerical description of each site and building situation. This type of description will be useful in future studies of radon occurrences.

REFERENCES

1. Nero, A. V., and W. M. Lowder, Eds. Indoor Radon, special issue of Health Phys., Vol. 45, No. 2, (Aug. 1983).

2. Rama and W. S. Moore. "Mechanism of Transport of U-Th Series Isotopes from Solids into Ground Water," Geochim. Cosmochin. Acta, Vol. 48, (1984), pp. 395-399.

3. Lowder, W. M. "Part One: Overview," (of Radon), in Indoor Air and Human Health, G. M. Gammage and S. V. Kaye, Eds. (Chelsea, MI: Lewis Publishers, Inc., 1985), pp. 39-42.

4. LeGrand, H. E.   "Radon and Radium Emanations from Fractured Crystalline Rocks -- A Conceptual Hydrogeological Model," Ground Water, Vol. 25, No. 1, (1987), pp. 59-69.

5. Tanner, A. B.   "Geologic Factors that Influence Radon Availability," in Radon -- Proceedings of Conference Sponsored by the Air Pollution Control Association, (Feb. 1986), pp. 1-11.

6. Nero, A. V.   "Indoor Concentrations of Radon-222 and Its Daughters: Sources, Range, and Environmental Influences," in Indoor Air and Human Health, G. M. Gammage and S. V. Kaye, Eds.   (Chelsea, MI: Lewis Publishers, Inc., 1985), pp. 43-67.

7. U.S. Environmental Protection Agency.   "A Citizen's Guide to Radon," Office of Air and Radiation, OPA-86-004, (1986), 14 p.

Session VII: Remedial Actions for Radon, Radium and Other Radioactivity

Moderator: Henry F. Lucas, Argonne National Laboratory

A METHOD OF RADON REDUCTION FOR NEW BUILDINGS

Hormoz Pazwash, Ph.D., P.E.
Woodward-Clyde Consultants, Wayne, New Jersey

INTRODUCTION

Radon is a radioactive gas and a natural product of the radioactive decay of radium (226 Ra) which occurs in uranium-bearing soils and rocks and, therefore, is present throughout the earth's crust. Radon is unstable, and generates a series of short-lived radioactive decay products (called radon daughters). This decay process continues until lead 206, a stable isotope is produced. In this country, naturally-occurring radon is found in highest concentrations in granitic rock areas including many parts of New England, New Jersey, Pennsylvania and California. The most acute radon problem in the northeast has been identified to be occurring along Reading Prong geologic formation. Radon is a deadly gas, exposure to which has been identified as a cause of lung cancer. According to recent estimates by EPA and private organizations 10,000 to 20,000 cases of the approximately 140,000 annual deaths due to lung cancer is attributed to exposure to radon [7,8].

Outdoor concentration of radon emitting from soil is small. But the soil gas can build up and occur at high concentrations indoors. There are two other sources of radon in houses; building materials and domestic water supplies. These latter sources are, however, less important than soil gas emanation, accounting for less than 10 percent of total exposure to radon.

The infiltration of radon from soil into houses generally occurs through cracks and openings in the foundations and under crawl space. Basements appear to be the most susceptable to high rates of radon entry from soil. The radon entry is caused by two effects. One is the movement of radon by molecular diffusion through air pores in the soil. A second, more significant air flow occurs due to the indoor-outdoor pressure differential. This pressure-driven flow which is the main component of radon entry into homes is created by the thermal "stack effect" and wind loading on the building. The "stack effect" relates to the pressure difference across any wall separating air masses of different temperature. In cold season, this effect causes a net inward flow of soil gas into a heated building. Wind loading produces differential pressure across different walls of a building. This also causes an exchange of air between the house and the soil but is less important than the stack effect. More detailed information about occurrences of radon in houses may be found in general references cited at the end of this paper [1,3,5,6].

For given conditions, concentration of radon in a home is affected by the natural ventilation rate. Based on limited data it appears that the ventilation rate in most conventional houses varies from 0.5 to 1.0 air exchanges per hour, averaged over cold and warm seasons, respectively. However, for new homes with tight, energy-efficient construction the rate is expected to be substantially lower. As a result, the average level of indoor radon has been increasing in the past few years and is expected to continue to increase in the future. A comparison of radon occurrence in energy saving and conventional houses may be found in references [2, 9].

Radon concentration is measured by units of radioactivity rather than mass because the actual mass involved is so small. Customarily, the curie has been used as the unit of radioactivity and the prefix "pico" (10^{-12}) normally attached. One picocurie per liter (pCi/L) is equivalent to the emission of 0.037 atoms of radon per second per liter. In new SI units, Becquerels (Bq) are used as the unit $(25 \text{ Bq/m}^3 = 0.7 \text{ pCi/l})$. But the picocurie unit is dominantly used in this country. Levels as high as 1000 pCi/l airborne radon concentration in homes are extremely rare. Even levels of 20 pCi/l are considered high by EPA experts. The EPA recommends 4 pCi/l as the mitigation action level. It is estimated that more than 12 percent of homes in radon-prone areas in this country have radon levels greater than 4 pCi/l and, therefore, are in need of remediation [8].

As noted, one source of indoor radon is domestic water. Since radon is highly soluble it may accumulate at high concentrations in ground water after being generated in the earth's crust. This can be a significant source of indoor radon in houses supplied by private wells where the storage time is usually too short to allow for radon to decay. While radon concentrations of 100 pCi/l is rare for indoor air, water-borne radon has been found in excess of 100,000 pCi/l. However, cancer deaths caused by inhaling water-borne radon appears to account for less than 5 percent of deaths from exposure to radon emanating directly from soil [7].

METHODS OF RADON REDUCTION IN HOMES

A variety of methods have been employed to reduce the level of radon in houses. In principal high radon concentrations in soil gas can be lowered, based on the following two concepts: 1) removing the source or reducing its transport indoors by sealing openings and holes in the floor and walls and 2) replacing contaminated indoor air by either ventilation or air cleaning. While the former approach is more practicable for new homes, the latter approach had been customarily employed for mitigation of radon in existing structures. However, experience has shown that the air treatment is less successful than originally anticipated. As a result methods based on source control are now being widely practiced for remediation of radon.

The EPA outlines nine different methods for controlling soil gas radon in existing homes. These include covering exposed earth, sealing cracks and openings, natural ventilation, forced ventilation, heat recovery ventilation, sub-slab ventilation, blockwall ventilation, air-supply systems and drain tile suction. These methods and their applications are briefly discussed in a 1986 publication by the EPA [10]. A more detailed description of radon reduction methods can be found in some of the papers contained in general references [3, 5, 6]. For example see [4] for initial results from tests involving application of three different methods of active ventilation for mitigation of radon under homes basements in eastern Pennsylvania.

The effectiveness of any one of the above methods depends upon several parameters: the unique characteristics of each house, the routes of radon entry, the level of radon, and how well the job is done. Houses are not alike - even two houses that look similar have small differences in construction. These differences can affect radon entry and the effectiveness of radon reduction methods. Underlying soils are often widely variable. Even for houses in close proximity, the soil may have different characteristics that can effect the result obtained from using a radon reduction technique. The importance of how a radon reduction method is implemented can not be overemphasized. Most remedies require the skilled services of a professional contractor with experience in radon reduction techniques. Therefore it is important for homeowners to carefully select a contractor. Checking with business agencies such as Chamber of Commerce, getting a second opinion from another contractor or seeking advice from one of the state's radiological health officials can help a homeowner in making a right decision.

For new structures, it is generally more cost effective to incorporate radon reduction construction features and ventilation during construction than to apply remedial techniques later. Construction of concrete foundation and slab with no openings, cracks or other low resistance pathways for soil gas migration is the simplest and the most cost-effective means of excluding radon from indoor air. However, the possibility exists that, due to differential settlement or other reasons, cracks may develop in well constructed foundations or walls in the future. Therefore, it is a good policy to implement an additional method of radon control. A good control strategy should contain a combination of different control measures.

High radon concentration in household ground-water supplies are effectively reduced by using either air strippering towers or granular activated carbon filters (GAC). The choice between the two depends on the size of the application. GAC has been successfully applied for treating private well water containing radon of the order of several thousand pCi/l. This system is much more cost effective than air stripping for such small applications. For larger public water supplies with demands in excess of 15 gpm (nearly 1 l/s), air strippers are more cost effective. Indeed, for larger than 300 gpm (nearly 20 l/s) applications air stripping is many times less expensive than GAC.

DESCRIPTION OF THE METHOD

The method of radon reduction that is described in this section was developed for control of radon for a new structure that was going to be constructed as headquarters of a company in Huntington County, New Jersey. Because the soil at the structure site had been suspected of having the potential for emitting radon, Woodward-Clyde Consultants was hired to design a ventilation system that would be installed during the construction. The purpose of ventilation system was to evacuate radon gas, if present. from the subbase so that it would not accumulate under the building foundation. Therefore, the system would serve as a preventive measure that could be used as a method of radon control, in case of radon emission from the soil.

The building included four blocks of attached structures, each with plan dimensions of 150 feet by 150 feet. We proposed that the ventilation system be installed in a granular subbase composed of 8 inch bed of washed rock gravel (min. 3/4 in diameter grain size). Our proposed ventilation system was composed of four identical units, each for ventilation of a single structure subbase. Figure 1 shows a general layout of the entire system and details of each unit. As shown, each unit consisted of an exhaust pipe network for the evacuation of air at the gravel bed and an open-to air inlet duct for the supply of fresh air to the subbase.

The outlet portion of the system consisted of five horizontal perforated PVC pipes, a collection pipe, a riser pipe and a ventilation fan (or a passive venturi) on the roof. The perforated PVC pipe had a 3 in. inside diameter and the collector and the stack are 4 in. and 5 in. pipes, respectively. Standard perforated 3 in. PVC pipes come in two perforation settings, one with two rows of holes around the pipe circumference and the other with three rows of holes. In either case, perforations are circular holes 5/8 in. in diameter, spaced 5 in. on center. We proposed using perforated pipes with three rows of perforation, providing a total opening of over 2 square inches per linear foot of pipe. Four inch perforated pipes may be used in place of 3 in. pipes. However, 4 in. pipes cost slightly more than 3 in. pipes while providing exactly the same breathing area. It was also noted that square air ducts of equivalent sizes can be used in place of the recommended 4 in. collector pipe and 5 in. stack pipe, if preferred for esthetic reasons. We suggested that the collector pipe (or duct) be placed outside the building foundation so that it would be easily accessible in the future.

The inlet portion of the system consistent of a 5 in. diameter inlet pipe, a 4 in. distributor pipe and 4 in. eject nipples in the subbase of each building. We noted that, similar to the collector pipe, the distributor pipe should be installed outside of the foundation for ease of future accessibility. We also noted that an equivalent air duct could be used instead of 4 in. air distributor pipe, if preferred.

The ventilation of air may be provided by a passive system using a turbine type venturi, or a powered fan. In the former case, the operation of the system is controlled by the prevailing winds; in the latter case the system can be operated according to a scheduled plan (in terms of air flow

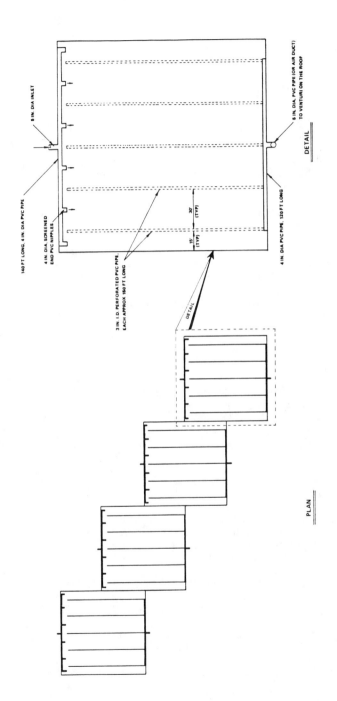

Figure 1. Basics features of the sub-slab ventilation system.

rat∍, period and frequency of operation). For passive ventilation, the venturi should be capable of providing a lift equivalent to at least 5 in. of water. With a forced (or active) ventilation, the fan should be capable of having the following characteristics:

- o   5 in. diameter inlet;
- o   200-400 CFM flow rate; and
- o   10 in. maximum pressure.

We proposed that a passive system be installed initially, and suggested that the stack should be installed near a heat source, if possible, to enhance upward draft. If the passive system was found to be inadequate for ventilating the subbase in the future, the original system could be modified by simply replacing the venturi with a powered fan.

In terms of construction features, the following suggestions were made:

> In constructing the basement, wall joints should be sealed with an elastic compound. Footings, slabs, and foundation walls should be constructed by well-vibrated, dense concrete.

The proposed ventilation system was approved by NJDEP and was installed in the headquarters building last summer.

REFERENCES

1.   "A Citizen's Guide To Radon," USEPA, office of air and Radiation, Report OPA 86-004, (Aug. 1986).

2.   Fleischer, R.L., A. Mogro-Campero and L.G. Turner,. "Indoor Radon Levels in the Northeastern U.S.: Effects of Energy-Efficiency in Homes," Health Physics, Special Issue on Indoor Radon, 45 (2): 407-412 (1983).

3.   Health Physics, Official Journal of the Health Physics Society, Pergamon Press Special Issue, on "Indoor Radon" 45(2) 1983.

4.   Henschel, D.B. and A.G. Scott. "The EPA Program to Demonstrate Mitigation Measures for Indoor Radon: Initial Results," Proceedings of APCA International Specialty Conference entitled "Indoor Radon,", Feb. 1986, pp. 110-121.

5.   "Indoor Air Quality Environmental Information Handbook: Radon," prepared by Mueller Associates, Inc., Syseon Corporation, Brookhaven National Laboratory, prepared for U.S. Dept. of Energy, Jan. 1986.

6.   Indoor Radon, Proceedings of APCA International Specialty Conference, Philadelphia, Penn. Feb. 1986.

7.   Kosowatz, J.J. "Radon Groundwater Alarm Rung," ENR News, p.10 July 31, 1986.

8.   Lowry, J.D. "Radon At Home," Civil Engineering 57(2), 44-47 (1987)

9.   Nero, A.V., M.L. Boegel, C.D. Hollowell, J.G. Ingersoll, and W.W. Nazaroff. "Radon Concentration and Infilteration Rates Measured in Conventional and Energy-Efficient Houses," Health Physics 45 (2) 401-405 (1983)

10. "Radon Reduction Methods, A Homeowner's Guide," USEPA Report OPA-86-005, (Aug. 1986).

TREATMENT SCHEME FOR CONTROLLING
THE MIGRATION OF RADIUM FROM
A TAILINGS IMPOUNDMENT

B. E. Opitz,[1]
Pacific Northwest Laboratories, Richland, Washington 99352

ABSTRACT

Under sponsorship of the Nuclear Regulatory Commission's
Uranium Research and Recovery Program, Pacific Northwest Laboratory
(PNL) has investigated the use of various neutralizing reagents and
techniques to attenuate the movement of contaminants associated
with acidic uranium mill tailings. The objective of this study was
to identify those contaminants which are not effectively attenuated
by common neutralization methods and to develop alternative control
measures. Of those contaminants associated with uranium mill tailings which were identified as not being effectively immobilized by
tailings neutralization, radium imposes an important environmental
concern in terms of potential groundwater contamination.

Control or attenuation of radium is of special concern primarily due to its radiological health implications. For that reason,
the Environmental Protection Agency (EPA) has implemented strict
guidelines governing the maximum allowable concentration in drinking waters. Current EPA guidelines call for total radium activities not to exceed 5 pCi/L. Due to the high activity of soluble
radium in the acidic uranium mill tailings environment (several
hundred to several thousand pCi/L), specific ion removal procedures

[1] Brian E. Opitz, Research Scientist, Pacific Northwest
Laboratory, Richland, Washington 99352.

were investigated for use in attenuating radium in order to prevent future groundwater contamination. Results of these investigations led to the development of a tailings additive comprised of a mixture of hydrated lime and barium chloride, which, when added to acidic tailings, can reduce the amount of leachable radium escaping a designated tailings impoundment.

In laboratory verification tests, this radium specific tailings treatment reduced the effluent solution activity of radium by three orders of magnitude, from >3500 pCi/L to 1.7 pCi/L, in comparison with untreated acidic tailings.

INTRODUCTION

In September of 1981, the Nuclear Regulatory Commission expanded the scope of the Uranium Research and Recovery Program at Pacific Northwest Laboratory (PNL) to address many technical issues of concern in the licensing, operation, and decommissioning of uranium recovery facilities. One of the more technical issues to be investigated was the use of tailings neutralization and other alternatives for immobilizing toxic materials in acidic uranium mill wastes.

The objective of this project was to assess viable alternatives for reducing contaminant mobility under a full range of site and environmental conditions. Initially, the project focused on a review of treatment techniques utilized by other acid waste generating industries and their potential applicability to acidic uranium mill waste (Sherwood and Serne 1983a [10]). Based on this review it appeared that neutralization of acidic uranium mill tailings solution would be a viable technique for reducing contaminant mobility. Laboratory experiments were then performed to evaluate specific neutralization reagents performance and to optimize treatment conditions for contaminant immobilization in acidic tailings and tailings solutions (Sherwood and Serne 1983b [11], Opitz et al. 1983 [8], Opitz et al. 1985 [9]). The results of these treatment experiments indicated that calcium alkalies ($CaCO_3$ and $Ca(OH)_2$) provide the highest quality neutralized effluent at the lowest cost.

The neutralization process is very effective in attenuating a high percentage of the dissolved constituents contained originally in acidic uranium mill tailings solutions. However, certain constituents commonly associated with tailings solution are not strictly dependent on solution pH for their solubility in uranium tailings liquors. Therefore, the activities or concentrations of these constituents show only slight and sometimes no attenuation after solution neutralization. One of these constituents not ideally controlled by neutralization is radium.

Control or attenuation of radium is of special concern primarily due to its radiological health implications. For that reason, the Environmental Protection Agency (EPA) has implemented strict guidelines governing the maximum allowable concentration in drinking waters. Current EPA guidelines calls for total radium activities not to exceed 5 pCi/L. Due to the high activity of soluble radium in the acidic uranium mill tailings environment (several hundred to several thousand pCi/L), specific ion removal procedures were investigated for use in attenuating radium in order to prevent future groundwater contamination.

In this study, we performed further laboratory tests using lime neutralization plus barium addition in the form of barium chloride. In these tests various amounts of lime and barium chloride were mixed as additives to solid acidic tailings from the Pathfinder (Lucky Mc) Gas Hills Mill outside of Riverton, Wyoming. The study's objective was to test a method of amending solid acidic tailings in order to attenuate the migration of radium and reduce the potential for future groundwater contamination.

MATERIALS AND METHODS

Our experimental methodology to assess the performance of the various amendments considers three components: the materials used, the test methods employed, and the analytical measurements performed. Descriptions of these components follow.

MATERIALS

Solid acidic tailings collected from the Pathfinder (Lucky Mc) Gas Hills Mill site were used in the radium attenuation studies. The tailings were collected from material near the edge of the embankment along the dam face on the west end of the the tailings impoundment as shown in Figure 1. The material used in these tests was a random tailings sample taken at one location and may vary greatly throughout the site.

The barium chloride ($BaCl_2$) used for amending the solid tailings was a reagent grade chemical obtained from a local supply house. Commercial grade lime ($Ca(OH)_2$) was used in the experiments and was finely ground to <200 mesh.

TEST METHODS

The tailings were oven dried at 105°C for 24 hours prior to use to facilitate handling. The dried tailings used in the experiments were passed through a number 10 sieve and subdivided into two 1000-g subsamples. One tailings subsample was amended with barium chloride and calcium hydroxide. The other subsample was untreated and to be used as a blank or base line value for comparison of attenuation results. The description of the amended and untreated tailings is shown in Table 1. The calcium hydroxide amendment was mixed into the tailings as an oven dried solid. Barium chloride was added to the tailings in liquid form after dissolving 3.286 g of $BaCl_2$-$2H_2O$ per liter of distilled, deionized water (DD H_2O). Distilled deionized water was added to the untreated tailings not containing barium chloride.

Upon addition of the various reagents and solutions to the acidic Lucky Mc tailings, the composite samples were thoroughly

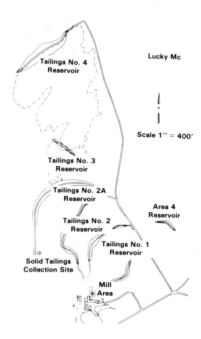

Figure 1.   Lucky Mc Mill site and sampling location.

Table 1.   Sample identification and description.

Sample ID	Description
1	1000 g tailings, 7.3 g $Ca(OH)_2$, 151 mL $BaCl_2$ solution.
2	1000 g tailings, 151 mL DD H_2O

mixed and allowed to stand overnight in sealed plastic bags to ensure a uniform moisture distribution. Subsamples of each mixture were collected for moisture content determination. The method used for moisture content determination was microwave drying (Gee and Dodson 1981 [3]).

Two polyacrylic cylindrical columns were compacted in accordance with the sliding weight tamper method (ASTM 1982 [1]), with the composite mixture from each sample bag. Sample 1 was compacted in column 1, sample 2 in column 2. The physical compaction data for each column are listed in Table 2. Every effort was made to keep all column compaction parameters equal so a fair comparison could be made at the completion of the experiment.

Table 2.   Radium attenuation column compaction data.

Column ID	Cell Diameter (cm)	Length (cm)	Volume (cm^3)	Dry Sample wt (g)	Bulk Density (g/cm^3)	Porosity	Pore Volume (cm^3)
1	8.89	8.89	552	833	1.51	0.44	244
2	8.89	8.89	552	833	1.51	0.44	244

Once compacted, the columns containing the amended tailings mixtures and the untreated tailings were saturated with approximately one pore volume of local groundwater. The method used to leach the compacted tailings columns followed the ASTM method for measuring permeability of granular soils under constant head conditions (ASTM 1982 [1]) except that permeability values were not calculated. The chemical composition of the local groundwater is shown in Table 3. Flow through the columns was from the bottom to top to enhance saturated flow conditions.

Flow rates through the columns were maintained at approximately 0.1 mL/min with Buchler multistatic constant flow pumps. Column effluent samples were collected at approximately one pore volume intervals (~250 mL) throughout the duration of the experiment. The pH and Eh of each sample was measured after collection. Subsamples were then preserved appropriately for subsequent anion, cation, and total radium determinations.

ANALYTICAL METHODS

Solid tailings characterization, including particle size distribution and saturated paste pH (treated and untreated tailings), was performed as described by Black (1965 [2]).

Radium-226 activities were determined by radium coprecipitation with barium sulfate followed by alpha scintillation counting (APHA-AWWA-WPCF 1980 [4]). The counters were calibrated using $BaSO_4$ precipitates of varying weights, each containing a known activity of ^{226}Ra.

Chemical characterization of the macro cations and some trace metals as appropriate (Al, Ca, Fe, Mg, Mn, Si, Sr, and Zn) in the leaching column effluents were performed with a Jarrell Ash spectrometer with an inductively coupled plasma source (ICP). A Perkin-Elmer Model 5000 with a HGA-500 graphite furnace was used to measure the remaining trace metals. The anions presented were determined using a Dionex Model 14 ion chromatograph (IC). Solution pH was measured using a Corning Model 130 pH meter and a combination electrode.

Table 3. Local groundwater chemical composition.

Parameter	Concentration (mg/L)
Al	0.1
As	<0.02
Ba	0.03
Ca	25.1
Fe	0.04
K	0.9
Mg	4.4
Mn	<0.3
Na	2.9
Si	3.6
Sr	0.1
Zn	0.2
Cl	1.5
NO_3	<0.5
SO_4	19.8
pH (units)	8.2

RESULTS AND DISCUSSION

SEDIMENT CHARACTERIZATION

The particle size distribution for the untreated acidic tailing and the saturated paste pH for both the untreated and amended
tailings are presented in Table 4. The particle size data suggest
that the tailings material can be classified as a sandy loam. No
significant size distribution difference was expected after addition of enough $Ca(OH)_2$ to raise the saturated paste pH of the
amended tailings to pH 6.8 so particle size measurements were not
performed. The saturated paste pH data for the untreated tailings,
pH 2.5, indicates the as-collected tailings remain very acidic
after they are discharged into the tailings impoundment. Low pH,
acidic conditions found in the untreated tailings enhance the
mobility of many contaminants.

RADIUM ATTENUATION EXPERIMENTS

The results of the radium attenuation experiments comparing
the untreated tailings with the amended tailings are shown in Figure 2. The first sample collected from the untreated acidic tailings at an adjusted pore volume of 0.51 contained 3345 pCi/L of
soluble radium. As local groundwater contact continued, the activity of the radium leached from the acidic tailings decreased to
1218 pCi/L after 1.6 pore volumes of leaching. The final sample
collected in these experiments contained 570 pCi/L at an adjusted
pore volume of 2.8, still over two orders of magnitude higher than

Table 4.   Tailings Characterization Data.

	Untreated	Amended
Particle size - clay	5	ND
% silt	32	ND
sand	63	ND
Saturated paste pH (pH units)	2.5	6.8

ND = Not determined

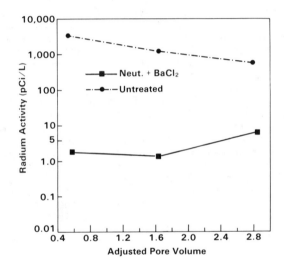

Figure 2.   Results of treated and untreated
             tailings experiments.

EPA's drinking water limit of 5 pCi/L total radium. The amended
tailings column (neutralized with $Ca(OH)_2$ plus $BaCl_2$ added) displayed a high degree of radium attenuation throughout the duration
of the experiment. The initial radium activity in the sample
collected at an adjusted pore volume of 0.6 was 1.7 pCi/L. This
represents a reduction in total radium activity of greater than
three orders of magnitude compared to the first sample collected
from the untreated acidic tailings. With further leachant contact,
the radium activity showed a further reduction to 1.1 pCi/L at 1.72
adjusted pore volumes. The last sample collected from the amended
tailings column displayed a slight increase in radium activity. At
an adjusted pore volume of 2.8, 6.1 pCi/L of soluble radium were

detected in the column effluent. Although the final sample collected from the amended tailings column was 1.1 pCi/L higher than
EPA's drinking water limit, a value of 6.1 pCi/L represents a
decrease of nearly two orders of magnitude over the untreated tailings column at similar leaching volumes.

The mechanism which accounts for the removal of soluble radium
from the tailings effluent solution is assumed to be coprecipitation. The reactants in the amended tailings material are $BaCl_2$ and
radium from the acidic tailings. The reaction is shown in Equation 1 and is probably the most common method for radium removal
from solution (Moffett 1976 [6]).

$$BaCl_2 + Ra^{2+} + SO_4^{2-} = 2 Cl^- + (Ra,Ba)SO_4 \qquad (1)$$

As barium sulfate is precipitated, radium is effectively
removed from solution by adsorption onto the precipitate surface or
is incorporated into the lattice structure of the solid precipitate. Radium removal is very much dependent on the solution parameters associated with the precipitation reactions. For example,
there must be an excess of sulfate in solution and the kinetics of
the coprecipitation reaction are sometimes slow so adequate column
residence time or reaction time (ranging from hours to days dependent on conditions) is necessary (Moffett 1976 [6]; Levins
1983 [5]). Since most U.S. mills utilize a sulfuric acid leach
cycle, excess sulfate is typically present. Furthermore, by neutralizing the acidic tailings, the reactants ($BaCl_2$ and radium) are
retained in the column long enough for the reaction to occur.
Tailings neutralization causes the pH dependent ions to precipitate
in the pore spaces within the column. The solids formed plug the
pathways through which solution normally travels, reduces the flow
rate of leachant through the neutral tailings, and therefore
increases the residence time for reactions. Further specifics of
the effects of tailings neutralization are discussed in the
solution chemistry section.

SOLUTION CHEMISTRY

The chemical composition data from the effluents collected
from both the untreated acidic tailings and the amended tailings
columns are shown in Table 5.

The data indicate that in addition to a substantial reduction
in total radium activity, similar reductions were detected for several of the macro cations, trace metals and sulfate in the column
effluents. Elements such as Al, As, Cr, Cu, Co, Fe, Si, Zn, and
SO_4 all exhibited large decreases relative to their original concentration in the untreated tailings column effluents. These
reductions are attributed to the change in solution pH (acidic to
neutral) during which time the solubility of the above elements is
exceeded resulting in precipitation reactions. The reaction products that form are likely metal hydroxides from the $Ca(OH)_2$ added
and sulfate salts as indicated by the large decrease in soluble SO_4
detected in the effluents as reported in Table 5. As mentioned
previously, the metal hydroxide and sulfate precipitates plug the

Table 5. Solution composition of untreated and amended tailings effluents.

Parameter	Untreated[a]			Amended[a]		
	1	2	3	1	2	3
pH	2.40	2.65	2.77	6.52	7.31	7.66
Adjusted pore volume	0.52	1.64	2.77	0.60	1.72	2.84
Total dissolved solids (g/L)	7.22	2.46	1.70	3.25	1.35	1.04
Total radium (pCi/L)	3345	1218	570	1.7	1.1	6.1
Macroions (mg/L)						
Al	2100	570	300	<10	<10	<10
Ca	610	560	610	1020	660	660
Fe	1400	390	240	<10	<10	<10
Mg	1740	472	254	1120	391	202
Mn	197	50	26	36	19	14
Na	1000	360	230	1070	280	160
Si	60	34	30	<10	5	6
Sr	2		0.9	<1	<1	<1
Zn	32	7.8	<5.0	<5	<5	<5
Cl	510	170	90	3860	565	140
NO_3	<20	<20	<20	<20	<20	<20
SO_4	25600	8100	5000	3430	2780	2630
Trace metals (mg/L)						
As	2.82	0.76	0.43	<0.02	<0.05	<0.02
Ba	<0.06	<0.02	<0.02	<0.04	<0.02	<0.02
Cd	0.27	0.08	0.04	0.02	<0.01	<0.01
Co	5.33	1.22	0.83	0.06	<0.02	<0.02
Cr	2.70	0.89	0.55	<0.01	<0.01	<0.01
Cu	2.69	0.76	0.34	<0.02	<0.02	<0.02
Pb	<0.02	<0.02	<0.02	<0.02	<0.02	<0.02
Se	0.24	<0.03	<0.02	0.22	<0.05	<0.05

[a] Effluent numbers

solution pathways and reduce the flow rate of leachant through the neutral tailings. The flow reduction allows a longer reaction time for the barium and radium reactions to occur and also limits the total volume of leachant to contact the tailings. By limiting or reducing the solution flow through tailings, the leaching of contaminants not attenuated by either neutralization or barium addition is also reduced (Opitz et al. 1985 [9]).

CONCLUSIONS

The results of these tests indicate that barium chloride addition to tailings can be an effective means of limiting the amount of soluble radium leached from acidic uranium mill tailings. Initial radium activities in the untreated acidic tailings (0.52 pore volumes) were greater than 3300 pCi/L. After 2.8 pore volumes of leaching the radium activities were 570 pCi/L, still greater than two orders of magnitude higher than EPA s water standard of 5 pCi/L. At similar pore volumes the radium activity of the barium treated tailings was 1.7 pCi/L (0.60 pore volumes) equivalent to a reduction by a factor of greater than 1900. At 2.8 pore volumes the radium activity was 6.1 pCi/L or 1.1 pCi/L higher than EPA's water standard but still a reduction of nearly two orders of magnitude.

Macro cation and trace metal reductions also occurred in the effluents from the neutralized and barium treated column. This was indicated in general by the reduction in the effluent total dissolved solids content in the amended tailings column, 7.22 g/L in the untreated tailings column effluent (0.52 pore volumes), compared with 3.25 g/L in the amended tailings column effluent (0.60 pore volume).

A tailings amendment technique such as this barium chloride/ hydrated lime treatment effectively reduces the amount of soluble radium, macro ions and trace metals that can be leached from acidic tailings. Utilization of tailings treatment techniques, such as the one described herein, appears to be an effective method for preventing the migration of radium acidic tailings.

ACKNOWLEDGMENTS

This work was sponsored and supported by the Office of Nuclear Regulatory Research of the Nuclear Regulatory Commission under DE-AC06-76RLO 1830, NRC FIN B2370. The authors acknowledge the interest and encouragement of NRC staff, particularly Mr. Frank Swanberg, who was the technical monitor of this project. Special appreciation is expressed to Mr. Robert Poyser at Pathfinder Gas Hills Mill who assisted in the collection of the test materials and Mr. R. J. Serne for reviewing this document prior to publication.

REFERENCES

1. ASTM. Annual Book of ASTM Standards Part 19, American Society for Testing and Materials, Philadelphia, Pennsylvania. 1982.

2. Black, C. A. Methods of Soil Analysis. American Society of Agronomy Monograph 9, Madison, Wisconsin. 1965.

3. Gee, G. W. and M. E. Dodson. "Soil Water Content by Microwave Drying: A Routine Procedure." Soil Science Society of America Journal 45:1234-1237. 1981.

4. APHA-AWWA-WPCF. Standard Methods for the Examination of Water and Wastewater. American Public Health Association, American Water Works Association, Water Pollution Control Federation, 15th edition, Washington, D. C. 1980.

5. Levins, D. M. "Mobilization of Radionuclides and Heavy Metals in Uranium Mill and Tailings Dam Circuits." Paper presented at The Office of the Supervising Scientist Workshop, May 19-20, 1983, Jabiru, Australia. 1983.

6. Moffett, D. 1976. The Disposal of Solid Waste and Liquid Effluents from the Milling of Uranium Ores. CANMET Report 76-19, Canada Centre for Mineral and Energy Technology, Ottawa, Canada.

7. Effluents from the Milling of Uranium Ores. CANMET Report 76-19, Canada Centre for Mineral and Energy Technology, Ottawa, Canada.

8. Opitz, B. E., M. E. Dodson, and R. J. Serne. Laboratory Evaluation of Limestone and Lime Neutralization of Acidic Uranium Mill Tailings Solution: Laboratory Progress Report. NUREG/CR-3449 (PNL-4809), U.S. Nuclear Regulatory Commission, Washington, D.C. 1983.

9. Opitz, B. E., M. E. Dodson and R. J. Serne. Uranium Mill Tailings Neutralization: Contaminant Complexation and Tailings Leaching Studies. NUREG/CR-3906 (PNL-5179), U.S. Nuclear Regulatory Commission, Washington, D.C. 1985.

10. Sherwood, D. R. and R. J. Serne. Tailings Treatment Techniques for Uranium Mill Waste: A Review of Existing Information. NUREG/CR-2938 (PNL-4453), U.S. Nuclear Regulatory Commission, Washington, D.C. 1983a.

11. Sherwood, D. R. and R. J. Serne. Evaluation of Selected Neutralizing Agents for the Treatment of Uranium Tailings Leachates: Laboratory Progress Report. NUREG/CR-3030 (PNL-4524), U.S. Nuclear Regulatory Commission, Washington, D.C. 1983b.

ASSESSING THE POTENTIAL WATER QUALITY HAZARDS CAUSED BY DISPOSAL OF RADIUM-CONTAINING WASTE SOLIDS BY SOIL BLENDING

G. Fred Lee, Ph.D., P.E., and R. Anne Jones, Ph.D.,
Department of Civil and Environmental Engineering
New Jersey Institute of Technology, Newark, New Jersey

INTRODUCTION

Soil blending has recently been proposed as a method for disposal of radium-containing waste solids. This approach is basically the dilution of the waste solids with "soils" in order to reduce the concentration of radium-226 to designated levels. While in principle this approach may be satisfactory, in practice appropriate environmental and public health protection will be difficult to achieve with this approach because of the potential for leaching of radium-226 which could contaminate surface and groundwaters, increasing the cancer risk of those using the waters. This paper reviews the factors that should be considered in developing a technically valid program for the disposal of radium-containing waste solids by soil blending that is protective of public health and the environment.

Particular attention is given to potential problems of surface and groundwater contamination with radium-226 that could arise from the disposal of radium-containing waste solids by blending with soil. However, it should be understood that an evaluation of such a waste/soil blending operation must also consider a wide variety of other radioactive as well as non-radioactive contaminants that can have adverse effects on surface and groundwater quality, and public health. For example, radon-222 is a daughter product of radium-226. Cothern [1] recently reviewed the potential hazards associated with the presence of naturally-occurring radon-222 in drinking water. He estimated that in the US between 2,000 and 40,000 excess cases of lung cancer per 70-year lifetime are due to radon in drinking water. Also, a recent National Research Council report entitled, "Scientific Basis for Risk Assessment and

Management of Uranium Mill Tailings," discusses the potential public
health hazards associated with radon-222 emanations from uranium
mill tailings [2]. These reports should be consulted for further
information on the problems of radon contamination.

The development of this review was prompted by a situation that
has arisen in the state of New Jersey with disposal of radium-
contaminated solids. There it has been proposed to dispose of
about 15,000 drums of soil contaminated with 0.5 g radium-226 by
blending-diluting the waste solid with a silty material to reduce
the concentration of radium-226 to 3 pCi/gram, and spreading the
mixture over about 40 acres of a sand and gravel quarry near
Vernon, NJ.

SELECTION OF ACCEPTABLE DILUTION OF RADIUM-226

One of the most important aspects that must be addressed in
developing a radium-containing waste/soil blending operation is the
acceptable radium-226 concentration in the blended mixture.
Proponents of the radium waste/soil blending approach in the New
Jersey situation have suggested a level of 3 pCi of radium-226 per
gram of blended soil as an acceptable concentration at the Vernon,
New Jersey site. This concentration was selected based on the fact
that the natural geological strata near the proposed disposal area
contain this concentration of radium-226. Since radium-226 has not
been found to be a problem in surface and groundwaters in the
proposed disposal area, it was concluded that achieving a dilution
of the radium-containing wastes to that concentration would also
not cause any water quality problems. This approach, however, is
not technically valid since it presumes, without proper evaluation,
that the leaching of radium-226 from the waste/soil mixture will be
the same as that from the natural strata of the region. It would
indeed be rare that that would be found. As discussed below, there
is a wide variety of factors that control the leaching of radium-
226 from solids. Furthermore, a disposal area containing blended
waste/soil mixtures containing 3 pCi/gram represents a tremendous
reservoir of radium which could lead to surface and groundwater
contamination.

The current US EPA drinking water MCL for radium-226 is 5
pCi/L. This concentration is associated with a cancer risk level
of one additional cancer per 50,000 people over a 70-year lifetime
[3]. This risk is substantially greater than the 10^{-6} risk that is
typically set as a desired maximum goal in groundwater clean-ups
near hazardous waste disposal sites. This leads to an important
philosophical aspect to the blending of radium-containing wastes
with soil to achieve a concentration of radium equivalent to
background levels, or so the concentration of radium-226 in the
groundwater does not exceed the drinking water MCL. It would
appear that soil blending for disposal of radium-contaminated waste
is being justified based on the fact that groundwaters in areas not
contaminated with radium wastes can be found to have elevated
radium-226 levels, and that there are areas that have high levels

of naturally-occurring radium in the soil. In the opinion of these authors, in light of the cancer risks associated with allowed levels of radium-226 in drinking water, these findings are not in themselves adequate justification for allowing cheaper waste disposal by soil blending. Allowing cheaper disposal of radium-contaminated soils at the expense (increased cancer risk) of other individuals or their heirs cannot be justified by indicating that there are people exposed to these risks from natural sources. Any elevation of the background concentration in surface waters or groundwaters represents an increased hazard to those who use this water for domestic purposes. Residents near waste disposal sites should not be exposed to elevated concentrations of a carcinogenic contaminant; any such exposure represents an attempt to develop a cheaper waste disposal practice than is necessary to fully protect the health and welfare of the residents near the disposal site.

Site-specific and waste-specific evaluations must be made to determine whether environmental or public health problems could result from the leaching of radium-226 from an area proposed to receive radium-containing wastes blended/diluted with soil. Factors that need to be considered in such an evaluation program are discussed subsequently.

With respect to the potential for groundwater contamination, it is important to consider not only the concentration of radium-226 in the blended soil mixture that could be leached as a result of percolation of water through the mixture, but also the depth of the blended soil mixture through which the percolate must pass. A relatively thin bed of radium-226-contaminated waste/soil mixture spread out over a very large area could cause less potential hazard for groundwater quality than a deep bed of the blended soil. While these two scenarios may result in the same average concentration in the waters of an aquifer, this "average" value may have little bearing on potential public health and environmental safety of the operation. For example, the leaching of radium from the thin bed may result in concentrations of radium in the groundwater that are elevated but still at levels within more accepted cancer risk, while leaching from the deep bed may result in pockets of groundwater that contain radium at levels that represent significantly elevated cancer risks compared to those which would be experienced if the waste had not been disposed in that area. Basically, the leaching tests must prove that no leaching of radium-226 can occur from the blended mixture. This is virtually impossible to do.

With respect to the potential for surface water contamination, consideration must be given to waterborne and airborne transport of the blended soils, and to groundwater and surface water transport of leachate to surface waters. Also, consideration must be given to the translocation of radium-226 by terrestrial plants which revegetate the disposal area. It is important to point out that the leaching characteristics of radium-226 from airborne, waterborne, or plant tissue-borne particulates at off-disposal-site locations are likely to be significantly different from the leaching characteristics of the radium-226 from the waste/soil mixture at the disposal site. This is discussed further below.

With the 1622-year half-life of radium-226, it is important to include in the assessment of acceptability of a disposal operation the long-term period (thousands of years) over which leaching and environmental contamination may take place. The normal short-term leaching tests would not typically be adequate to properly define the hazards that a radium-waste/soil blending operation could represent to surface and groundwater quality and public health due to the fact that leaching can occur at a very slow rate and result in concentrations of radium-226 associated with increased cancer risk above that generally accepted for a very long period of time - thousands of years.

It is clear that a detailed, properly-conducted, laboratory and field, site-specific investigation in a hazard assessment framework must be conducted before an "acceptable" radium-226 concentration-dilution can be selected for a blended waste/soil disposal operation. The concentration selected would be highly dependent on the leachability of radium-226 from all media of potential importance, surface and groundwater hydrology, thickness of the blended soil mixture, the characteristics of the soils used for dilution, the characteristics of the radium-containing wastes, the characteristics of the strata through which the water percolates below the disposal area, and the characteristics of the final surface cover and vegetation over the site, as well as other factors. The potential significance of many of these factors is discussed below.

Leaching of Radium-226

There have been several studies directed toward evaluating the leaching of radium-226 from natural strata and wastes. It is known that some natural geological strata leach radium-226 to the groundwaters resulting in concentrations of radium-226 well-above the drinking water standard. Hursh [4] reported that natural sources of radium-226 resulted in groundwaters´ in southern Wisconsin and northern Illinois having concentrations of radium-226 that were in excess of drinking water standards. He also reported that radium-226 concentrations in deep well water from Claremont, Oklahoma were several hundred pCi/L; sources of this radium were also natural. Based on the US EPA [3] reportings, the radium-226 average content of all surface water supplies was estimated to range from 0.1 to 0.5 pCi/L; the average concentration in large groundwater systems in individual states was estimated to range from 0.1 to 3 pCi/L. The US EPA [3] reported that of the East Coast aquifer systems on which compliance monitoring was done, about 500 exceeded the radium-226 MCL of 5 pCi/L. It was also reported that while the highest concentration reported for any groundwater drinking water system was about 200 pCi/L, few supplies exceeded 50 pCi/L. It is clear that while radium-226 is not typically leached from natural strata to groundwaters to result in concentrations above the drinking water standard, there are situations in which this does occur. The US EPA [3] indicated that aquifers that had low concentration of radium-228 are carbonate, metamorphic rock, quartzose sand, sandstone and basic igneous rock aquifers. Those with high levels of radium-228 include granite, arkosic sand, and quartzose sandstone aquifers with high total dissolved solids [3].

In the 1950´s the US Public Health Service [5] began intensive studies on the contamination of surface waters in the Colorado Plateau area by radium-226 from uranium milling operations. It was found that the average dissolved radium-226 concentration in river water upstream from the mill discharges (non-contaminated areas) was 0.3 pCi/L. River waters below the uranium mill discharges were reported to contain radium-226 at levels ranging from 0.3 to 86 pCi/L. Radium-226 was clearly being leached from uranium mill tailings in concentrations which represented potential hazards. Further, studies on biota (algae, insects, and fish) showed that the radium-226 in the water accumulated within aquatic organisms [5]. Anderson et al. [6] reported bioconcentration factors for radium-226 in fish, algae, and aquatic insects of 500 to 1000. Such bioaccumulation has potential significance in judging the environmental safety of disposing radium-contaminated waste solids by dilution-blending with soil.

In the early 1960´s Shearer [7] conducted a detailed study of the leachability of radium-226 from uranium mill tailings and river sediment from the Colorado Plateau area; a summary of this work was published by Shearer and Lee [8]. They [8] found that the leachability of radium-226 depended on a variety of factors including duration of contact with leaching solution, composition of leaching solution such as its barium and sulfate concentrations, characteristics of the uranium mill tailings especially whether they were derived from an alkaline or acid leach mill circuit and particle size, and the liquid to solid ratio in the leaching system. Shearer [7,8] reported that the dominant factors affecting radium-226 solubilization from uranium mill tailings was the solubilization of barite (radium-barium-sulfate). Huck and Anderson [9] reported that radium-226 was strongly sorbed onto barite; the sorbed and precipitated radium were extractable with EDTA. Yagnik et al. [10] found that radium-226 could be extracted from acid leach process tailings with HNO_3, DTPA, EDTA, or $CaCl_2$. While not specifically investigated, it appears that acid rain could potentially enhance the leachability of radium-226 if the pH were lowered sufficiently. However, it is possible that the increased sulfate concentration of acid rain could, in some instances, decrease the leachability of radium-226.

It is evident that radium-226 can be leached from uranium mill wastes, and that a wide variety of factors can influence the amount leached and the rate of leaching. It is also clear from the studies that have been done that it is inappropriate to assume that the amount and rate of leaching of radium-226 from any solid in a particular water or situation is similar to the that from natural strata or other solids. Site-specific leaching tests must be done to determine whether radium-226 can leach from a waste/soil blended mixture at a sufficient rate to represent a hazard to public health or the environment. It is important that conditions of these leaching tests closely resemble the conditions of the areas in which there is potential concern about radium-226 leaching. For example, if there is a potential for airborne, waterborne, or plant tissue-borne radium-226 to be transported to a wetland area where leaching could occur, the leaching test should be conducted under the same pH and redox (reducing) conditions as those that occur in the wetland, and with waters of high humic/organic content typical of the wetland. Similarly, in non-calcareous areas (soils with low $CaCO_3$ content) the characteristics of the leaching solution should take into account the pH and sulfate content of the precipitation in order to simulate the conditions that could occur as the waters percolate through the blended soils.

A number of investigators have found that radium-226 tends to become strongly sorbed onto clays and other surfaces. It appears that while radium-226 may be leachable from a particular solid waste, if the blending soil or surrounding geological strata has a high clay content. the radium would be sorbed onto the clays and thereby rendered potentially significantly less mobile. It is, therefore, important to use soils with a high clay content in a waste/soil blending operation. Silts or sands with low clay content would not likely sorb the radium to a sufficient extent to prevent surface and/or groundwater contamination. Site-specific evaluation of the particular types of clay that might be available should be conducted to determine their potential for sorption of radium-226 under the conditions that could be found at and near the disposal site. If it is proposed that the clays of the system be relied on for attenuation of radium-226, comprehensive studies need to be done to understand the clay-radium interactions.

The characteristics of the subsurface geology at the disposal site could be very important in determining whether any leached radium-226 contaminates groundwater of the region. Areas low in clay content (e.g., sand, silt) should be avoided because of their typically expected, low capacity for sorption of radium-226 and higher permeabilities. As with any hazardous waste disposal operation, the placement of the disposal area above cavernous limestone systems should be avoided. It is important to properly characterize the subsurface geology-hydrology, and the tendency for transport of radium-226 to existing, as well as potential, groundwater supplies and surface waters. This characterization should include evaluation of samples of the geological strata for their tendency to sorb radium-226.

Rate of Leaching

Short-term (a few hours´ to a few days´ duration) leaching tests of the type conducted by Shearer and others showed that appreciable concentrations of radium-226 were leachable from certain types of waste solids. Shearer [7] found that an apparent equilibrium in the amount of radium-226 leached was achieved within a few days. After leaching, he allowed the leached solids to remain wet for up to 21 days and conducted leaching tests again. He did not detect any additional leaching of radium-226 during these subsequent leaches. These results should not be interpreted to mean that radium-226 associated with all other solids would behave in the same manner. They also cannot be interpreted to mean that if he had allowed additional time for intraparticle diffusion to occur, additional leaching would not have been found.

As noted above, it has been proposed to dispose of 15,000 drums of contaminated wastes containing a total of 0.5 g of radium-226 by blending with soil and depositing the mixture in a sand and gravel quarry near Vernon, NJ. One proposed plan called for dilution to achieve 3 pCi/gram of blended soil; this mixture was to be mixed over a 40-acre area within the quarry. Calculations made by the authors showed that infiltration of precipitation through the blended waste/soil mixture could result in radium-226 concentrations in the percolate through the 40-acre area in excess of the drinking water standard for a period of more than a thousand years. These calculations demonstrate the importance of properly considering the fact that a very slow rate of radium leaching can result in contamination of groundwaters such that the radium-226 concentration represents a significant increase in the cancer risk associated with the drinking water. Lysimeter studies should be conducted to simulate the proposed blended soil column. These columns should be hydraulically loaded in a pattern resembling the normal precipitation pattern of the area, and should be studied for no less than a year, but preferably several years to determine whether sufficient leaching would likely occur in the immediate future to cause groundwater contamination. Further, any radium-contaminated waste/soil blending operation must include a commitment to monitor the groundwater under the disposal area ad infinitum to ensure that radium leaching is not occurring.

One possible way to try to avoid contamination of groundwater in a waste/soil blending operation is to follow the approach that is being followed for solid and hazardous waste disposal sites, of capping the site with a layer of low-permeability clay. This approach is designed to minimize the amount of water that enters the site and thereby minimize leachate generation. While this approach will work during the time that the cap is properly maintained, it is questionable that regulatory agencies - society - will continue to maintain the cap for the thousands of years during which radium-226 leaching could occur. It would seem more appropriate to treat the radium-226-containing wastes to prevent radium-226 migration, such as with barium sulfate, and then to place the treated waste in a double-lined disposal area with a leachate collection system of the type used for hazardous waste disposal. It is important to emphasize, however, that as with essentially all hazardous wastes, a 30-year post-closure monitoring period is grossly inadequate to provide for long-term public health and environmental protection.

There are situations in which the concentrations of contaminants in leachates below waste disposal areas that are initially below a water quality standard may build up to exceed the standard in groundwater systems. During periods of low precipitation, moisture flux in the unsaturated zone may be to the atmosphere. When this occurs, contaminants in leachate migrating through the unsaturated zone can precipitate and accumulate at some stratum below the soil surface. With the next major inflow of moisture (e.g., following a rainfall event) the accumulated contaminants can be rapidly transported to the groundwaters, yielding a localized area in the saturated zone that exceeds the drinking water standards. Because of the relatively poor mixing that typically occurs in groundwaters, these localized areas may persist for considerable distances and periods of time in an aquifer system.

MONITORING

Any radium-226 waste/soil blending operation should include a long-term (ad infinitum) commitment to monitor the surface water, groundwater, and terrestrial vegetation in the vicinity of the disposal site. Monitoring of the surface waters receiving drainage from the site should include monitoring water, fine-grain sediments where they accumulate downstream of the disposal area, fish and other aquatic life, and aquatic vegetation. The groundwater monitoring should include monitoring of both saturated and unsaturated flow. Initially, the monitoring should be done quarterly for surface waters, and monthly for groundwater. After a year or two, it may be possible to reduce the monitoring frequency to once or twice per year. It is important to emphasize that just because the monitoring program does not detect any problems in 10, 50, or 100 years, the program cannot be terminated. Depending on how well the site was prepared and maintained, it is possible that problems may not become apparent until 50 to several hundred years after the disposal took place. To ensure that the residents of the region of the disposal site are in fact provided with long-term public health protection, a trust fund of sufficient magnitude should be established to ensure that funds are available ad infinitum for monitoring, maintenance, and remediation should contamination be found.

The monitoring program should include not only radium-226, but also other contaminants present in the waste solids that could have an adverse effect on surface water or groundwater quality.

CONCLUSIONS

The disposal of radium-226-contaminated waste solids by soil blending appears to have limited utility. The potential for long-term, slow leaching of radium-226 from the blended soils, coupled with the very low drinking water standard for radium-226 and the increased cancer risk associated with that standard, mandates that great caution be exercised in utilizing such an approach. Radium-226 has been found to be leachable from some natural strata, as well as from uranium mill tailings and similar materials. The rate and extent of leaching depends on a wide variety of factors, with liquid to solid ratio, and barium and sulfate content of the water and matrix being the most important. The pH, organic content, and clay type and content may also be important. The leached radium-226 can be readily sorbed on and immobilized by clays and possibly other solids. Site-specific, long-term leaching and sorption studies should be done to evaluate the potential for surface and groundwater contamination at the proposed disposal site for the blended soil. Any soil blending system should be accompanied by a long-term monitoring of groundwaters and surface water, vegetation, and organisms to ensure that potentially significant environmental contamination with radium-226 or other contaminants from the site does not occur. As part of establishing the soil blending system, funds should be provided to ensure that monitoring, site maintenance, and remedial measures will be undertaken as long as there is any radium-226 present at the site in sufficient amounts to potentially cause environmental contamination.

ACKNOWLEDGMENT

The preparation of this paper was supported by the Department of Civil and Environmental Engineering, and the NSF/Industry/University Cooperative Center for Research in Hazardous and Toxic Substances at the New Jersey Institute of Technology. The assistance of C. Timm, Jacobs Engineering, Washington, D.C., and S. Law of the Uranium Mill Tailings Remedial Action Project, DOE, Albuquerque, NM in acquisition of background literature is greatly appreciated. Also the assistance of Mr. C. Ray, graduate student in Environmental Engineering at the New Jersey Institute of Technology, and Dr. D. Watts, Director of Operations for the NSF/Industry/University Cooperative Center for Research in Hazardous and Toxic Substances is appreciated.

REFERENCES

1. Cothern, C. R., "Radon in Drinking Water," To be Published in Journ. American Water Works Association April (1987).

2. Uranium Mill Tailings Study Panel, Board on Radioactive Waste Management, National Research Council, Scientific Basis for Risk Assessment and Management of Uranium Mill Tailings, National Academy Press, Washington, D.C. (1986).

3. US EPA, "Water Pollution Control; National Primary Drinking Water Regulations; Radionuclides," 40 CFR Part 141; Federal Register 51:34836-34862, September 30 (1986).

4. Hursh, J. B., "Natural Occurrence of Radium in Man and in Waters and in Food," Health Physics, Vol 2. (1960).

5.   Tsivoglou, E.C., et al., "Report of Survey of Contamination of Surface Waters by Uranium Recovery Plants," US Public Health Service, R.A. Taft Sanitary Engineering Center, Cincinnati, OH (1956).

6.   Anderson, J. B., et al., "Effects of Uranium Mill Wastes on Biological Fauna of the Animas River," National Symposium of Radioecology, Colorado State University, Fort Collins, CO (1961).

7.   Shearer, S. D., The Leachability of Radium-226 from Uranium Mill Waste Solids and River Sediments, Ph.D. dissertation, University of Wisconsin-Madison (1962).

8.   Shearer, S. D., and Lee, G. F., "Leachability of Radium 226 from Uranium Mill Solids and River Sediments," Health Physics 10:217-227 (1964).

9.   Huck, P. M., and Anderson, W. B., "Deposition of 226-Ra on Surfaces during Precipitation and Leaching of $(Ba, Ra)SO_4$," Water Research 17:1403-1406 (1983).

10.   Yagnik, S. K., Baird, M. H., and Banerjee, S., "An Investigation of Radium Extraction from Sulfuric Acid Leach Uranium Mill Tailings," Presented at AIChE Annual Meeting, Miami, FL (1978).

Radon in Water Supply Wells:
Treatment Facility Requirements and Costs

David J. Hiltebrand*

John E. Dyksen**

Kalyan Raman***

Introduction

Over the past several years, the United States Environmental Protection Agency has been preparing revised National Primary Drinking Water Regulations. The revisions may change the maximum contaminant level (MCL) for some currently regulated contaminants and develop MCLs for additional contaminants. Radon has been included in the list of contaminants for which the EPA is considering an MCL.

Radon-222 is an inert, noble gas that is formed by the radioactive decay of the element radium-236 (1). Radon can enter the drinking water supply from radioactive mining operations, industrial discharge or the decay of naturally occurring radium within the local bedrock. Most often, radon occurs naturally in ground water as a result of the decay of radium in water and in the rock and soil matrix surrounding the water (2). The

* Project Environmental Scientist, Malcolm Pirnie, Inc., 100 Eisenhower Drive, Paramus, New Jersey

** Senior Project Manager, Water Quality and Treatment, Malcolm Pirnie, Inc., 100 Eisenhower Drive, Paramas, New Jersey 07652

*** Engineer, Malcolm Pirnie, Inc., 100 Eisenhower Drive, Paramus, New Jersey 07652

concentration of radon in water supplies ranges from the background levels of less than 50 picocuries per year (pCi/L) which is normally found in the surface supplies, to over 1,000,000 pCi/L in drilled wells (3).

The health risk concern for radon is from human exposure to radiation through ingestion and primarily inhalation, since the radon gas is released into the household from faucets, bathtubs, showerheads and other water-using appliances. Most of the radon contained in a water supply is lost to the air. High airborne concentrations can build up inside the home particularly when the household air exchange rate is low (4).

Presented in this paper is a discussion of the feasible treatment methods to remove radon from drinking water and the associated costs. Because acceptable radon levels have not as yet been proposed by the USEPA, treatment methods were studied for the following influent and effluent combinations and the resulting required percent removals:

Radon (pCi/L)		Required
Influent	Effluent	Percent Removal
750,000	20,000	97
	1,000	99.9
100,000	20,000	80
	10,000	90
	5,000	95
	1,000	99
10,000	5,000	50
	1,000	90

The influent levels were chosen in order to include radon levels which have been found to occur and the effluent levels were chosen in order to present a broad range of percent removal requirements.

The most effective technologies for the removal of radon from drinking water have been identified as:

- Granular Activated Carbon (GAC) Adsorption
- Packed Tower Aeration

GAC Adsorption

In drinking water treatment the use of granular activated carbon (GAC) in the United States has been limited primarily to applications for the control of synthetic organic chemicals and taste and odor compounds. GAC adsorption has been used to remove radon from water to concentrate it for analytical measurements. However, there are very few documented evidences of using GAC as a continuous process to remove radon from drinking water (5). Since the detention of radon in drinking water supplies, a number of research and pilot-scale studies have been undertaken to evaluate the effectiveness of GAC for controlling radon. Because of the decay of radon and radon progeny accumulated within the GAC contractor, concern has been expressed about the gamma radiation hazard. Another potential concern regarding the use of the GAC unit is the fate of radon progeny in the GAC bed. However, preliminary tests with commercial units indicate that even after four months of operation, no radon progeny were detected in effluent water (5). Due to their simplicity, use of GAC units to remove radon in small water supply systems is attractive. No moving parts are required to operate the units. In areas where turbidity is present, occasional backwashing of the GAC units may be required.

The application of granular activated carbon adsorption to the removal of radon from drinking water supplies involves the following major design considerations:

- Adsorption/Decay Steady State

- Empty Bed contact Time

- Contractor configuration: Downflow versus upflow, pressure versus gravity, single-stage versus multi-stage or parallel

Adsorption/Decay Steady State - In the classic application of granular activated carbon adsorption to the removal of organic compounds, adsorption isotherms have been found to be useful screening tools for determining preliminary carbon usage rages, and for evaluating the effectiveness of different types of GAC. However, while adsorption isotherms may give an indication of the potential of a giver carbon for radon removal, they do not yield sufficient data to develop design criteria for GAC treatment systems. Radon has a short half-life of 3.82 days, and it has been found that large portions of the adsorbed radon decay within the GAC bed before breakthrough. Thus, in effect, the GAC bed is extended many times over the life indicated by the adsorption isotherm.

For a typical nondecaying adsorbate, a breakthrough point occurs when the effluent concentration equals the influent concentration. In contrast, the radon removal relationship shows an initial period where the adsorption is maximum and decay is minimum. During this period the radon concentration in the treated water gradually increases, after which time it levels off at a steady state value. A steady state is established within the carbon bed when the adsorption rate equals the rate of decay. The decay of radon perpetuates the life of the bed because the bed is effectively continuously self-regenerating.

Empty Bed Contact Time - The empty bed contact time (EBCT) provides an indication of the quantity of carbon which will on-line at any one time, and thus reflects the capital cost for the system. In general, increasing the empty bed contact time results in lower carbon usage rates. Typical EBCTs for SOC removal range from 15 to 30 minutes. Test data regarding the removal of radon from ground water appear to indicate that radon requires a longer EBCT than other adsorbable materials.

The relationship between empty bed contact time and the percent of radon remaining based on pilot studies, is shown on Figure 1. As illustrated, an EBCT of 130 minutes is required to reduce a radon influent level from 100,000 pCi/L to 1,000 pCi/L (99 percent removal).

Contactor Configuration - Based on the estimated contact times for radon removal, one can determine a conceptual process design by evaluating the various contactor configurations. The two basic modes of contactor operation are downflow and upflow. Upflow beds typically have been applied to situations where suspended solids removal also is required.

Downflow fixed bed contactors offer the simplest and most common contactor configuration for radon removal in ground water. The contactors can be operated either under pressure or by gravity. Pressure contactors may be more applicable to ground water systems because of the nature of these systems. The use of gravity contactors for most ground water systems would involve repumping the treated water to the distribution system. On the other hand, pressure contactors might be used without repumping, thus reducing both capital and operating costs. A diagram of a pressure system is presented on Figure 2.

Packed Tower Aeration

Air stripping has been used effectively in water treatment to reduce the concentration of taste and odor producing compounds and certain organic compounds. However, the use of air stripping for the purpose of controlling radon is a relatively new concept in the drinking water industry. To date, few pilot-scale packed tower aeration studies have been conducted for radon removal. However, as shown in Table 1, radon has a larger Henry's Law constant than tetrachloroethylene and trichloroethylene which are known to be easily removed by air stripping. Since radon acts similarly to some highly volatile organic compounds (VOCs) and packed columns have been shown to be the most efficient form of aeration for VOC removal, packed tower aeration has been considered as a most effective technology (7,8).

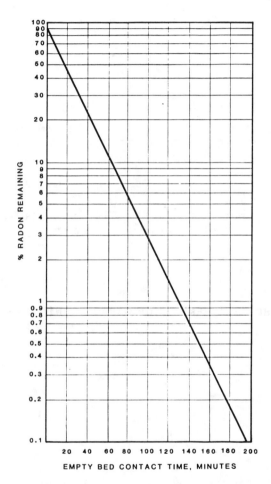

FIGURE 1 – IMPACT OF EMPTY BED CONTACT TIME
ON PERCENT RADON REMAINING

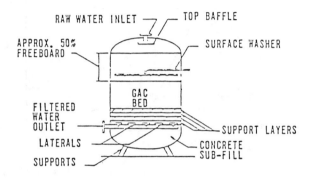

RAW WATER INLET — TOP BAFFLE

APPROX. 50%
FREEBOARD — SURFACE WASHER

GAC
BED

FILTERED
WATER
OUTLET —

— SUPPORT LAYERS

LATERALS —

— CONCRETE
SUB-FILL

SUPPORTS —

FIGURE 2 - SCHEMATIC OF CARBON CONTACTOR

TABLE   1

HENRY'S LAW CONSTANTS FOR SELECTED COMPOUNDS

Compound	Henry's Law Constant[1] (atm)
Vinyl Chloride	3.5×10^5
Radon	2.26×10^3
Carbon Dioxide	1.51×10^3
Tetrachloroethylene	1.11×10^3
Trichloroethylene	5.5×10^2

Note:

1.   Temperature 20 C.

The rate at which radon is removed from water by aeration depends on several factors:

- Air:water ratio
- Contact time
- Available area for mass transfer
- Temperature of the water and the air
- Physical chemistry of the contaminant

The first three factors may be controlled in the design of an air stripping unit, while the other two factors are set for a specific water supply.

Air:water Ratio - The air flow requirements for a packed column depend on the Henry's Law constant for the particular compound(s) to be removed from the water. On a perfect aeration system, the minimum air water ratio of the Henry's Law constant. The greater the Henry's Law constant, the less air is required to remove the compound from water. Because aeration systems are not perfect, the actual air:water ratios to achieve a given removal efficiency are greater than the minimum.

Contact Time - The contact time is a function of the depth of the packing material. An increase in the depth of packing material results in a greater contact time between the air and the water, and consequently, higher removals are achieved.

Mass Transfer Area - The available area for mass transfer is a function of the packing material. Various sizes and types of packing material are available including 1/4 inch to 3-inch sizes and in metal, ceramic and plastic materials. In general, the smaller packing material provide a greater available area for mass transfer per volume of material, thus increasing the mass of contaminant removed.

Equipment Required - A diagram of a typical packed tower installation is shown on Figure 3 and consists of the following:

- Packed Tower: Either metal (steel or aluminum), plastic, or concrete construction. Internals (packing, supports, distributors, must eliminators) are generally made of metal or plastic.

- Blower: Typically centrifugal type, either metal or plastic construction. Noise control may be required depending on the size and system location.

- Effluent Storage: Generally provided as a concrete clearwell below the packed tower.

- Effluent Pumping: Generally required since effluent is at atmospheric pressure. Vertical turbine pumps mounted on clearwell are typical.

In ground water applications, water is generally pumped directly from wells to the top of the packed tower. The effluent flows from the bottom of the tower into a clearwell from where it is usually pumped into the distribution system.

Treatment Costs - Treatment costs have been developed for radon removal by GAC adsorption and packed column aeration, for the various influent/effluent concentrations listed earlier. Costs for GAC adsorption were developed using a computer model written for the USEPA by Culp/Wesner/Culp, Inc. (9). Packed column aeration costs were estimated by the USEPA, using a computer model developed by the Office of Drinking Water (10). Cost factors used in developing the treatment costs were:

Electric Power	$ 0.086/Kwh
Labor - Small System Sizes (<150,000 gpd)	$ 5.90/hr
- Large System Sizes (>150,000 gpd)	$14.30/hr
Diesel Fuel	$ 0.80/gal
Natural Gas	$ 0.0027/scf
Siatework	15% of construction costs
Contractor's Overhead & Profit	12% of construction costs (including siatework)
Contingencies	15% of construction costs
Engineering & Technical Fee	15% of construction costs (including siatework & contractor's O&P)
Interest Rate	10%
Number of Years	20

Provided below is a brief discussion on the design assumptions that were used in sizing the treatment systems, and the treatment costs.

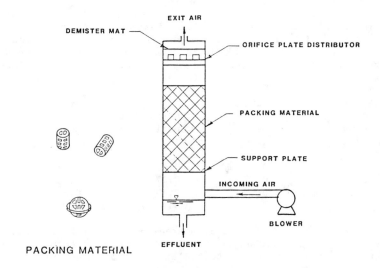

PACKED COLUMN

FIGURE 3 - SCHEMATIC OF PACKED COLUMN AERATION

GAC Adsorption - The costs are for a complete carbon contacting facility, including vessels, piping and valves, instrumentation, control, backwaste pumping facilities, housing and purchase and placement of virgin carbon in the contactors. Skid mounted package GAC units are used for systems requiring a total volume of less than 350 ft^3. Larger systems are field constructed and are equipped with shop-fabricated steel pressure vessels. A bed depth of 20 feet provides an EBCT of 30 minutes per contactor at a hydraulic loading rate of 5 gpm/sq. ft. To achieve the long detention times required for radon removal, it is assumed that multiple units would be operated series.

The results of surveys of small treatment systems performed in connection with this cost estimation indicate that the plants generally satisfy their peak water demand through a combination of treatment and storage capacities. On this basis, the following flow capacities were assumed for systems less than 1 MGD.

Average Daily Flow, (mgd)	Treatment Design Capacity (mgd)	Storage Capacity, (mg)	System Supply Capacity, (mgd)
0.01	0.03	0.03	0.06
0.04	0.07	0.07	0.14
0.13	0.17	0.15	0.32
0.40	0.50	0.46	0.96

The capital, operating and maintenance costs (O&M), and the costs for producing 1,000 gallons of water by GAC adsorption for various system sizes are presented in Table 2. The high EBCTs required for removing radon result in total production costs that range from 57 to 1,577 cents per 1,000 gallons for systems with design flows of 0.06 to 15.44 mgd. In addition to the costs, the need to use multiple contactors to achieve the required EBCTs can result in the need for more contactors than would be practical. For example, in order to achieve a 99 percent removal of radon, an EBCT of 100 minutes is required. For a system with a design capacity of 7.52 mgd, this would require the use of 40 contactors, each with a diameter of 12 feet, a bed depth of 29 feet and an overall height of 33 feet.

Packed Column Aeration - The design parameters for packed tower aeration were developed utilizing the results of pilot studies conducted by Dr. L. Lowry of the University of Maine at Orono. Based upon these results, the following packing heights were required for a liquid loading rate of 30 gpm/sf and an air:water ratio of 15:1:

TABLE 2

ESTIMATED COST FOR REMOVING RADON
BY GAC ADSORPTION

System Size Category Population Range Design Flow (MGD) Average Daily Flow (MGD)	Influent (pCi/L)									
	750,000				100,000				10,000	
	Effluent (pCi/L)				Effluent (pCi/L)				Effluent (pCi/L)	
	20,000	10,000	5,000	1,000	20,000	10,000	5,000	1,000	5,000	1,000
Percent Removed	97.30	98.70	99.30	99.90	80.00	90.00	95.00	99.00	50.00	90.00
25-100 Total Capital Cost (K$)	140.00	158.80	172.60	362.20	93.70	112.30	129.30	164.70	63.70	112.30
0.06 O&M Cost (K$/year)	7.48	8.46	9.40	32.31	5.02	6.10	6.93	8.86	3.09	6.10
0.01 Total Production Cost (cents/1,000 gal)	504.20	571.39	625.37	1577.53	337.75	406.55	466.12	594.43	222.81	406.55
101-500 Total Capital Cost (K$)	428.60	496.50	554.00	495.90	153.30	190.90	388.50	516.20	98.90	190.90
0.14 O&M Cost (K$/year)	49.28	52.93	55.55	66.40	11.40	14.57	49.63	54.11	7.35	14.57
0.04 Total Production Cost (cents/1,000 gal)	606.53	677.31	734.39	758.89	179.04	225.22	579.99	698.59	115.48	225.22
501-1,000 Total Capital Cost (K$)	579.90	585.10	747.50	952.90	462.60	600.90	520.60	710.90	160.30	600.90
0.32 O&M Cost (K$/year)	69.53	77.57	83.55	101.03	53.77	60.95	64.62	79.27	13.16	60.95
0.13 Total Production Cost (cents/1,000 gal)	283.54	301.36	352.97	438.68	222.69	270.95	259.08	335.30	65.90	270.95
1,001-3,300 Total Capital Cost (K$)	1131.00	1530.00	1493.40	2018.40	709.00	813.00	998.20	1407.20	553.20	813.00
0.96 O&M Cost (K$/year)	129.84	164.90	181.34	258.29	80.01	99.78	115.21	173.04	58.15	99.78
0.40 Total Production Cost (cents/1,000 gal)	179.92	236.04	244.35	339.30	111.84	133.75	159.22	231.73	84.33	133.75
3,301-10,000 Total Capital Cost (K$)	5505.70	6824.50	9183.50	10642.80	2614.30	3745.80	4836.70	7205.80	1288.10	3745.80
3.00 O&M Cost (K$/year)	770.49	966.84	1115.86	1587.18	342.04	490.02	657.38	1032.91	150.65	490.02
1.30 Total Production Cost (cents/1,000 gal)	298.67	372.70	462.50	597.95	136.80	196.00	258.27	396.06	83.64	196.00
10,001-25,000 Total Capital Cost (K$)	12977.50	16126.90	18453.50	25785.20	6174.40	8645.30	11183.90	17159.20	2757.90	8645.30
7.50 O&M Cost (K$/year)	1971.88	2486.81	2879.48	4137.67	828.91	1260.09	1676.11	2660.65	362.79	1260.09
3.25 Total Production Cost (cents/1,000 gal)	294.73	369.32	425.46	604.12	131.01	191.83	252.04	394.20	57.89	191.83

TABLE 3

ESTIMATED COST FOR REMOVING RADON
BY PACKED COLUMN AERATION

System Size Category / Population Range / Design Flow (MGD) / Average Daily Flow (MGD)	Influent (pCi/L)										
	750,000				100,000				10,000		
	Effluent (pCi/L)				Effluent (pCi/L)				Effluent (pCi/L)		
	20,000	10,000	5,000	1,000	20,000	10,000	5,000	1,000	5,000	1,000	100
Percent Removed	97.30	98.70	99.30	99.90	80.00	90.00	95.00	99.00	50.00	90.00	99.00
25-100 (0.06 / 0.01)											
Total Capital Cost (K$)	62.30	63.28	65.41	69.65	58.90	58.90	59.90	63.60	58.90	58.90	58.90
O&M Cost (K$/year)	0.64	0.68	0.70	0.81	0.56	0.56	0.60	0.69	0.56	0.56	0.56
Total Production Cost (cents/1,000 gal)	167.72	170.99	176.66	189.49	157.55	157.55	160.80	172.00	157.55	157.55	157.55
101-500 (0.14 / 0.04)											
Total Capital Cost (K$)	70.80	74.06	76.51	83.21	66.40	66.40	69.65	74.60	66.40	66.40	66.40
O&M Cost (K$/year)	1.10	1.20	1.20	1.40	0.91	0.91	1.00	1.20	0.91	0.91	0.91
Total Production Cost (cents/1,000 gal)	57.33	60.27	62.02	68.03	53.00	53.00	55.90	60.60	53.00	53.00	53.00
501-1,000 (0.32 / 0.13)											
Total Capital Cost (K$)	88.12	93.02	94.66	106.11	79.30	79.30	83.20	94.70	79.30	79.30	79.30
O&M Cost (K$/year)	2.20	2.40	2.50	3.00	1.90	1.90	2.10	2.50	1.90	1.90	1.90
Total Production Cost (cents/1,000 gal)	25.85	27.45	28.05	31.85	23.10	23.10	24.46	28.00	23.10	23.10	23.10
1,001-3,300 (0.96 / 0.40)											
Total Capital Cost (K$)	137.23	147.06	151.98	176.60	120.85	120.85	129.00	148.70	120.85	120.85	120.85
O&M Cost (K$/year)	5.80	6.30	6.60	7.80	4.90	4.90	5.40	6.40	4.90	4.90	4.90
Total Production Cost (cents/1,000 gal)	15.00	16.15	16.75	19.55	13.00	13.00	14.00	16.35	13.00	13.00	13.00
3,301-10,000 (3.00 / 1.30)											
Total Capital Cost (K$)	258.79	290.08	298.67	364.30	224.24	224.24	248.92	293.40	224.24	224.24	224.24
O&M Cost (K$/year)	18.00	19.00	20.00	24.00	15.00	15.00	16.00	20.00	15.00	15.00	15.00
Total Production Cost (cents/1,000 gal)	10.20	11.19	11.56	14.08	8.70	8.70	9.53	11.50	8.70	8.70	8.70
10,001-25,000 (7.50 / 3.25)											
Total Capital Cost (K$)	576.26	626.17	659.52	810.45	460.20	460.20	526.50	642.84	460.20	460.20	460.20
O&M Cost (K$/year)	45.00	49.00	50.00	60.00	38.00	38.00	42.00	49.00	38.00	38.00	38.00
Total Production Cost (cents/1,000 gal)	9.50	10.33	10.75	13.08	7.76	7.76	8.75	10.50	7.76	7.76	7.76

Percent Removal	Packing Height (ft)
50	3.5
80	7.5
90	10
95	14
97	17
99	22
99.9	33

Since packing heights of less than 10 feet are not practical for packed towers costs estimates are based upon a minimal packing height of 10 feet. Also, for small systems (less than 1 mgd) it is most cost effective to treat the entire flow rather than using a combination of treatment and storage.

The capital costs of the packed column include: PVC column, packing, intervals, blowers, piping, clearwell, instrumentation, air ducts and electrical wiring.

The capital, O&M, and the costs for producing 1,000 gallons of water by packed column aeration for various system sizes are presented in Table 3. As indicated in the table, total production costs for removing radon by packed tower aeration range from 9 to 190 cents per thousand gallons.

Conclusion

Although both GAC and packed tower aeration have been identified as the most effective technologies for the removal of radon from drinking water, the long EBCTs required for removing radon by GAC results in substantially greater production costs that are required for packed tower aeration. In addition to the costs, the long EBCTs can result in the need for a larger number of contactors than would be practical. Therefore, base upon the facility requirements and costs it would appear that packed tower aeration will become the most viable treatment alternative. However, the selection of a particular treatment method to remove radon will need to include an evaluation of th potential air quality concerns created by the exhaust gases from the aeration system which will contain radon.

Bibliography

1. Radon in Water and Air-health Risks and Control Measures. The land and Water Resources Center, University of Maine at Orono, February, 1983.

2. Partridge, J.E.; Horton, T.R.; Sensintaffar, E. L., 1979. A study of Radon-222 Released from Water During Typical Household Activities. Report for Office of Radiation Programs Technical Note CRP/EERF-79-1.

3. Personal Communication. September 18, 1983, from Dr. Jerry Lowry of the University of Maine to Alan Hess of Malcolm Pirnie.

4. Castren, O., 1977. "The Contribution of Bored Wells to Respirating Radon Daughter Exposure in Finland, TRPA, Paris.

5. Lowry, J. D. and Brandown, J. E., 1981. "Removal of Radon from Ground Water Supplies using Granular Activated Carbon or Diffused Aeration," University of Maine, Department of Civil Engineering, Orono, Maine 06469.

6. Personal Communication. May 2, 1984 from Dr. Jerry Lowry of University of Maine to Kathleen Beyer of Malcolm Pirnie. Unpublished data from EPA project R-810829-C1-0.

7. Hess, A. F., Dyksen, J. E., and Dunn, H. J., 1983. "Control Strategy - Aeration Treatment Technique in Occurrence and Removal of Volatile Organic Chemicals From Drinking Water," American Water Works Association Research Foundation, Denver.

8. Treybal, R.E., 1980. Mass Transfer Operations. McGraw-Hill Book Co., New York, New York. pp.187-219.

9. WATER CO$T. A Computer Program for Estimating Water and Wastewater Treatment Costs, by Culp.Wesner.Culp, Santa Ana, California 92707.

10. Personal Communication. May 9, 1986 from Michael Cummins of Office of Drinking Water, USEPA, to John Dyksen of Malcolm Pirnie.

Frederick T. Varani, P.E.
EDC, Inc., Lakewood, Colorado

Robert T. Jelinek, P.E.
Richard P. Arber Associates, Denver, Colorado

Randy J. Correll
U.S. Public Health Service, Seattle, Washington

INTRODUCTION

The U.S. Environmental Protection Agency (EPA) recently announced its intention to propose national interim primary drinking water regulations (NIPDWR) for several radionuclides not currently regulated, including uranium [1] . Current regulations for radionuclides in drinking water include a limits of 5 pCi/L for radium and 15 pCi/L for gross alpha particle activity, excluding both uranium and radon. As mandated by the recently amended Safe Drinking Water Act, EPA will proposed health - based maximum contaminant level goals (MCLGs) and enforceable maximum contaminant levels (MCLs) for these radionuclides that will apply to all community water systems in this country.

Health effects research has indicated that uranium is known to be toxic to kidneys. An additional health threat from ingestion of uranium is that of bone cancer. Since uranium is a known or probable human carcinogen, under EPA rules, it requires a MCLG in drinking water of zero. The proposed MCL for uranium is 10 pCi/L [2]. Of the approximately 60,000 community water systems in the U.S., it is estimated that between 100 and 2,000 would exceed this proposed MCL [3].

High levels of naturally occurring uranium in water are most prevalent in the Colorado Plateau and Rocky Mountain region. Gross alpha concentrations, uranium concentrations, and U_{234}/U_{238} ratios in the Rocky Mountain states have been documented in a nation-wide study of public groundwater supplies [4]. With respect to small public systems and private wells in the Front Range area of Colorado west of Denver, two sampling programs have identified gross alpha and uranium levels of concern [5,6]. The general location of the wells sampled in West Jefferson County, Colorado during these two sampling programs are shown in Figure 1. These wells are owned by the Jefferson County School District which has operated schools in this mountain area for the past decade.

Removal of naturally occurring uranium in drinking water by the ion exchange process using a strong base anionic resin has been documented in several laboratory and field investigations [7,8], most notably a two year field study which included 12 field units, in the Denver, Colorado and Gallup, New Mexico areas [9]. Several wells operated by the School District were included in this two year investigation.

Each ion exchange system tested during the two year field investigation consisted of a fiberglass tank packed with 0.25 cu.ft. of Sybron Chemical Ionac A641 resin. The loading rate used was 0.25 gpm. None of the units were regnerated nor was the treated water consumed. Some of the field units became saturated with uranium at 21,000, 29,000, and 12,000 bed volume through put. over 62,000 bed volumes were treated at a School District installation. The remaining columns removed uranium to non-detectable levels until the sampling program was terminated.

The focus of this paper is to summarize the radiochemical quality data for two of the School District wells and discuss the U_{234}/U_{238} ratio observed in the area, present loading and regeneration results from pilot and full-scale ion exchange systems, and discuss techniques for disposal of radioactive regenerent brines.

QUALITY DATA AND U_{234}/U_{238} RATIO

Over the past several years, radiochemical quality data have been collected by the School District for wells of concern in West Jefferson County shown in Table 1. Included as Figure 2 is the flow chart for gross alpha particle activity monitoring commonly used to determine compliance with NIPDWR for radionuclides. Using this evaluation procedure it was determined that although the water quality complied with the NIPDWR, significant gross alpha activity still existed.

The fluorometric method is routinely used for uranium analysis in which a mass measurement for uranium is converted to an activity level using a conversion factor of 0.7 ug/pCi. The conversion factor assumes U_{234} and U_{238} are present in equal amounts. Using the more expensive alpha-spectroscopic analysis to determine the individual activity contributions from U_{234} and U_{238}, for the District well which was being examined, the activity ratio of U_{234}/U_{238} was determined to be 2.4:1. Therefore, the majority of the gross alpha activity was attributable to these species of uranium. Additionally, two different reference materials, Americium-241 (AM-241) or Uranium-

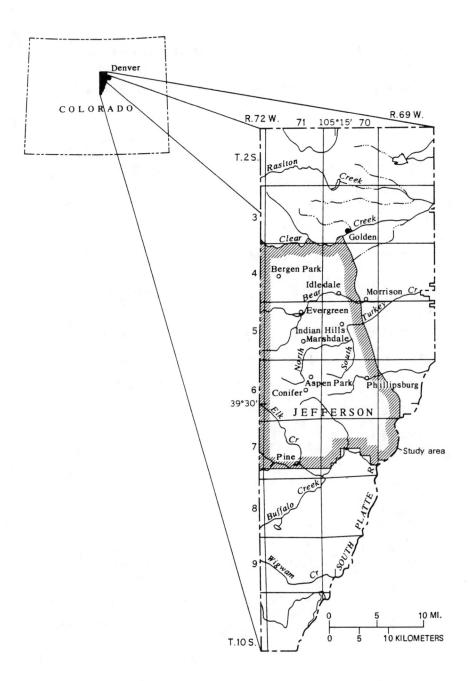

Figure 1. Location of wells in W. Jefferson County, Colorado

Table 1. Historical summary of District well water quality.

Parameter	Marshdale Well	Coal Creek Well
TDS	130-170 mg/L	220 mg/L
Ca	20-22 mg/L as $CaCo_3$	31 mg/L as $CaCo_3$
Na	---	61 mg/L
Bicarbonate Alkalinity	---	215 mg/L as $CaCo_3$
pH	7.3-7.2 units	---
Hardness	82 mg/L as $CaCO_3$	78 mg/L as $CaCo_3$
Iron (total)	0.12 - 3.1 mg/L	0.22 mg/L
Iron (dissolved)	0.07 - 0.01 mg/L	<0.05 mg/L
Manganese	0.033 - .045 mg/L	<0.05 mg/L
Magnesium	3.9 - 4.6 mg/L	1.0 mg/L
Sulfate	36 mg/L	6.8 mg/L
Gross Alpha Activity	80 - 170 pCi/L	50 - 60 pCi/L
Gross Beta Activity	0-30 pCi/L	
Uranium Activity	40-50 pCi/L	16 pCi/L
Uranium Concentration	0.047 - 0.09/mg/L	0.024 mg/L
Radium 226 Activity	0.6 - 1.3 pCi/L	1.9 pCi/L
Radium 228 Activity	1.0 - 1.8 pCi/L	

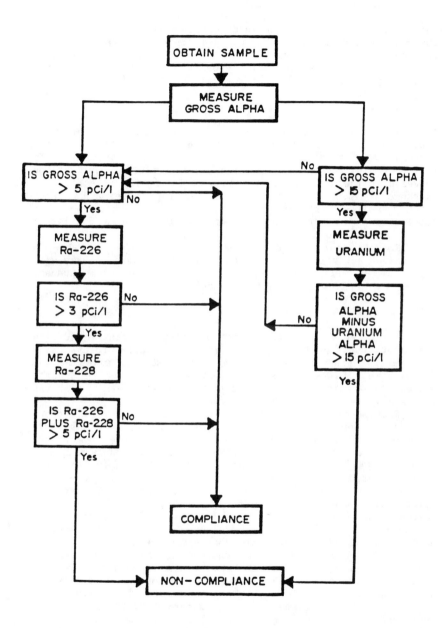

Figure 2. Flow chart for gross alpha particle activity monitoring

Natural (U-nat), may be used in determining gross alpha activity. The use of the AM-241 reference was found to more accurately represent the gross alpha levels. Use of the U-nat reference would indicate the existence of gross alpha emitters other than uranium species, which was not the case in the District well.

PILOT STUDY

As mentioned previously, alpha-spectroscopic analysis indicated that virtually all of the gross alpha activity in the District well was being contributed by uranium species. To verify uranium removal results from previous field investigations on wells in the area, a pilot column employing the selective ion exchange resin IONAC A641 was operated for three weeks using water from the District well. Analysis of weekly composite samples of column influent and effluent indicated a consistent removal of 99 percent of gross alpha activity.

Results from this pilot testing are included in Table 2. Gross alpha activity in the column influent appeared to vary with the volume of well water pumped. The larger the volume pumped, the higher the influent gross alpha activity.

FULL SCALE SYSTEM

After removal of gross alpha activity was verified during the pilot investigation, a full-scale ion exchange process using a commercial grade water softening system was installed at the District well in May, 1986. The system was designed to treat a maximum flow rate of eight gallons per minute (gpm), although five gpm is typically produced by the well.

Prefilters in parallel, with pore openings of one micron, precede the ion exchange columns to avoid plugging of the resin by large particulate material. Two 16-inch diameter by 52-inch in height Hydromax pressure vessels, each filled with 3 cu. feet. of Ionac A642 (potable water grade) strong-base anion exchange resin are used for uranium removal. The columns are operated in series, with sample valves located upstream of the first column, between the columns, and in the effluent piping of the second column. The location of these sample points allows monitoring of the level of exhaustion of each column. Treated water flows to an existing 30,000 gallon buried concrete storage tank.

As gross alpha analysis requires several weeks, two ion exchange vessels were installed to provide a factor of safety against uranium breakthrough while waiting for analysis results. A process flow schematic is included as Figure 3. Capital costs for the installed system were approximately $17,000, including engineering. The system was housed in an existing building.

To meet the requirements established by the Hazardous Wastes section of the Colorado Department of Health, the resin must be regenerated monthly. Control of the regeneration is accomplished with a timer and automatic valves and is performed

Table 2.   Results from pilot study at District well.

Sample Dates	Column Influent Gross Alpha (pCi/L)	Column Effluent Gross Alpha (pCi/L)	% Removal of Gross Alpha
11/1/85 to 11/8/85	110	1.6	99
11/8/85 to 11/15/85	120	1.4	99
11/15/85 to 11/22/85	27	0.3	90

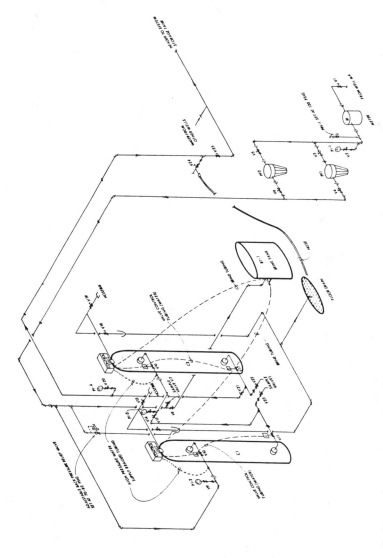

Figure 3.   Full scale process schematic.

in an upflow mode. Five bed volumes of salt brine are used in regeneration.

Results from eight months of full-scale operation are summarized in Table 3. Through mid-January, 1987 approximately 800,000 gallons, or 18,000 bed volumes, were treated. The system was regenerated five times. The data indicate consistent removal of gross alpha activity from the water of greater than 89 percent. Gross alpha activity in the raw well water ranged from 24 to 130 pCi/L. With the frequency of regeneration required, it appears that a single ion exchange column is adequate for removal of uranium contamination from the District well.

During August 1986, the mass balance of uranium loaded into the resin was compared to the amount recovered through regeneration and indicated that essentially complete recovery was achieved.

Operation and maintenance (O&M) costs for uranium removal only for the District system including labor for operation, regeneration and sample collection, sample analysis, resin replacement, salt for brine and electrical requirements are estimated to be $2.30 per 1,000 gallons of water treated. These O&M costs do not include costs for brine disposal.

DISCUSSION OF BRINE DISPOSAL

Removal of uranium from well water is a relatively proven technology when compared to the uncertainties and unknowns associated with final disposal of uranium-laden regenerant brines.

Safety and regulatory requirements were considered in identifying the various options for disposal that were available. Several options for disposal of the rinse waters and brines from the ion exchange system were proposed to the Colorado Department of Health for their consideration. They included:

1. Hauling and discharge into the equalization basin at a nearby wastewater treatment facility. Since no sanitary sewer is located near the District well, all domestic wastes are hauled to the nearby wastewater treatment plant. It was proposed that the brine be hauled along with the sanitary wastes.

2. Use of a licensed hauler to transport and dispose of the brines at an acceptable uranium disposal facility in another state. Estimates for this option ranged from $250 to $350 per 55 gallon drum.

3. Injection into a newly-drilled disposal well on the property or into an existing unused well. Costs associated with the depth of such a well and concerns of regulating officials at the Division of Radiation Control of the Colorado Department of Health were a significant

Table 3. Summary of gross alpha activity associated with uranium removal system.

Sample Date	Total Volume Treated To-date (Gal)	Raw Well Water (pCi/L)	Effluent of first Column (pCi/L)	Effluent of second Column (pCi/L)	Gross Alpha Removal (%)	Wastewater Treatment Plant where Regenerent Brine was Discharged Plant Effluent (pCi/L)	Sludge (pCi/L)
5/30/86	0	32 ± 6	1.9 ± 2.1	3.4 ± 2.5	89		
6/19/86	68,030	130 ± 10	1.2 ± 1.9	1.5 ± 2.0	99		
6/26/86	91,840	99 ± 11	1.5 ± 2.0	3.8 ± 2.6	96		
7/15/86	156,470	110 ± 10	0.9 ± 1.9	1.8 ± 2.1	98		
7/22/86	180,280	84 ± 10	2.3 ± 2.1	2.6 ± 2.2	97		
7/28/86	200,690	66 ± 9	1.9 ± 2.0	1.0 ± 1.7	98		
8/8/86	238,100	80 ± 10		3.2 ± 2.4	96		
8/13/86	225,110						
10/9/86	485,210	24 ± 6	1.7 ± 2.0	2.0 ± 2.1	92		
11/6/86	602,490						19±7
11/13/86	630,020	110 ± 10	1.9 ± 2.3	1.9 ± 2.3	98		
12/15/86	750,887					18±7	3.5±1.8 pCi/gm
1/14/87	799,210					34±14	1.0±1.1 pCi/gm

consideration in determining the viability of this alternative.

4. Transportation of the brine to a relatively nearby uranium mill for treatment. After it was realized that permit amendments would be required by the mill, this alternative was eliminated from further study.

Following discussions with the Drinking Water, Hazardous Wastes, and Permit sections of the Colorado Department of Health, and consideration of costs and safety, Alternative 1, hauling to the nearby wastewater treatment facility was selected for implementation.

As a requirement of the Colorado Department of Health, gross alpha activity in the wastewater treatment plant effluent and sludge must be monitored by the District on a monthly basis. Results from several samples of plant effluent and sludge are also shown in Table 3. O&M costs for disposal of brines from the District system are estimated to be $1.10 per 1000 gallons of water treated.

CONCLUSIONS

Based upon the work performed at the Jefferson County School District, the following conclusions have been identified:

1. Although MCLs for uranium have not been adopted at this time, the School District decided to implement a system to remove uranium from water used for drinking and other purposes at several facilities. As it appears that drinking water regulations for uranium will be adopted, many communities will be required to remove the radionuclide from their water.

2. If all the contributors to the gross alpha activity of a water cannot be accounted for using common techniques for uranium analysis, the alpha-spectroscopic analysis should be used to determine the individual contributions from U_{234} and U_{238}.

3. Consistent removal of high levels of gross alpha activity from well water to levels acceptable for drinking water may be accomplished at reasonable cost for small systems with conventional water softening equipment employing a strong base anionic ion exchange resin.

4. Not unlike the nuclear power industry, disposal of uranium-laden regenerant brines is the most complex task involved in a project for removal of uranium from water. Disposal of brines is site specific. Costs and methods of disposal are driven by safety and regulations.

REFERENCES

1.   American Water Works Association. "Mainstream", 20(10):1-2 (1986).

2.   Hanson, S.W., D. Wilson, and N. Gunaji. "Removal of Uranium from Drinking Water by Ion Exchange and Chemical Clarification" unpublished report, U.S. EPA MERL contract CR-810453-01-0 (1985).

3.   Reid, G.W., P. Lassovszky, and S.W. Hathaway. "Treatment, Waste Management and Cost for Removal of Radioactivity from Drinking Water", J. Health Physics 48(3): 671-694 (1985).

4.   Horton, T.R. "Nationwide Occurence of Radon and Other Natural Radioactivity in Public Water Supplies", U.S. EPA Report 520/5-85-008, (1985).

5.   Hall, D.C., and C.J. Johnson. "Drinking Water Quality and Variations in Water Levels in the Fractured Crystalline – Rock Aquifer, W. Central Jefferson County, CO", U.S. Geological Survey Water Resources Investigations 79-94 (1979).

6.   Johnson, C.J., unpublished results in letter from Jefferson County Health Department to residents of homes relying on wells for drinking water (August 20, 1979).

7.   Bondietti, E.A., S.K. White, and S.Y. Lee. "Methods of Removing Uranium from Drinking Water: II, Present Municipal Treatment and Potential Treatment Methods", U.S. EPA Report - 570/9-82-003, ORNL EIS-194 (1982).

8.   Reid, G.W., P. Lassovszky, and S.W. Hathaway. "Treatment, Waste Management and cost for Removal of Radioactivity from Drinking Water", J. Health Physics 48(3): 671-694 (1985).

9.   Hathaway, S.W. "Ion Exchange Technology for Removing Naturally Occurring Uranium from Drinking Water.", unpublished results (1986).